State Space and Input-Output Linear Systems

DAVID F. DELCHAMPS

State Space and Input-Output Linear Systems

With 12 Illustrations

Springer-Verlag
New York Berlin Heidelberg
London Paris Tokyo

David F. Delchamps
School of Electrical Engineering
Cornell University
Ithaca, NY 14853-5401
USA

Library of Congress Cataloging-in-Publication Data
Delchamps, David F.
State space and input-output linear systems.
Bibliography: p.
Includes index.
1. System analysis. 2. State-space methods.
3. Linear systems. I. Title.
QA402.D39 1988 003 87-32061

Camera ready copy prepared by the author using Troff.
Printed and bound by Edward Brothers Inc., Ann Arbor, Michigan.
Printed in the United States of America.

9 8 7 6 5 4 3 2 1

ISBN 0-387-96659-5 Springer-Verlag New York Berlin Heidelberg
ISBN 3-540-96659-5 Springer-Verlag Berlin Heidelberg New York

this book is dedicated to
the memory of my grandfather

Julian Montgomery West

and

to four fellow travelers
who've taught me a lot

Stephen West Delchamps
Arthur Lawrence Gruen
Jeffrey Arthur Tice
Geoffrey Alan Williamson

PREFACE

It is difficult for me to forget the mild sense of betrayal I felt some ten years ago when I discovered, with considerable dismay, that my two favorite books on linear system theory — Desoer's *Notes for a Second Course on Linear Systems* and Brockett's *Finite Dimensional Linear Systems* — were both out of print. Since that time, of course, linear system theory has undergone a transformation of the sort which always attends the maturation of a theory whose range of applicability is expanding in a fashion governed by technological developments and by the rate at which such advances become a part of engineering practice. The growth of the field has inspired the publication of some excellent books; the encyclopedic treatises by Kailath and Chen, in particular, come immediately to mind.

Nonetheless, I was inspired to write this book primarily by my practical needs as a teacher and researcher in the field. For the past five years, I have taught a one semester first year graduate level linear system theory course in the School of Electrical Engineering at Cornell. The members of the class have always come from a variety of departments and backgrounds, and consequently have entered the class with levels of preparation ranging from first year calculus and a taste of transform theory on the one extreme to senior level real analysis and abstract algebra on the other. I have therefore found it desirable, if not necessary, to teach the course in a manner which permits one group of students to catch up on their linear algebra and differential equations, while at the same time giving better prepared students a rigorous introduction to more advanced topics. The composition and tone of the book reflect these objectives; the exercises, which are scattered throughout, vary in difficulty but are all of central importance. The entire book can be covered easily in two semesters at the first year graduate level, although much of the purely mathematical material can be skipped if the background of the class permits.

Furthermore, in the course of doing research in the systems area, it has always seemed to me that it would be nice to have available a *concise*, rigorous, and fairly readable reference which treated not only the central results from continuous- and discrete-time state space and input-output linear system theory but also some of the theory's mathematical foundations and recent developments, such as the polynomial matrix fraction approach to input-output systems. In writing this book, I have sought to provide such an account. To be sure, the selection of topics indicates my own taste, and I am certain that most readers will find some omission or another particularly annoying. Still, the book is intended only to trace one path through a complicated network of results; the reader can judge whether the path I've selected is a logically coherent or informative one.

It goes without saying that the book owes much of its existence to the help I've received from co-workers and friends too numerous to mention. I'd like first of all to thank all of my colleagues from the School of Electrical Engineering at Cornell for their encouragement and support over the last five years. The second half of the book was written while I was visiting the Mathematical Sciences Institute at Cornell, and I would therefore also like to thank the M.S.I. for its support. Linda Struzinsky deserves a great deal of credit for her careful typing of the earlier part of the manuscript, and for helping me understand the vagaries of our word processing system.

Truly indispensable to me has been the assistance of Geoff Williamson and Scott English, each of whom has read every word of the manuscript several times. Geoff and Scott have caught more typographical errors than I'd care to remember, and have given me some valuable advice on improving the exposition. Needless to say, any remaining typographical errors and flawed writing are my responsibility alone. I would also like to thank Andrew Nobel and Ian Dobson for reviewing earlier versions of some of the sections in Part I.

Finally, I'd like to thank the people from whom I've learned essentially everything I know about linear system theory: Roger Brockett, Chris Byrnes, Peter Caines, Bradley Dickinson, and the late Michael Murray. These five fabulous teachers have inspired me (and many others) more than they realize.

David F. Delchamps
Ithaca, New York
18 September 1987

TABLE OF CONTENTS

PART I
MATHEMATICAL PRELIMINARIES

PART II
STATE SPACE LINEAR SYSTEMS

PART III
INPUT-OUTPUT LINEAR SYSTEMS

PART IV
STABILITY AND FEEDBACK

Introduction

The theory of linear systems has grown so rapidly and in so many different directions during the last quarter century that it has become difficult, if not impossible, to describe its boundaries in any definitive or straightforward fashion. The pioneers whose research inspired the early development of the theory, most notably Hendrick Bode, Harry Nyquist, and Norbert Wiener, would probably be somewhat surprised at the number of pure algebraists and geometers, mathematical physicists, econometricians, coding theorists, and others who today count linear system theory among their principal areas of research. It is fair to say, however, that linear system theory has been nurtured to its current state of maturity largely by engineers and applied mathematicians whose interests have centered on promoting *mathematical modeling* as a valuable resource whose potential contribution to enhancing one's understanding of practical scientific and engineering problems should not be underestimated.

This is not, however, a book about mathematical modeling. Rather, it is a book about a specific class of mathematical models whose usefulness in a variety of engineering and applied scientific contexts has been amply demonstrated. The range of processes in the "real world" for which linear systems serve as effective models encompasses elementary electrical networks; communications, signal processing, and control systems; complicated mechanical processes, most notably those which arise in aerospace applications; and even economic and demographic processes, not to mention many others.

The subtleties and pitfalls inherent in the very act of attempting to describe an observed phenomenon analytically make the field of mathematical modeling a rich and demanding area of study in itself, complete with its own techniques and its own class of experts. Unfortunately, space constraints prevent us from delving too deeply into modeling issues here. Instead, our objective will be to explore the theory of the models themselves, in this case of linear systems; we propose to develop many of the theory's central results, most of which which should arguably be a part of the toolkit of every engineer and applied mathematician, from their mathematical and phenomenological foundations.

A *system*, at least in the context of this book, is a *mathematical model* for a *real-world process* which accepts a certain number of *inputs* and give rise to a certain number of *outputs*. The inputs may be regarded as *driving* the process, while the outputs may be interpreted as observable manifestations of what is going on inside the process and how this *internal behavior* is affected by the inputs.

An intelligent outsider — say, someone who is interested in *analyzing* such a process — may or may not be able to manipulate its inputs at will; indeed, some or all of the inputs may be wholly or partially determined by external factors such as the weather, gravity, or even the laws of supply and demand. Nonetheless, the systems considered in this book should be regarded as models for processes whose inputs may be specified, at least to a certain extent, by people who wish either to understand the processes better or to make them behave in a prescribed fashion. It is perhaps this *prescriptive* orientation which makes system theory, as a field, much more than just an elaborately axiomatized branch of classical dynamics in the presence of forcing functions. It is hoped that the approach taken here toward the development of the the central results of linear system theory, along with the selection of material to be covered, will illuminate some of the points at which system theory departs from such pure *descriptive* science.

Let us consider now the question which this book attempts to answer: what is a *linear* system? Roughly speaking, a linear system is a special sort of mathematical model for an input-output process of the kind described above. What makes the model special is its *linearity*. What is linearity? The word is full of meaning, in a mathematical sense, and using it to describe something as nebulously specified as a "mathematical model for something going on in the real world" presupposes a certain amount of structure for the various building blocks which comprise the model. Moreover, if the model is to reflect at least approximately what is really happening in the modeled process, then the process itself must have structural features which parallel those of the model.

More specifically, a linear system serves as a model for a process which may be viewed macroscopically as a *box* which accept a certain number of input *time functions* and emits a certain number of output *time functions* which depend, in some linear fashion, on the input functions. This box may have associated with it some internal structure; the settings and configurations possible for the situation which obtains inside the box may or may not be accessible to outside measurement or manipulation, but will also depend *linearly* on the input functions.

In order to specify what is meant by a time function in a given context, we must first make clear what is meant by *time*. In this book, time may be either *continuous* or *discrete*. We view a continuous-time function as a mapping which has as its domain some subset, usually an interval, of the real line **R**; a discrete-time function, on the other hand, is defined on a subset of the integers Z. Specifying the "time axis," of course, is only half the battle; we need to identify the sets in which time functions are to take their *values*. If the models we're considering are to be linear, then these sets of values must have associated with them some linear structure. Accordingly, our time functions will always take values in *vector spaces*. These vector spaces might be familiar ones such as $\mathbf{R}^n$, the set of all n-tuples of real numbers; we will, however, have occasion to deal with time functions which assume values in *abstract* vector spaces whose specification is far less concrete.

Having made some sense of time functions, it is now possible to give a preliminary description of the linear systems which we'll be studying. A *continuous-time state space linear system* will be a model for an input-output process which has associated with it some well-defined notion of *internal situation*, or *state*. The evolution of the state of such a process, like its inputs and outputs, will be modeled as a time function which takes values in a vector space of *states* for the process. The value of this function at a given instant t_0 may be viewed as describing an *initialization*, or *setting*, for various quantities associated with the process whose future evolution summarizes the time development of the process's internal situation under the influence of the inputs. We'll assume that a great deal of information about the process is contained in the value of its state at t_o; more precisely, this value will tell us *at least* enough so that we can figure out the entire future evolution of the state *and* of the output functions once the input functions are specified for times greater than t_o.

There will be certain rules which govern the evolution of the state under the influence of the inputs and of the outputs under the influence of the inputs and the state. Linearity makes its first appearance in the specification of how these *state transitions* and *readouts* occur. The idea is that the state at some time t_1 should depend *linearly* not only on the state at each "earlier" time $t_o < t_1$ but also on the input function which is applied between t_o and t_1. Similarly, the value of the output at any time t should depend linearly on the state at time t and the value of the input at time t.

The intuitive characterization we've just given of a continuous-time state space linear system is formalized in §6 below, as is the obviously analogous notion of a *discrete-time* state space linear system. In §18, we introduce formally the notion of *input-output linear systems*. Input-output systems serve as models for input-output processes which have associated with them no well-defined concept of state or internal situation. Such processes are well thought of as *black boxes* which accept input time functions and give rise to output time functions in a linear fashion. Historically, the study and application of linear input-output models predates most research on state space models, which didn't become popular among engineers and applied scientists until the 1950's. Input-output linear system theory has its roots in electrical engineering, and it was electrical engineers such as Bode and Nyquist who first promulgated system theoretic approaches, particularly those involving *frequency domain techniques*, to problems in engineering analysis and design.

It is not surprising that electrical networks have played a pivotal role in the development of linear system theory. Passive electrical components such as resistors, capacitors, and inductors are very well approximated by linear models; even active nonlinear devices, such as transistors, are usually operated in their *linear ranges*. One fundamental example of a physical process which admits a state space linear system model is a simple network which contains only resistors, capacitors, and inductors, all modeled as linear elements, along with a certain number of independent sources of voltage and current. There are assumed to be certain specific points in the network at which voltage and current measurements can be made. If the voltages across all the capacitors and the currents through all the inductors are known at a given time t_o, then knowledge of the behavior of the sources (i.e., the inputs to the system) for $t \geqslant t_o$ enables one to determine the behavior of *all* the voltages and currents in the network for $t \geqslant t_o$. For this reason, the vector of capacitor voltages and inductor currents is a reasonable candidate for the *state* of the system; the adoption of linear models for the various circuit elements forces the time evolution of the state and the measurements (i.e., outputs) to depend linearly on the inputs and initial state, and a continuous-time state space linear system results.

Real-world manifestations of input-output linear systems are somewhat more difficult to characterize, since one usually visualizes a black box as having something inside it. Imagine, however, a black box which contains an electrical network and is to be used as a signal amplifier. In some sense, what's really inside the box is less important from a design standpoint than the behavior of the box as a processor of incoming time functions. Consequently, we are tempted to model such a box in terms only of its *input-output behavior*, and to adopt the view that this box is, as far as we are concerned, essentially equivalent to any other box which has the same input-output properties. If the dependence of the outputs on the inputs is linear, then the signal amplifier may be modeled effectively as an input-output linear system.

Having established the general class of models which this book aims to describe, it is probably worthwhile to say a few words about some the specific topics that we'll be encountering along the way. Linear system theory has a lexicon all its own which enhances the theory's descriptive power and inspires a particular set of attitudes toward problems to which the theory may be applied. This language has its roots for the most part in the language of linear algebra and linear differential equations, and most of Part I is devoted to a fairly brisk tabulation of some central results from these two areas.

§1 covers the elementary arithmetic of real and complex matrices; although the treatment is self-contained, the material is presented in a manner which assumes that the reader has had at least some previous exposure to the subject. Probably the most important results from §1, at least from the standpoint of linear system theory, are those involving the the computation of the rank, range, and nullspace of a matrix by means of Gauss Elimination. §2 focuses on the basic Existence and Uniqueness Theorem for first order linear vector differential equations; §3 summarizes the corresponding results for linear difference equations.

In §4, some fundamental notions from *abstract* linear algebra are introduced. Again, although the presentation begins at at "square one," it will probably be more easily accessible to readers who have at least a nodding acquaintance with the idea of an abstract vector space. The results of §4 may be regarded as *abstract generalizations* of the results of §1; on the other hand, it is equally valid to view the material in §1 as a collection of *concrete manifestations* of the more primitive constructions in §4. In any case, §1 and §4 are intended to illuminate at least a few of these fundamental correspondences. The topics covered in §5, similarly, are generalizations of some geometric notions which should be more or less familiar to everyone. In particular, the material on norms of linear mappings will play a crucial role in the development of stability theory for linear systems in Part IV.

Parts II and III are devoted to a careful analysis of the many important features of state space and input-output linear systems, as defined in §6 and §18, respectively. It should be emphasized that although the general definitions serve as umbrellas under which we develop the properties of such systems, the exposition in Parts II and III does *not*, for the most part, adhere to an axiomatic, general-to-specific format. It is arguable that such an approach would not only run counter to the historical evolution of the subject, but would also obscure the meaning and intuitive appeal of most of the fundamental principles which make linear system theory an exciting and relevant branch of applied mathematical modeling.

In this spirit, §7 brings the abstract definitions in §6 down to earth in the form of *realizations* for state space linear systems. The four sections which follow are dedicated to some further results from linear algebra on which the analysis of realizations for *time-invariant* state space systems depends. All of these results are of independent mathematical interest, as well; §§8-10 present a good portion of the theory of eigenvalues, eigenvectors, and generalized eigenvectors for linear transformations on finite-dimensional vector spaces, and in §11 the Jordan canonical form is derived.

The remainder of Part II addresses some fundamental linear system theoretic notions which have no direct analogues in other fields of mathematical modeling. The properties of *reachability* and *controllability* reflect the degree to which inputs to a state space linear system may be used to affect the evolution of the system's state; *observability* and *constructibility* reflect the degree to which the state evolution influences the system's output. §17 presents a proof of R. E. Kalman's Canonical Structure Theorem, which says that a given state space linear system may be *decomposed* into interconnected subsystems which are exhibited in such a way that their reachability and observability properties are easier to assess than those of the original system.

Part II is concerned with the theory of input-output linear systems. An input-output linear system may be viewed mathematically as defining a linear mapping (or mappings) between vector spaces of time functions. More concretely, the study of input-output linear systems provides an excellent forum for careful consideration of the words *analysis* and *synthesis* and their importance in the context of engineering modeling. More precisely, it might be one's objective to *analyze* a given black box by observing its response (i.e., output) corresponding to certain special input time functions, and drawing conclusions about its behavior when more general input functions are applied. On the other hand, one might wish to *synthesize* a given input-output behavior by *building* a box which responds to inputs in a prescribed fashion. The *frequency domain approach* to input-output linear systems, which is addressed in §§19-20, has a decidedly analytical flavor, while realization theory, which occupies most of §§21-22, lies more in the realm of synthesis.

The development of frequency domain theory in §§19-20 is predicated on the reader's having had at least some prior acquaintance with continuous- and discrete-time Fourier transforms, although definitions of these transforms are presented and some of their properties are proven. The Laplace transform and the z-transform are also defined, and their relevance to the the *transfer function analysis* of input-output linear systems is discussed. §21 is concerned the realization theory of time-invariant input-output linear systems. Realization theory establishes a strong connection between input-output linear system theory and the state space theory of Part I. Specifying a *minimal realization* for a given input-output system is seen to be in some sense equivalent to describing how to "build" a *state space system* of minimum complexity whose input-output behavior is the same as that of the given input-output system. §22 approaches transfer function analysis and realization theory from the viewpoint of *matrix fraction descriptions;* this approach has become increasingly important, especially in applications of linear system theory, during the last ten to fifteen years. Many of the central constructions and results concerning polynomial matrix factorization of rational matrix-valued functions are derived, including the Smith and Smith-McMillan canonical forms and the equivalence of irreducibility, minimality, and relative primeness of polynomial matrix factorizations.

In the first two sections of Part IV, the important subject of *stability theory* for state space and input-output linear systems is addressed. The stability of state space linear system, which is the subject of §23, is defined in the usual *Lyapunov* sense, and conditions are derived which assess a system's stability in terms of properties of its state transition matrix. §24 gives similar criteria for the *bounded-input bounded-output* stability of input-output linear systems in terms of the system's weighting pattern; these conditions reduce to "classical" assertions about the system's *impulse response* and *transfer function* if the system is time-invariant.

The material covered in §25 represents another collection of results all of which are specific to linear system theory; the presentation of these results has a decidedly *control-theoretic* slant. The Nyquist Criterion for the stability of a single-input, single-output feedback system and Wonham's famous result on pole placement by means of state feedback are both derived, as is Luenberger's technique for the implementation of state feedback controllers for time-invariant systems by means of observers.

§§26-27 derive the solution to the *linear quadratic regulator problem* for time-invariant discrete- and continuous-time systems. The solution to this problem, although it is almost thirty years old, stands as one of the enduring triumphs of modern linear system theory. The problem marks a point at which essentially *all* of the system-theoretic constructions described in this book, most notably reachability and observability, feedback, and stability theory, must be brought to bear in order to solve a single problem which has proven to be of great practical importance in applications.

PART I
MATHEMATICAL PRELIMINARIES

1. *Some Linear Algebra*

In this section, we present a rapid review of certain fundamental results from linear algebra; most of these results will be of paramount importance later on. A good presentation of similar material at a comparable level is given in [Strang]. It should be emphasized that this section is intended mainly as a *review*, and not as a detailed exposition; at the very least, it should serve to familiarize the reader with certain notational conventions and terminology which we shall be using in the sections which follow.

By a *scalar* we mean a real or complex number; scalars are usually denoted by lower-case Greek letters or lower-case subscripted Roman letters. We usually denote by **F** the set of scalars with which we are working; thus, **F** stands either for **R** (real numbers) or for **C** (complex numbers). An $(m \times n)$ (real or complex) *matrix* A is a rectangular array of (real or complex) scalars having m rows and n columns. The scalar lying at the junction of the ith row and jth column of an $(m \times n)$ matrix A is denoted either by a_{ij} or by $[A]_{ij}$, $1 \leqslant i \leqslant m$, $1 \leqslant j \leqslant n$. Specifying an $(m \times n)$ matrix A is equivalent to specifying each a_{ij}.

Given an $(m \times n)$ matrix A and a scalar α, we may define a new $(m \times n)$ matrix αA by the formula $[\alpha A]_{ij} = \alpha a_{ij}$, $1 \leqslant i \leqslant m$, $1 \leqslant j \leqslant n$. Given two $(m \times n)$ matrices A and B, we define their *sum* $A + B$ as the $(m \times n)$ matrix given by

$$[A+B]_{ij} = a_{ij} + b_{ij}\ , \quad 1 \leqslant i \leqslant m\ ,\ 1 \leqslant j \leqslant n\ .$$

It is clear that matrix addition is commutative and associative, and that multiplication of matrices by scalars distributes over matrix addition. The *transpose* of an $(m \times n)$ matrix A is the $(n \times m)$ matrix A^T given by

$$[A^T]_{ij} = a_{ji}\,, \quad 1 \leqslant i \leqslant n\,,\ 1 \leqslant j \leqslant m\,.$$

There are certain matrices which deserve special mention.

1.1 Special Matrices

(a) A *k-vector* is simply a $(k \times 1)$ matrix. A k-vector is usually denoted by a lower-case Roman letter; the single number in the ith row of a k-vector is denoted x_i.

(b) The $(m \times n)$ *zero matrix* is the $(m \times n)$ matrix 0 all of whose entries are zero. Observe that $A + 0 = 0 + A = A$ for every $(m \times n)$ matrix A.

(c) A *symmetric matrix* is an $(n \times n)$ (i.e., square) matrix A satisfying $A^T = A$. A *skew symmetric matrix* is an $(n \times n)$ matrix A which satisfies $A^T = -A$.

(d) A *diagonal matrix* is an $(n \times n)$ matrix A satisfying $a_{ij} = 0$ when $i \neq j$.

(e) The $(n \times n)$ *identity matrix* I_n has entries given by

$$[I_n]_{ij} = \delta_{ij} = \begin{cases} 1 & i = j \\ 0 & i \neq j \end{cases}.$$

(f) A *Hermitian matrix* is an $(n \times n)$ matrix A which satisfies $A^\dagger = A$, where $A^\dagger$ denotes the *conjugate transpose* of A; i.e.,

$$[A^\dagger]_{ij} = ([A]_{ji})^*\,,$$

where α^* denotes the complex conjugate of α. A *skew-Hermitian* matrix A satisfies $A^\dagger = -A$. Observe that every real (skew-) symmetric matrix is (skew-) Hermitian. □

Given an $(m \times n)$ matrix A and an $(n \times r)$ matrix B, define the *product* AB (or, "*B (A) multiplied on the left (right) by A (B)*") as the $(m \times r)$ matrix whose (i, j) element is given, for $1 \leqslant i \leqslant m$, $1 \leqslant j \leqslant n$, by

$$[AB]_{ij} = \sum_{k=1}^{n} a_{ik} b_{kj}\,.$$

Observe that the number of columns in A must be the same as the number of rows in B for the product AB to make sense. The following examples serve to illustrate various situations which may arise.

1.2 Examples Of Matrix Multiplication

(a) AB is defined; BA is undefined.

$$A = \begin{bmatrix} 1 & 2 \\ 3 & 4 \end{bmatrix} \qquad B = \begin{bmatrix} 1 & 0 & 1 \\ 2 & 1 & 2 \end{bmatrix}$$

$$AB = \begin{bmatrix} 5 & 2 & 5 \\ 11 & 4 & 11 \end{bmatrix}$$

(b) AB and BA are both defined.

$$A = [1 \;\; 2 \;\; 3] \qquad B = \begin{bmatrix} 1 \\ 2 \\ 3 \end{bmatrix}$$

$$AB = 14 \qquad BA = \begin{bmatrix} 1 & 2 & 3 \\ 2 & 4 & 6 \\ 3 & 6 & 9 \end{bmatrix}$$

(By convention, a (1×1) matrix is the same as a scalar.)

(c) AB and BA are both defined and are the same size.

$$A = \begin{bmatrix} 0 & 1 \\ 0 & 0 \end{bmatrix} \qquad B = \begin{bmatrix} 0 & 0 \\ 1 & 0 \end{bmatrix}$$

$$AB = \begin{bmatrix} 1 & 0 \\ 0 & 0 \end{bmatrix} \qquad BA = \begin{bmatrix} 0 & 0 \\ 0 & 1 \end{bmatrix}$$ □

It is important to notice (c.f. 1.2(a)) that the product BA may be undefined even if AB *is* defined. Moreover, when both AB and BA are defined, they may be of different sizes (1.2(b)); and, even if of the same size, AB and BA may be unequal (1.2(c)). It is evident that matrix multiplication distributes over matrix addition and commutes with multiplication of matrices by scalars. Another important property of matrix multiplication is the identity

$$(AB)^T = B^T A^T$$

which is an immediate consequence of the definitions.

Observe also that if A is an arbitrary $(m \times n)$ matrix, then the identity matrices serve as identity elements for matrix multiplication in the following sense:

$$I_m A = A I_n = A .$$

To see this, note that for $1 \leqslant i \leqslant m$, $1 \leqslant j \leqslant n$, we have

$$\begin{aligned} [I_m A]_{ij} &= \sum_{k=1}^{m} [I_m]_{ik} a_{kj} \\ &= \sum_{k=1}^{m} \delta_{ik} a_{kj} = a_{ij} . \end{aligned}$$

A similar argument shows that $AI_n = A$.

We now focus our attention temporarily on square matrices.

1.3 Definition: An $(n \times n)$ matrix A is said to be *invertible* if and only if there exists an $(n \times n)$ matrix B such that

$$AB = BA = I_n .$$

In this case, the matrix B is called the *inverse of* A, and is denoted by A^{-1}. □

The reader should check that there can be at most one $(n \times n)$ matrix B satisfying Definition 1.3 for a given A; indeed, if B and D are two such matrices, then $AB = I_n$ implies that $(DA)B = D$, which in turn means that $B = D$ since $DA = I_n$. A trivial example of an invertible matrix is I_n itself, which is its own inverse. The $(n \times n)$ zero matrix is evidently not invertible.

1.4 Facts: (*a*) *If A and B are $(n \times n)$ invertible matrices, then AB is also invertible; in this case,*

$$(AB)^{-1} = B^{-1}A^{-1} .$$

(*b*) *If A is an invertible $(n \times n)$ matrix, then so is A^T; in this case, $(A^T)^{-1} = (A^{-1})^T$* .

(*c*) *If A, B, and D are $(n \times n)$ matrices satisfying $AB = I_n = BD$*, then $A = D$.

Proofs: (a) and (b) are elementary consequences of the definitions. As for (c), which says essentially that a "right inverse" and a "left inverse" for a given $(n \times n)$ matrix must be equal, simply multiply both sides of the equation $AB = I_n$ on the right by D, use $BD = I_n$, and obtain $A = D$. □

It is often of great importance to determine whether a given $(n \times n)$ matrix A is invertible; invertibility or non-invertibility of an arbitrary A may seldom be determined by simple inspection. Important criteria for invertibility are given by the following central result.

1.5 Theorem: *Let A be an $(n \times n)$ matrix. The following three conditions are equivalent:*

(a) *A is invertible.*

(b) *The only n-vector x satisfying $Ax = 0$ is $x = 0$.*

(c) *For every n-vector y, there exists a unique n-vector x satisfying $Ax = y$.*

We postpone the complete proof of Theorem 1.5 until after Theorem 1.10; for now, it is easy to see that 1.5(a) implies both 1.5(b) and 1.5(c): given arbitrary y and invertible A, multiply both sides of the equation $Ax = 0$ or $Ax = y$ by A^{-1} to obtain $x = 0$ or $x = A^{-1}y$. In order to prove the remainder of Theorem 1.5, we shall require some additional machinery; as it happens, the technique we are about to discuss is of independent interest, and is useful in a variety of situations. We begin by defining *elementary matrices* as follows.

1.6 Definition: An *elementary matrix* is any of the following three types of matrix:

(a) Given integers $k \neq l$ between 1 and n and a scalar α, define $E_n(k, l, \alpha)$ as the $(n \times n)$ matrix obtained from I_n by replacing the zero at the (k, l) position by α. Formally,

$$[E_n(k, l, \alpha)]_{ij} = \delta_{ij} + \alpha\delta_{ik}\delta_{jl} .$$

(b) Given integers $k \neq l$ between 1 and n, define $P_n(k, l)$ as the $(n \times n)$ matrix obtained from I_n by interchanging the k th row and the l th row.

(c) Given a nonzero scalar α, define $\Gamma_n(k, \alpha)$ as the $(n \times n)$ matrix obtained from I_n by replacing the 1 at the (k, k) position with α. □

The reader should check that each of the above elementary matrices is invertible; in fact, we have the following easily verified formulas for the inverses of the elementary matrices:

$$[E_n(k, l, \alpha)]^{-1} = E_n(k, l, -\alpha)$$

$$[P_n(k, l)]^{-1} = P_n(k, l)$$

$$[\Gamma_n(k, \alpha)]^{-1} = \Gamma_n(k, \alpha^{-1}) .$$

We shall be concerned primarily with the effect on a given $(n \times n)$ matrix A which is produced when A is multiplied on the left by an elementary matrix.

1.7 Exercise: Show formally that, given an $(n \times n)$ matrix A:

(a) $E_n(k, l, \alpha)A$ is the matrix obtained from A by replacing the k th row with the sum of itself and α times the original l th row; i.e.,

$$[E_n(k,l,\alpha)A]_{ij} = \begin{cases} a_{ij} & i \neq k \\ a_{ij} + \alpha a_{lj} & i = k . \end{cases}$$

(b) $P_n(k, l)A$ is the matrix obtained from A by interchanging the k th and l th rows.

(c) $\Gamma_n(k, \alpha)A$ is the matrix obtained from A by replacing the k th row with its multiple by α; i.e.:

$$[\Gamma_n(k, \alpha)A]_{ij} = \begin{cases} a_{ij} & i \neq k \\ \alpha a_{ij} & i = k . \end{cases}$$ □

The procedure which we now describe is called *Gauss elimination*. It is an algorithm which enables one to "transform" a given $(n \times n)$ matrix A into a standard form which illuminates certain basic properties of A. We shall see later that the same procedure may also be applied to the analysis of arbitrary non-square matrices. Gauss elimination arose (and may still be viewed) as a systematic method for solving systems of simultaneous linear equations; the ramifications of this viewpoint should be revealed in what follows.

It is far easier to illustrate Gauss elimination by example than to describe it in words and symbols. Nonetheless, we present the algorithm formally before explaining it.

1.8 Gauss Elimination Algorithm: *Given an arbitrary $(n \times n)$ matrix A:*

Step 0: Set $i = 1$, $j = 1, M = A$.

Step 1: If $m_{ij} = 0$, *find the smallest* $k \geqslant i$ *such that* $m_{kj} \neq 0$; *multiply* M *on the left by* $P_n(k, i)$ *and rename the resulting matrix* M. *If no such* k *exists, increase* j *by 1 and repeat this step. The algorithm terminates here if* $j = n+1$.

Step 2: Multiply M *on the left, in any order you choose, by the succession of matrices*

$$E_n(k, i, -\frac{m_{kj}}{m_{ij}}),$$

for $i < k \leqslant n$. *Rename the resulting matrix* M.

Step 3: Increase i *and* j *by 1 and return to step 1. The algorithm terminates here when* $i = n$. □

1.9 Examples: The matrices shown are the intermediate M's; the algorithm steps used in obtaining them are shown above the arrows.

(a)

$$A = \begin{bmatrix} 1 & 2 \\ 3 & 4 \end{bmatrix} \xrightarrow{2} \begin{bmatrix} 1 & 2 \\ 0 & -2 \end{bmatrix}$$

(b)

$$A = \begin{bmatrix} 0 & 1 & 4 \\ 2 & 3 & 5 \\ 4 & 1 & 1 \end{bmatrix} \xrightarrow{1} \begin{bmatrix} 2 & 3 & 5 \\ 0 & 1 & 4 \\ 4 & 1 & 1 \end{bmatrix}$$

$$\xrightarrow{2} \begin{bmatrix} 2 & 3 & 5 \\ 0 & 1 & 4 \\ 0 & -5 & -9 \end{bmatrix} \xrightarrow{2} \begin{bmatrix} 2 & 3 & 5 \\ 0 & 1 & 4 \\ 0 & 0 & 11 \end{bmatrix}$$

(c)

$$A = \begin{bmatrix} 1 & 2 & 3 \\ 2 & 4 & 6 \\ 3 & 6 & 2 \end{bmatrix} \xrightarrow{2} \begin{bmatrix} 1 & 2 & 3 \\ 0 & 0 & 0 \\ 0 & 0 & -7 \end{bmatrix}$$

$$\xrightarrow{1} \begin{bmatrix} 1 & 2 & 3 \\ 0 & 0 & -7 \\ 0 & 0 & 0 \end{bmatrix}$$

□

The operations performed on the matrix M in Steps 1 and 2 of 1.8 are called *elementary row operations;* in each case, the operation is accomplished (c.f. Exercise 1.7) by multiplying M on the left with an elementary matrix or sequence of elementary matrices. If these matrices are kept track of and indexed in order of their use, we obtain an equation of the form

$$CA = C_N C_{N-1} \cdots C_2 C_1 A = M \ ,$$

where N is the number of elementary matrices used, the C_i are the matrices themselves, C is their product, and M is the matrix arrived at by means of 1.8. M has a special structure and a special name; it is called the *echelon form* of the matrix A. The word echelon is used because the echelon form of A contains a "staircase" (i.e. echelon) of zeros supporting an "inverted staircase" of (possibly) nonzero numbers.

The echelon form M of an $(n \times n)$ matrix A contains a certain number $r(A)$ of nonzero rows and $(n - r(A))$ zero rows; the zero rows, if there are any, lie at the bottom of M. The *pivot indices* j_k, $1 \leqslant k \leqslant r$, of A are the indices of the columns in which one finds the leftmost nonzero entries of the nonzero rows in M; in Example 1.9(a) we have $r(A) = 2$, $j_1 = 1$, $j_2 = 2$; in 1.9(b), $r(A) = 3$, $j_1 = 1$, $j_2 = 2$, $j_3 = 3$; in 1.9(c) $r(A) = 2$, $j_1 = 1$, $j_2 = 3$. Observe that if $r(A) = n$, then the pivot indices must be $j_1 = 1, j_2 = 2, \ldots, j_n = n$. Vestigial doubts in the mind of the reader concerning the importance of Gauss elimination should be dispelled by the following central result.

1.10 Theorem: *Let A be an $(n \times n)$ matrix. The following three conditions are equivalent.*

(a) $r(A) = n$.

(b) The only n-vector x satisfying $Ax = 0$ is $x = 0$.

(c) For any n-vector y, there exists a unique n-vector x such that $Ax = y$.

Proof: We show first that (a) implies (c). If $r(A) = n$, each "diagonal element" m_{ii} in A's echelon form $M = CA$ is nonzero. Given $y = [\, y_1 \ldots y_n \,]^T$, define the n-vector z by $z = C^{-1}y = [\, z_1 \ldots z_n \,]^T$; observe that C is invertible by Fact 1.4(a) since it is the product of (invertible) elementary matrices. Next, solve the equation $Mx = z$ for x as follows: obviously, $x_n = z_n / m_{nn}$; next solve for x_{n-1} via

$m_{n-1,n-1}x_{n-1} + m_{n-1,n}x_n = z_{n-1}$. Continue to solve for x_i, $1 \leqslant i \leqslant n-1$, in this fashion; the procedure works since each m_{ii} is nonzero. The vector x then satisfies $Mx = z$, and is obviously uniquely determined; multiplying both sides of the equation $Mx = z$ by C reveals x as the unique n-vector satisfying $Ax = y$.

It is clear that (c) implies (b); $x = 0$ always satisfies $Ax = 0$, and is unique in this regard if (c) is assumed.

It remains for us to show that (b) implies (a). Let M be the echelon form of A; suppose $r(A) < n$. We'll construct in this case a nonzero n-vector x satisfying $Ax = 0$. Observe that $Ax = 0$ if and only $Mx = 0$; we therefore seek an $x \neq 0$ satisfying $Mx = 0$.

Construct such an x as follows: since $r < n$, there is a smallest integer i_0, $1 \leqslant i_0 \leqslant n$, which is *not* a pivot index for A. Thus $1, 2, 3, \ldots, i_0-1$ are pivot indices. Set

$$x_{i_0} = 1, \quad x_n = x_{n-1} = \cdots = x_{i_0+1} = 0.$$

The remaining x's, i.e. $x_1, x_2, \cdots, x_{i_0-1}$, may then be solved for uniquely so as to satisfy $Mx = 0$; to see this, note that $Mx = 0$, with $x_{i_0}, \cdots, x_n$ chosen as indicated, if and only if

$$\begin{bmatrix} m_{11} & m_{12} & \ldots & m_{1,i_0-1} \\ 0 & m_{22} & & \cdot \\ \cdot & 0 & & \cdot \\ \cdot & \cdot & & \cdot \\ \cdot & \cdot & & \cdot \\ 0 & 0 & & m_{i_0-1,i_0-1} \end{bmatrix} \begin{bmatrix} x_1 \\ x_2 \\ \cdot \\ \cdot \\ \cdot \\ x_{i_0-1} \end{bmatrix} = - \begin{bmatrix} m_{1,i_0} \\ \cdot \\ \cdot \\ \cdot \\ \cdot \\ m_{i_0-1,i_0} \end{bmatrix}.$$

The matrix in this equation, however, is already in echelon form; the same technique used above in showing that (a) implies (c) may now be used to solve for $x_1, \cdots, x_{i_0-1}$. □

We are now equipped to prove Theorem 1.5.

Proof Of Theorem 1.5: Suppose A is invertible; we've seen already that $Ax = 0$ is satisfied only by $x = 0$. By Theorem 1.10, we conclude that, given an n-vector y, $Ax = y$ is satisfied by a unique n-vector x. Thus we have (a) implies (b) implies (c). We prove now that (c) implies the invertibility of A.

Assuming (c), let $b^1, b^2, \cdots, b^n$ be the unique n-vector solutions to the equations

$$Ab^1 = e^1, \ldots, Ab^n = e^n,$$

where e^j, $1 \leqslant j \leqslant n$, is the n-vector with a 1 in the jth position and zeroes elsewhere; that is, $[e^j]_i = \delta_{ij}$. It is then a simple matter to verify that

$$AB = I_n,$$

where B is the $(n \times n)$ matrix whose jth column is b^j.

Now, property (b) holds for B; if $Bx = 0$, then $ABx = I_n x = x = 0$. Thus property (c) also holds for B; we conclude, using the same logic as we have just used for A, that there exists an $(n \times n)$ matrix D satisfying $BD = I_n$. By Fact 1.4(c), $A = D$; we have $AB = BA = I_n$, meaning that A is invertible. □

It is important to note that if every "n" in the Gauss Elimination Algorithm 1.8 (except the termination criterion in Step 1) is replaced by "m," one obtains an algorithm which may be used in reducing an arbitrary $(m \times n)$ (i.e. non-square) matrix A to echelon form. The pivot indices and $r(A)$ are defined in the same way. Observe that $r(A) \leqslant m$ and $r(A) \leqslant n$, as well.

1.11 Examples: (same notation as in Examples 1.9)

(a)

$$A = \begin{bmatrix} 1 & 2 & 3 & 4 \\ 2 & 4 & 6 & 8 \\ 0 & 6 & 3 & 9 \end{bmatrix} \overset{2}{\rightarrow} \begin{bmatrix} 1 & 2 & 3 & 4 \\ 0 & 0 & 0 & 0 \\ 0 & 6 & 3 & 9 \end{bmatrix} \overset{1}{\rightarrow} \begin{bmatrix} 1 & 2 & 3 & 4 \\ 0 & 6 & 3 & 9 \\ 0 & 0 & 0 & 0 \end{bmatrix}$$

(b)

$$A = \begin{bmatrix} 1 & 2 & 3 \\ 4 & 5 & 6 \\ 7 & 8 & 10 \\ 10 & 11 & 12 \end{bmatrix} \overset{2}{\rightarrow} \begin{bmatrix} 1 & 2 & 3 \\ 0 & -3 & -6 \\ 0 & -6 & -11 \\ 0 & -9 & -18 \end{bmatrix} \overset{2}{\rightarrow} \begin{bmatrix} 1 & 2 & 3 \\ 0 & -3 & -6 \\ 0 & 0 & -1 \\ 0 & 0 & 0 \end{bmatrix}$$ □

The following assertion is an immediate consequence of the manipulation in the proof of Theorem 1.10.

1.12 Lemma: *If A is an $(m \times n)$ matrix with $m < n$, then there exists a nonzero n-vector x satisfying $Ax = 0$.*

Proof: Find an $(m \times m)$ matrix C via Gauss Elimination such that $CA = M$ is the echelon form for A. Once again, let $r(A)$ be the number of nonzero rows in M and define the pivot indices of A as before. Since $r(A) \leqslant m < n$, there exists a smallest i_0 which is not a pivot index for A. Set $x_{i_0} = 1$, and set $x_{i_0+1} = \cdots = x_n = 0$. Solve now for $x_1, \cdots, x_{i_0-1}$ exactly as in the proof of Theorem 1.10. □

1.13 Example: (A, M as in Example 1.11(a)) Set $x_3 = 1$, and $x_4 = 0$; solve for x_1, x_2 via

$$\begin{bmatrix} 1 & 2 \\ 0 & 6 \end{bmatrix} \begin{bmatrix} x_1 \\ x_2 \end{bmatrix} = \begin{bmatrix} -3 \\ -3 \end{bmatrix},$$

which yields $x_2 = -1/2$, $x_1 = -2$. Thus, $x = [\,-2\ -1/2\ 1\ 0\,]^T$ satisfies $Ax = 0$. □

It is worth emphasizing the important role that the Gauss Elimination Algorithm 1.8 has played in the development of linear algebra, particularly over the last forty years. The presence of high-speed digital computers has made it advantageous to have efficient *algorithms* for solving problems in linear algebra; neat, closed-form solutions to such problems are, more often than not, computationally wasteful. A very readable account of the computational utility of Gauss Elimination is given in [Strang]. In the present context, Gauss Elimination has provided us already with a convenient way of proving and perhaps better understanding Theorem 1.5. It will also serve to help us "compute" three very important "objects" associated with any $(m \times n)$ matrix A, namely, the rank, range, and nullspace of A. These objects may be regarded as manifestations of certain fundamental notions from "abstract" linear algebra which we shall encounter in later sections. It seems conceptually worthwhile to introduce them at this point in the more concrete setting of matrices and n-vectors.

As usual, we denote by $\mathbf{F}$ the set of scalars with which we are working; thus, F stands either for $\mathbf{R}$ (real numbers) or for $\mathbf{C}$ (complex numbers). By $\mathbf{F}^n$, we mean the set of all n-vectors x whose components x_i, $1 \leqslant i \leqslant n$, lie in $\mathbf{F}$. Formally, then,

$$F^n = \{x = \begin{bmatrix} x_1 \\ \cdot \\ \cdot \\ \cdot \\ x_n \end{bmatrix} : x_i \in F,\ 1 \leqslant i \leqslant n\}.$$

Observe that vectors in F^n may be added and multiplied by scalars to generate new vectors in F^n. These two operations give F^n some additional structure which is of central importance.

1.14 Definition: A *subspace* W of F^n is a sub*set* of F^n which is *closed* under the operations of addition and scalar multiplication. □

1.15 Examples:

(a) F^n and $\{0\}$ are both (somewhat trivial) subspaces of F^n.

(b) $\{\begin{bmatrix} \alpha \\ 0 \end{bmatrix} : \alpha \in F\}$ is a subspace of F^2.

(c) Given $x^1, \ldots, x^k \in F^n$, the set W of all vectors of the form $\alpha_1 x^1 + \ldots + \alpha_k x^k$, with $\alpha_i \in F$, $1 \leqslant i \leqslant k$, is a subspace of F^n. □

Example 1.15(c) is particularly important; W in that example is called the subspace of F^n *spanned* by $x^1, \ldots, x^k$. The set $\{x^1, \ldots, x^k\}$ is called a *spanning set* for W. W contains precisely all the *linear combinations* of $x^1, \ldots, x^k$. Note that W itself is also a "spanning set" for W, albeit a large one. By finding the smallest spanning set possible for a given subspace W of F^n, one may give a precise characterization of how "large" the subspace W actually is. To this end, we make a definition.

1.16 Definition: A set $\{x^1, \ldots, x^k\}$ of vectors in F^n is *linearly independent* if and only if the only scalars $\alpha_1, \ldots, \alpha_k$ for which $\alpha_1 x^1 + \ldots + \alpha_k x^k = 0$ are $\alpha_i = 0$, $1 \leqslant i \leqslant k$. □

1.17 Exercise: Show that if $\{x^1, \ldots, x^k\}$ is a linearly independent set of n-vectors and A is an invertible $(n \times n)$

matrix, then $\{Ax^1, ..., Ax^k\}$ is a linearly independent set. [*Hint:* Theorem 1.5 along with the Definition 1.16.] □

1.18 Lemma: *A linearly independent subset of* $\mathbf{F}^n$ *contains at most n vectors.*

Proof: Let $\{x^1, \ldots, x^m\}$ be a set of n-vectors with $m > n$. Let A be the $(n \times m)$ matrix having x^j as its jth column, $1 \leqslant j \leqslant m$. By Lemma 1.12, there exists an m-vector $z \neq 0$ such that $Az = 0$; hence

$$z_1 x^1 + \ldots + z_m x^m = 0 .$$

Since $z \neq 0$, we conclude that $\{x^1, \ldots, x^m\}$ is not a linearly independent set. □

1.19 Definition: Let W be a subspace of $\mathbf{F}^n$. A *basis* for W is a linearly independent ordered set of vectors which is a spanning set for W. [N.b.: we have yet to show that a basis exists for any subspace W.] □

1.20 Theorem: *Every subspace W of* $\mathbf{F}^n$ *possesses a basis* $\{x^1, \ldots, x^k\}$.

Proof: Let x^1 be any nonzero vector in W. Let W_1 be the subspace of $\mathbf{F}^n$ spanned by x^1; i.e., $W_1 = \{\alpha_1 x^1 : \alpha_1 \in \mathbf{F}\}$. Clearly, W_1 is a subspace of W and $\{x^1\}$ is a basis for W_1; if $W_1 = W$, then $\{x^1\}$ is a basis for W. Suppose that a linearly independent set $\{x^1, \ldots, x^l\}$ of vectors in W has been chosen; let W_l be the subspace of W spanned by these vectors. Evidently, $\{x^1, \ldots, x^l\}$ is a basis for W_l; if $W_l \neq W$, choose $x^{l+1} \in W$ which does not lie in W_l. Since x^{l+1} is not in W_l, it is easily seen that $\{x^1, \ldots, x^{l+1}\}$ is a linearly independent set which forms a basis for W_{l+1}. Continue in this fashion until $W_k = W$; by Lemma 1.18, the procedure must terminate for some $k \leqslant n$. □

1.21 Theorem/Definition: *If W is a subspace of* $\mathbf{F}^n$, *then every basis for W contains the same number of vectors. This number is called the* dimension of W.

Proof: Let $\{x^1, \ldots, x^k\}$ and $\{y^1, \ldots, y^l\}$ be bases for W with $k \neq l$; assume $k < l$. Since the y^i form a basis for W, we may express each y^i as a linear combination of the x^j; i.e., there exist scalars a_{ij}, $1 \leqslant i \leqslant l$, $1 \leqslant j \leqslant k$ such that

$$y^i = \sum_{j=1}^{k} a_{ij} x^j , \quad 1 \leqslant i \leqslant l .$$

Let A be the (lxk) matrix satisfying $[A]_{ij} = a_{ij}$. By Lemma 1.12, since $k < l$, there exists an l-vector $c \neq 0$ satisfying $A^T c = 0$; i.e., $\sum_{i=1}^{l} c_i a_{ij} = 0$ for $1 \leqslant j \leqslant k$. But this implies that $\sum_{i=1}^{l} c_i y^i = 0$, contradicting the linear independence of $\{y^1, \ldots, y^l\}$. □

Observe that the ordered set $\{e^1, \ldots, e^n\}$ of n-vectors defined by $[e^i]_j = \delta_{ij}$ forms a basis for $\mathbf{F}^n$; thus, $\mathbf{F}^n$ has dimension n. By Lemma 1.18, every subspace of $\mathbf{F}^n$ has dimension at most n. It is also important to observe that any linearly independent set $\{x^1, \ldots, x^k\}$ of n-vectors which lie in a k-dimensional subspace W of $\mathbf{F}^n$ forms a basis for W. (If the set were not a basis for W, there would exist $x^{k+1} \in W$ such that $\{x^1, \ldots, x^{k+1}\}$ was a linearly independent set, contradicting k-dimensionality of W.) It follows that the only n-dimensional subspace of $\mathbf{F}^n$ is $\mathbf{F}^n$ itself.

Thus the term "basis for a subspace W of $\mathbf{F}^n$" may actually be *defined* as "a maximal linearly independent set of vectors in W." Note also that if V and W are both subspaces of $\mathbf{F}^n$, and W is contained in V, then the dimension of W is less than or equal to the dimension of V. To see this, let k be the dimension of W and let $\{x^1, \ldots, x^k\}$ be a basis for W; $\{x^1, \ldots, x^k\}$ is then a linearly independent set of vectors contained in V. If $W = V$, then $\{x^1, \ldots, x^k\}$ is also a basis for V; if $W \neq V$, then one may use the procedure in the proof of Theorem 1.20 to find $x^{k+1} \in V$ with $\{x^1, \ldots, x^{k+1}\}$ a linearly independent subset of V, implying that the dimension of V is at least $k+1$.

We turn now to a brief discussion of two important subspaces associated with an $(m \times n)$ matrix A.

1.22 Definition: Let A be an $(m \times n)$ matrix with entries in $\mathbf{F}$.

(a) The *nullspace* of A (written $\mathrm{N}(A)$) is the set of all $x \in \mathbf{F}^n$ such that $Ax = 0$.

(b) The *range* of A (written $\mathrm{R}(A)$) is the set of all $y \in \mathbf{F}^m$ y such that $y = Ax$ for some $x \in \mathbf{F}^n$. □

1.23 Exercises:

(a) Show that N(A) and R(A) (where A is an $(m \times n)$ matrix) are indeed *subspaces* of F^n and F^m, respectively.

(b) Let A be an arbitrary $(m \times n)$ matrix; show that R(A) is the subspace of F^m spanned by the columns of A viewed as m-vectors.

(c) If C is an invertible $(m \times m)$ matrix, show that $\mathrm{R}(C) = \mathrm{F}^m$.

(d) If C is any $(m \times m)$ matrix, show that, for every $(m \times n)$ matrix A, R(CA) is precisely the set of all Cy, where $y \in \mathrm{R}(A)$.

(e) Show that if C is invertible, then $\{a^{j_1}, \ldots, a^{j_k}\}$ (a subset of the columns of A) forms a basis for R(A) *if and only if* $\{Ca^{j_1}, \ldots, Ca^{j_k}\}$ forms a basis for R(CA). [*Hint*: Show that the second set is a spanning set for R(CA); then use Exercise 1.17.] □

1.24 Definition: Let A be an $(m \times n)$ matrix. The *rank* of A (written rk(A)) is the dimension of R(A) as a subspace of F^m. □

Observe that $\mathrm{rk}(A) \leqslant m$, by Lemma 1.18. It is also true, as we shall see, that $\mathrm{rk}(A) \leqslant n$; hence the rank of an $(m \times n)$ matrix is at most $\min(m, n)$. Note that if A is $(n \times n)$ and invertible, then, by Exercise 1.23(c), $\mathrm{rk}(A) = n$. An $(m \times n)$ matrix A is said to be of *full rank* if $\mathrm{rk}(A) = \min(m, n)$; in particular, an invertible matrix is of full rank. Our next theorem characterizes the rank, range, and nullspace of a matrix in terms of the echelon form for the matrix; the result may be viewed as giving a means whereby rk(A), N(A), and R(A) may be "computed".

1.25 Theorem: *Let A be an $(m \times n)$ matrix; let M be the echelon form for A. Let $r(A)$ denote the number of nonzero rows in M; let $j_1, \ldots, j_{r(A)}$ be the pivot indices for A. Then:*

(a) $\mathrm{rk}(A) = r(A)$; in fact, the columns of A indexed by $j_1, \ldots, j_{r(A)}$ form a basis for R(A). *(Consequently, $\mathrm{rk}(A) \leqslant \min(m, n)$.)*

(b) The dimension of N(A) *is $(n - r(A))$; i.e., the nullspace of A has dimension equal to the number of columns*

of A not indexed by the pivot indices.

Proof: (a) First, we show that the columns of M indexed by the pivot indices form a basis for R(M). By Exercise 1.23(e), since $A = C^{-1}M$, it may then be concluded that the columns of A indexed by the pivot indices form a basis for R(A), which therefore has dimension $r(A)$. The last statement proves $r(A) = \text{rk}(A)$.

The columns of M indexed by the pivot indices are linearly independent because of the nature of the echelon form. To see this, observe that the $(r(A) \times r(A))$ matrix $\hat{M}$ whose (i, k) element is given by

$$[\hat{M}]_{ik} = m_{i, j_k}, \quad 1 \leqslant k \leqslant r(A)$$

is an upper triangular matrix with nonzero (i, i) elements; saying that $c_1 m^{j_1} + \ldots + c_{r(A)} m^{j_{r(A)}} = 0$ is equivalent to saying $\hat{M}c = 0$, where m^l denotes the lth column of M and c denotes the $r(A)$-vector whose jth entry is c_j. But $\hat{M}c = 0$ readily implies $c = 0$; $\hat{M}$ is already in echelon form and has nonzero diagonal elements (use Theorem 1.10). Thus the columns of M indexed by the pivot indices are linearly independent.

Moreover, it may be seen that an arbitrary column of M not corresponding with a pivot index may be expressed as a linear combination of the columns to its left which *do* correspond with pivot indices. Thus the columns of M corresponding with pivot indices form a basis for the subspace of $\mathbf{F}^m$ spanned by the columns of M, which is R(M) by Exercise 1.23(b), and which in turn is what we wanted.

On, now, to the proof of (b). In the proof of Theorem 1.10, there was constructed an n-vector satisfying $Ax = 0$ by setting $x_{i_0} = 1$, $x_{i_0+1} = \ldots = x_n = 0$, and solving for $x_1, \ldots, x_{i_0-1}$, where i_o was the smallest integer which was *not* a pivot index for A. A similar procedure may be used to find other solutions to $Ax = 0$ by setting, individually, each x_{i_l} *not* corresponding with a pivot index equal to 1, setting other such x_{i_l}'s equal to zero, setting all x_i's with $i \geqslant i_l$ equal to zero, and solving for the unspecified x_i's, all of which correspond to pivot indices, and all of which satisfy $i < i_l$.

There are $(n - r(A))$ vectors so constructed; each lies in N(A), and together they form a linearly independent set, since each contains a 1 in a position where all the others contain zeros.

To show that these vectors form a basis for N(A), it suffices to show that the dimension of N(A) is *at most* $(n - r(A))$. The last assertion is a direct consequence of Lemma 1.18 along with

1.26 Lemma: *If* $\{x^1, \ldots, x^k\}$ *is a linearly independent set of* n*-vectors in* N(A), *and* $z^1, \ldots, z^l$ *are* n*-vectors such that* $\{Az^1, \ldots, Az^l\}$ *is a linearly independent set of* m*-vectors, then* $\{x^1, \ldots, x^k, z^1, \ldots, z^l\}$ *is a linearly independent set of* n*-vectors.*

Proof: (of Lemma 1.26): Assume that $\alpha_1 x^1 + \ldots + \alpha_k x^k + \beta_1 z^1 + \ldots + \beta_l z^l = 0$; multiply both sides by A; obtain $\sum \beta_i A z^i = 0$, implying $\beta_i = 0$. Conclude $\sum \alpha_i x^i = 0$, implying $\alpha_i = 0$ by linear independence of $\{x^1, \ldots, x^k\}$. □

We now complete the proof of Theorem 1.25. The linearly independent set of vectors in N(A) constructed above contains $(n - r(A))$ vectors. If the dimension of N(A) were greater than $(n - r(A))$, then there would exist a linearly independent set of $(n - r(A) + 1)$ vectors in N(A). These vectors, taken together with $e^{j_1}, e^{j_2}, \ldots, e^{j_{r(A)}}$, where $[e^l]_i = \delta_{il}$, as usual, would form a linearly independent set of $(n+1)$ vectors in F^n by Lemma 1.26; to see this, note that $Ae^{j_k} = a^{j_k}$, and $\{a^{j_1}, \ldots, a^{j_{r(A)}}\}$ is a basis for R(A), hence a linearly independent set. Lemma 1.18 asserts that this situation may not occur □

We continue by stating and proving two important results concerning rank. The first states that the rank of the product AB of two matrices A and B may not exceed the rank either of A or of B; the second, which is related to Exercise 1.23(c), asserts that a non-square *full-rank* matrix possesses a certain measure of "invertibility."

1.27 Theorem: *Let A and B be matrices having respective sizes $(m \times n)$ and $(n \times r)$. The rank of the product AB may not exceed the rank either of A or of B; i.e.*

$$\mathrm{rk}(AB) \leqslant \min\{\mathrm{rk}(A), \mathrm{rk}(B)\}.$$

Proof: Observe that the range of AB is contained in the range of A, since any vector of the form ABx is of the form Az, where $z = Bx$. Hence (c.f. the discussion which precedes

Definition 1.22) $\mathrm{rk}(AB)$, which is the dimension of $\mathrm{R}(AB)$, is at most $\mathrm{rk}(A)$, which is the dimension of $\mathrm{R}(A)$; i.e., $\mathrm{rk}(AB) \leqslant \mathrm{rk}(A)$. Moreover, if x is in the nullspace of B, then x is also in the nullspace of AB, since $(AB)x = A(Bx) = 0$. Thus the nullspace of B is contained in the nullspace of AB; hence the dimension of the nullspace of AB, which by Theorem 1.25(b) is $r - \mathrm{rk}(AB)$, is at least as large as $r - \mathrm{rk}(B)$, which is the dimension of the nullspace of B. We see that $\mathrm{rk}(AB) \leqslant \mathrm{rk}(B)$, and the result follows. □

1.28 Theorem: *Let A be an $(m \times n)$ matrix; suppose $m \leqslant n$ and A has (full) rank m. Then there exists an $(n \times m)$ matrix B such that $AB = I_m$.*

Proof: Let $\{e^1, \ldots, e^m\}$ be the "standard basis" for F^m; i.e., $[e^i]_j = \delta_{ij}$, $1 \leqslant i, j \leqslant m$. Since $\mathrm{rk}(A) = m$, we know that the range of A is F^m; hence there exist n-vectors $x^1, \ldots, x^m$ such that $Ax^i = e^i$, $1 \leqslant i \leqslant m$. Let B be the $(n \times m)$ matrix whose ith column is x^i, $1 \leqslant i \leqslant m$; it is easy to check that $AB = I_m$. □

The matrix B constructed in the proof of Theorem 1.28 is called a *right inverse* for A; as it happens, a given $(m \times n)$ matrix A with $\mathrm{rk}(A) = m$ has infinitely many right inverses if $m < n$. If $m = n$, then the right inverse B is uniquely determined, and is simply A^{-1}. Moreover, if $n \leqslant m$ and A is $(m \times n)$ with rank n, then an argument similar to that used in the proof of Theorem 1.28 shows the existence of a *left inverse* for A – namely, an $(n \times m)$ matrix B such that $BA = I_n$.

1.29 Exercise: Carry out the argument alluded to in the final sentence of the preceding paragraph; also, prove the statement about the nonuniqueness of B in Theorem 1.28 when m is less than n. □

We close this section by defining and discussing briefly the determinant of a square matrix A. There are many ways of approaching determinants; we choose to present an inductive definition.

1.30 Definition: The *determinant* of an $(n \times n)$ matrix A is the number $\det(A)$ defined inductively as follows:

(a) The determinant of a (1×1) matrix (i.e. a scalar) α is α itself.

(b) Given an $(n \times n)$ matrix A define Δ_{ij}, for each i, j with $1 \leqslant i, j \leqslant n$, to be the determinant of the $((n-1) \times (n-1))$ matrix obtained by removing from A the ith row and jth column. Then define $\det(A)$ by

$$\det(A) = \sum_{j=1}^{n} (-1)^{1+j} a_{1j} \Delta_{1j} . \qquad \square$$

The Δ_{ij} defined in 1.30(b) is called the "*i,j minor*" of the matrix A. The formula giving $\det(A)$ in terms of its $(1, j)$ minors corresponds to determinant evaluation by means of "expanding along the first row" of A. It may be shown (see, for example, [Strang]) that, for arbitrary i:

$$\det(A) = \sum_{j=1}^{n} (-1)^{i+j} a_{ij} \Delta_{ij} ,$$

and that, in fact, for arbitrary j:

$$\det(A) = \sum_{i=1}^{n} (-1)^{i+j} a_{ij} \Delta_{ij} .$$

These two formulas correspond, respectively, with "expanding along the ith row (respectively, jth column)" of A in order to evaluate $\det(A)$.

It is tedious but fairly straightforward (see [Strang]) to prove the following important property of the determinant.

1.31 Fact: *If A and B are $(n \times n)$ matrices, then* $\det(AB) = \det(A)\det(B)$. □

We may also readily infer from the definition of the determinant that the elementary matrices given in Definition 1.6 have determinants given by:

$$\det E_n(k, l, \alpha) = 1 ;$$

$$\det P_n(k, l) = -1 ;$$

and

$$\det\Gamma_n(k, \alpha) = \alpha .$$

1.32 Exercise: Show, using Definition 1.30, that

(a) For every A, $\det(A) = \det(A^T)$.

(b) If A is a diagonal matrix, then $\det(A)$ is equal to $a_{11}a_{22}\ldots a_{nn}$, the product of the diagonal elements of A.

(c) If A is an upper (lower) triangular matrix – i.e., if $a_{ij} = 0$ for $i > j$ $(i < j)$ – then $\det(A) = a_{11}a_{22}\ldots a_{nn}$.

(d) If M is the echelon form of an $(n \times n)$ matrix A, then for some k, $\det(A) = (-1)^k (\det M)$. □

Exercises 1.32(c) and (d) yield yet another criterion for the invertibility of an $(n \times n)$ matrix A. We have shown (Theorems 1.5 and 1.10) that an $(n \times n)$ matrix A is invertible precisely when M, the echelon form for A, has no zero rows (i.e., $r(A) = n$); this last property is equivalent, by construction of M, to the statement $m_{ii} \neq 0$, $1 \leqslant i \leqslant n$. Since M is upper triangular, $\det(M) = m_{11}m_{22}\ldots m_{nn}$; thus $\det(A) \neq 0$ precisely when $r(A) = n$. That is: *A is invertible if and only if* $\det(A) \neq 0$. A formula for A^{-1} involving determinants may, in fact, be derived from the definitions; see [Strang]. The formula is:

$$A^{-1} = \frac{1}{\det A}\,\mathrm{adj}(A),$$

where $\mathrm{adj}(A)$ is the $(n \times n)$ matrix given by

$$[\mathrm{adj}(A)]_{ij} = (-1)^{i+j}\Delta_{ji},$$

with Δ_{ij} as in Definition 1.30.

We call $\mathrm{adj}(A)$ the *adjugate matrix of A*; observe the order of indices on the right-hand side of the last equation. The quantity $(-1)^{i+j}\Delta_{ij}$ is often called the (i, j) *cofactor* of A; thus, some authors refer to the adjugate matrix as the "transpose of the matrix of cofactors." The formula given above for A^{-1} is an example of a "computationally inefficient" specification of A^{-1} (c.f. the discussion following Example 1.13, along with [Strang]). A more economical method for inverting an invertible $(n \times n)$ matrix is given in the following exercise.

1.33 Exercise: The *Gauss-Jordan method* for inverting an invertible $(n \times n)$ matrix A entails the following sequence of steps:

Step 1: Form the $(n \times 2n)$ matrix B in which the matrix A constitutes the first n columns, and the $(n \times n)$ identity matrix I_n appears as the last n columns.

Step 2: Perform Gauss Elimination on B so as to reduce it to echelon form; the matrix C such that CA is the echelon form for A now appears as the last n columns in the echelon form for B, while the echelon form for A appears as the first n columns in the echelon form for B. Why?

Step 3: Perform Gauss Elimination "from the bottom up" on the echelon form for B so as to make the first n columns a diagonal matrix with nonzero diagonal elements given by the pivots in the echelon form for A; why are these pivots necessarily nonzero?

Step 4: Multiply the matrix obtained in Step 3 on the left by the (easily found) inverse of the $(n \times n)$ diagonal matrix alluded to in Step 3; then the $(n \times n)$ identity matrix I_n appears as the first n columns of the newly obtained $(n \times 2n)$ matrix, and A^{-1} appears as the last n columns; why? □

2. *Linear Differential Equations: Existence and Uniqueness Theorems*

In this section, we prove the fundamental existence and uniqueness theorems for first-order linear ordinary differential equations. The existence proof presented here is constructive; although it does not generalize readily to nonlinear differential equations, it has its advantages over more abstract approaches to the existence theorem. (See, for example, [Vidyasagar], or any other book which uses the *Contraction Mapping Theorem* to prove existence.)

We shall be dealing with real and complex *vector-* and *matrix-valued* functions of a real variable. A real or complex n-vector function x is a mapping from the real numbers into the set of all real or complex n-vectors; we write $x : \mathbf{R} \to \mathbf{R}^n$ or $x : \mathbf{R} \to \mathbf{C}^n$ for such a function; another frequently used notation is $t \to x(t)$. A vector function x may be regarded as a vector each of whose components x_i is an ordinary scalar-valued function defined on R. Similarly, an $(n \times n)$ *matrix function* A is an $(n \times n)$ matrix each of whose elements a_{ij} is a scalar-valued function defined on **R**. A vector function x or matrix function A is said to be continuous or differentiable precisely when each of its component functions x_i or a_{ij} is continuous or differentiable in the usual sense. (See §5 below for an equivalent way of defining continuity of n-vector valued functions.)

In our development of linear system theory, the variable t plays the role of *time;* vector- and matrix-valued functions of t represent the *time-evolution* of important quantities associated with linear systems. Of primary interest to us are real or complex vector differential equations of the form

$$\frac{d}{dt}x(t) = A(t)x(t) + f(t), \quad t \in \mathbf{R} \tag{DE}$$

together with "initial conditions"

$$x(t_o) = x_o \in \mathbf{F}^n, \quad t_o \in \mathbf{R} \tag{IC}$$

where **F** is either **R** or **C**; A is a known $(n \times n)$ matrix function; f is a known n-vector function; $t_o \in \mathbf{R}$ and $x_o \in \mathbf{F}^n$ are given; and $t \to x(t)$ is an unknown differentiable vector function for which we wish to solve. Here, $\frac{d}{dt}x(t)$ denotes the vector function whose ith component is the ordinary derivative $\frac{dx_i}{dt}$; often, we write $\dot{x}(t)$ for $\frac{d}{dt}x(t)$. Observe that (DE) may be regarded as a *system* of n coupled first-order scalar linear differential equations:

$$\dot{x}_i(t) = \sum_{j=1}^{n} a_{ij}(t)x_j(t) + f_i(t), \quad t \in \mathbf{R}, \quad 1 \leqslant i \leqslant n.$$

A physicist or a mathematician might call (DE) a "linear system" for this reason; while our definition of *linear system* will be much more specific, equation (DE) will play a central role.

Let us first review the standard theory of first-order real *scalar* linear differential equations. We focus on

$$\dot{x}(t) = a(t)x(t) + f(t), \quad t \in \mathbf{R} \tag{de}$$

$$x(t_o) = x_o \in \mathbf{R}, \quad t_o \in \mathbf{R}. \tag{ic}$$

Our first result says that, under suitable regularity conditions, any solution x to (de) and (ic) is unique.

2.1 Theorem: *Let a and f be continuous real-valued functions; let $x_o \in \mathbf{R}$, $t_o \in \mathbf{R}$ be given. Pick any $s_1 < t_o$, $t_1 > t_o$.*

Then there exists at most one $x:[s_1, t_1] \to \mathbf{R}$ *which satisfies (de) and (ic) for all* $t \in [s_1, t_1]$.

Proof: Let x and $\hat{x}$ be two such solutions; we show that $x(t) = \hat{x}(t)$ for every $t \in [s_1, t_1]$. Define a function $z:[s_1, t_1] \to \mathbf{R}$ by

$$z(t) = \tfrac{1}{2}[x(t) - \hat{x}(t)]^2, \qquad t \in [s_1, t_1].$$

Then, since both x and $\hat{x}$ satisfy (de) and (ic),we have, for all $t \in [s_1, t_1]$,

$$\begin{aligned} \dot{z}(t) &= [x(t) - \hat{x}(t)][\dot{x}(t) - \tfrac{d}{dt}\hat{x}(t)] \\ &= 2a(t)z(t) \end{aligned}$$

and

$$z(t_o) = x(t_o) - \hat{x}(t_o) = 0.$$

Define a function ρ by

$$\rho(t) = \exp[-2\int_{t_o}^{t} a(\tau)d\tau], \qquad t \in \mathbf{R}.$$

Since a is continuous, the fundamental theorem of calculus implies that

$$\frac{d}{dt}\rho(t) = \frac{d}{dt}[-2\int_{t_o}^{t} a(\tau)d\tau]\rho(t) = -2a(t)\rho(t), \qquad t \in \mathrm{R}.$$

Thus for $t \in [s_1, t_1]$

$$\begin{aligned} \frac{d}{dt}[\rho(t)z(t)] &= \dot{\rho}(t)z(t) + \rho(t)\dot{z}(t) \\ &= -2a(t)\rho(t)z(t) + 2a(t)\rho(t)z(t) \\ &= 0, \end{aligned}$$

and

$$\rho(t_o)z(t_o) = 0.$$

Since $\rho(t)z(t)$ is differentiable on $[s_1, t_1]$ and has zero derivative everywhere, it is a constant; since $\rho(t_o)z(t_o) = 0$, $\rho(t)z(t) = 0$ for all $t \in [s_1, t_1]$. Moreover, $\rho(t) > 0$ for all t implies $z(t) = 0$ for all t, meaning $x(t) = \hat{x}(t)$ for all $t \in [s_1, t_1]$. □

2.2 Exercise: State and prove a complex version of Theorem 2.1. [*Suggestion:* let $z(t) = |x(t) - \hat{x}(t)|^2$; then $\dot{z}(t) = 2\,\mathrm{Re}\{a(t)\}z(t)$, $t \in \mathbf{R}$.] □

A by-product of the proof of Theorem 2.1 is a method for finding the necessarily unique solution to

$$\begin{aligned} \dot{m}(t) &= a(t)m(t), \qquad t \in \mathrm{R} \\ m(t_o) &= m_o \end{aligned} \tag{*}$$

where a is continuous, and $t_o \in \mathbf{R}$, $m_o \in \mathbf{R}$ are given. If $m : \mathrm{R} \to \mathrm{R}$ is a solution to (*), then

$$\frac{d}{dt}[\exp(-\int_{t_o}^{t} a(\tau)d\tau)\, m(t)] = 0, \qquad t \in \mathbf{R},$$

so

$$m(t) = K \exp[\int_{t_o}^{t} a(\tau)d\tau]$$

for some constant K ; since $m(t_o) = m_o$, we conclude that

$$m(t) = m_o \exp[\int_{t_o}^{t} a(\tau)d\tau], \qquad t \in \mathbf{R}.$$

It is readily verified that this function m solves (*).

We adopt the notation

$$\phi(t, s) = \exp[\int_{s}^{t} a(\tau)d\tau], \qquad s, t \in \mathbf{R}.$$

Then the function

$$t \to \phi(t, t_o)$$

may be viewed as the unique solution to (*) above when $m_o = 1$. $\phi(t, t_o)$ is often called an *integrating factor* for the equations (de) and (ic), since if x is a solution to (de) and (ic), one has

$$\begin{aligned} \frac{d}{dt}[\phi^{-1}(t, t_o)x(t)] &= \phi^{-2}(t, t_o)[\phi(t, t_o)\dot{x}(t) - x(t)\frac{d}{dt}\phi(t, t_o)] \\ &= \phi^{-1}(t, t_o)f(t), \qquad t \in \mathbf{R}. \end{aligned}$$

The bracketed quantity, whose value is x_o at $t = t_o$, may then be solved for by a simple *integration* from t_o to t :

$$\phi^{-1}(t, t_o)x(t) = x_o + \int_{t_o}^{t} \phi^{-1}(\tau, t_o)f(\tau)d\tau, \qquad t \in \mathbf{R}.$$

The form of the integrating factor $t \to \phi(t, t_o)$ enables us to obtain a nice formula for $t \to x(t)$ from the last equation. First, note that

$$\begin{aligned}\phi(t\,,t_o)\phi^{-1}(\tau,t_o) &= \phi(t\,,\tau)\\ &= \exp[\int_\tau^t a(\sigma)d\sigma]\end{aligned}$$

for every $t\,,\tau\,,t_o\in \mathbf{R}$; thus

$$x(t)=\phi(t\,,t_o)x_o+\int_{t_o}^{t}\phi(t\,,\tau)f(\tau)d\tau\,,\qquad t\in\mathbf{R}\,.$$

It is readily verified this last equation does indeed give a solution x for (de) and (ic). Observe that this solution is defined *for all* $t\in\mathbf{R}$. The preceding computation is summarized in

2.3 Theorem: *Let a and f be continuous real-valued functions of $t\in\mathbf{R}$. Then there exists, for every $x_o\in\mathbf{R}$, $t_o\in\mathbf{R}$, a (necessarily unique) solution $x:\mathbf{R}\to\mathbf{R}$ to (de) and (ic); it is given by*

$$x(t)=\phi(t\,,t_o)x_o+\int_{t_o}^{t}\phi(t\,,\tau)f(\tau)d\tau,\qquad t\in\mathbf{R},$$

where, for every real t and s, the function $t\to\phi(t\,,s)$ is the unique solution to

$$\begin{aligned}\dot m(t) &= a(t)m(t)\\ m(s) &= 1\,,\end{aligned}$$

and is given explicitly by

$$\phi(t\,,s)=\exp[\int_s^t a(\tau)d\tau]\,,\qquad t\,,s\in\mathbf{R}\,.$$ □

Now, we have a formula for the integrating factor $\phi(t\,,s)$; we arrived at this formula by way of a "trick." A somewhat "better" way of arriving at the last formula is by the method of *Picard Iteration;* we describe it here since it is the method which generalizes to the vector case.

Suppose we wish to solve

$$\begin{aligned}\dot m(t) &= a(t)m(t)\,,\qquad t\in\mathbf{R}\\ m(t_o) &= 1\,.\end{aligned}$$

If we integrate both sides between t_o and t, we obtain (since $m(t_o)=1$):

$$m(t) = 1 + \int_{t_0}^{t} a(\tau)m(\tau)d\tau\,, \qquad t \in \mathbf{R}\,.$$

Next, substitute *this* formula for m where $m(\tau)$ appears in the integral:

$$m(t) = 1 + \int_{t_0}^{t} a(\tau_1)\,[1 + \int_{t_0}^{\tau_1} a(\tau_2)m(\tau_2)d\tau_2]d\tau_1\,.$$

Next substitute *this* formula for m where $m(\tau_2)$ appears in the integral, and so on, and so on, and there appears an "infinite sum" of iterated integrals:

$$1 + \int_{t_0}^{t} a(\tau_1)d\tau_1 + \int_{t_0}^{t}\int_{t_0}^{\tau_1} a(\tau_1)a(\tau_2)d\tau_2 d\tau_1 + \ldots, \qquad t \in \mathbf{R}\,;$$

the k th term is

$$\int_{t_0}^{t}\int_{t_0}^{\tau_1}\cdots\int_{t_0}^{\tau_{k-1}} a(\tau_1)a(\tau_2)\ldots a(\tau_k)d\tau_k d\tau_{k-1}\ldots d\tau_1\,, \qquad t \in \mathbf{R}\,,$$

where the τ_i's are "dummy" variables of integration.

Now, this infinite series has no meaning unless it may be shown to "converge" in some sense. We invoke the following fundamental fact (see, for example, [Simmons]):

2.4 Fact: *(Weierstrass m-Test) Given an "infinite series" of functions* m_k, *defined on* $[s_1, t_1]$:

$$\sum_{k=0}^{\infty} m_k(t) \qquad s_1 \leqslant t \leqslant t_1,$$

suppose that there exist positive constants $\{c_k\}$, $k = 0, 1, \ldots$ *such that*

$$\sum_{k=0}^{\infty} c_k < \infty$$

and

$$|m_k(t)| \leqslant c_k\,, \qquad t \in [s_1, t_1]\,.$$

Then the series $\sum m_k(t)$ *converges uniformly on* $[s_1, t_1]$ *to a function* $m: [s_1, t_1] \to \mathbf{R}$. □

In the present context, the functions m_k are

$$m_k(t) = \int_{t_o}^{t} \cdots \int_{t_o}^{\tau_{k-1}} a(\tau_1)...a(\tau_k) d\tau_k \ldots d\tau_1 , \qquad t \in \mathrm{R} .$$

For the s_1 and t_1-values, just choose any $t_1 > t_o$ and $s_1 < t_o$. We'll now find constants c_k such that

$$|m_k(t)| \leqslant c_k , \qquad t \in [s_1, t_1],$$

and for which the sum of the c_k converges. Note that the multiple integral giving m_k is taken over a variant of the "standard k-simplex" in $(\tau_1, \ldots, \tau_k)$-space; for example, if $k = 2$, we have the following region of integration:

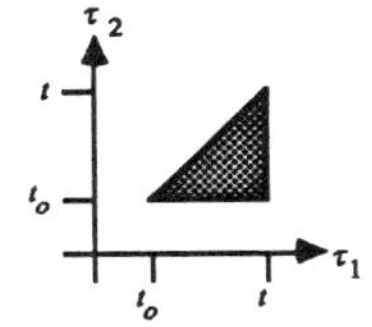

Figure 2.1 A region of integration

The "volume" of the region of integration, (a geometric fact which may be demonstrated via an easy induction) is

$$\frac{|t - t_o|^k}{k!}$$

(I.e., there are $k!$ such simplices in the "k-cube" of side $|t - t_o|$). Moreover, since a is continuous, the integrand

$$a(\tau_1)a(\tau_2) \ldots a(\tau_k)$$

is clearly bounded in magnitude for $t \in [s_1, t_1]$ as follows:

$$|a(\tau_1)a(\tau_2) \ldots a(\tau_k)| \leqslant [\max_{s_1 \leqslant t \leqslant t_1} |a(t)|]^k = \gamma^k ,$$

where γ is defined by the last equality. Thus, for $t \in [s_1, t_1]$:

$$\begin{aligned} |m_k(t)| &= |\int_{t_o}^{t} \cdots \int_{t_o}^{\tau_{k-1}} a(\tau_1) \ldots a(\tau_k) d\tau_k \ldots d\tau_1| \\ &\leqslant \int_{s_1}^{t_1} \cdots \int_{s_1}^{\tau_{k-1}} |a(\tau_1) \ldots a(\tau_k)| d\tau_k \ldots d\tau_1 \\ &\leqslant \int_{s_1}^{t} \cdots \int_{s_1}^{\tau_{k-1}} \gamma^k \, d\tau_k \ldots d\tau_1 \\ &\leqslant \gamma^k [\int_{s_1}^{t_1} \cdots \int_{s_1}^{\tau_{k-1}} d\tau_k \ldots d\tau_1] = \frac{\gamma^k}{k!}(t_1 - s_1)^k . \end{aligned}$$

Now, just set $c_k = \dfrac{\gamma^k (t_1 - s_1)^k}{k!}$; then

$$|m_k(t)| \leqslant c_k\,, \qquad t \in [s_1, t_1]\,,$$

and

$$\sum_{k=0}^{\infty} c_k = e^{\gamma(t_1 - s_1)} < \infty.$$

Thus the series $\sum_{k=0}^{\infty} m_k(t)$, $t \in [s_1, t_1]$, converges uniformly to some function m defined on $[s_1, t_1]$.

Furthermore, one may find a simple formula for the limiting function m. Since each integrand

$$a(\tau_1)a(\tau_2)\ldots a(\tau_k)$$

is *symmetric* in the variables $\tau_1, \ldots, \tau_k$, its integral over the indicated region is just $1/k!$ times its integral over the whole "cube" $[t_o, t] \times [t_o, t] \times \ldots \times [t_o, t]$ (k times); but for every $t \in \mathrm{R}$,

$$\begin{aligned}\int_{t_o}^{t} \cdots \int_{t_o}^{t} a(\tau_1)a(\tau_2)\ldots a(\tau_k)d\tau_k \ldots d\tau_1 \\ = [\int_{t_o}^{t} a(\tau_1)d\tau_1]\,[\int_{t_o}^{t} a(\tau_1)d\tau_2]\ldots[\int_{t_o}^{t} a(\tau_k)d\tau_k] \\ = [\int_{t_o}^{t} a(\tau)d\tau]^k\,.\end{aligned}$$

Hence

$$m_k(t) = \frac{1}{k!}[\int_{t_o}^{t} a(\tau)d\tau\,]^k\,, \qquad t \in \mathrm{R}\,.$$

Thus

$$m(t) = \sum_{k=0}^{\infty} \frac{1}{k!}\,[\int_{t_o}^{t} a(\tau)d\tau\,]^k = \exp[\int_{t_o}^{t} a(\tau)d\tau\,]\,, \qquad t \in \mathrm{R}\,,$$

which is just our old formula.

We turn now to the vector equations

$$\dot{x}(t) = A(t)x(t) + f(t)\,, \quad t \in \mathrm{R} \qquad \text{(DE)}$$

$$x(t_o) = x_o\,, \quad t_o \in \mathrm{R}, \quad x_o \in \mathrm{R}^n\,. \qquad \text{(IC)}$$

Here, A is an $(n \times n)$ matrix of functions and f is an n-vector of functions. We mimic the scalar strategy: first, we show that if a solution exists then it must be unique.

2.5 Theorem: *Let $t_o \in \mathbf{R}$, $x_o \in \mathbf{R}^n$ be given; let $s_1 < t_o$, $t_1 > t_o$ be arbitrary. Let A be a continuous real $(n \times n)$ matrix function and let f be a continuous real n-vector function, each defined on $[s_1, t_1]$. Then there exists at most one function $x : [s_1, t_1] \to \mathbf{R}^n$ which satisfies*

$$\dot{x}(t) = A(t)x(t) + f(t) \qquad (DE)$$

$$x(t_o) = x_o \qquad (IC)$$

for all $t \in [s_1, t_1]$.

Proof: Suppose x and $\hat{x}$ are two such solutions; define $z : [s_1, t_1] \to \mathbf{R}$ by

$$z(t) = [x(t) - \hat{x}(t)]^T [x(t) - \hat{x}(t)], \; t \in [s_1, t_1].$$

Then $z(t)$, for each t, is simply the sum of the squares of the component functions of the difference vector $x(t) - \hat{x}(t]$. Now, take $\dot{z}(t)$:

$$\dot{z}(t) = [x - \hat{x}]^T [\dot{x} - \tfrac{d}{dt}\hat{x}] + [\dot{x} - \tfrac{d}{dt}\hat{x}]^T [x - \hat{x}].$$

Since both terms are scalars, they are "symmetric matrices;" i.e.,

$$[x - \hat{x}]^T [\dot{x} - \tfrac{d}{dt}\hat{x}] = [\dot{x} - \tfrac{d}{dt}\hat{x}]^T [x - \hat{x}],$$

so that

$$\dot{z}(t) = 2[x(t) - \hat{x}(t)]^T [\dot{x}(t) - \tfrac{d}{dt}\hat{x}(t)].$$

That is:

$$\dot{z}(t) = 2[x(t) - \hat{x}(t)]^T A(t)[x(t) - \hat{x}(t)], \qquad t \in \mathbf{R},$$

and

$$z(t_o) = 0.$$

Now, the term on the right-hand side of the first equation is

$$2[x(t) - \hat{x}(t)]^T A(t)[x(t) - \hat{x}(t)] = 2 \sum_{i=1}^{n} \sum_{j=1}^{n} a_{ij}(x - \hat{x})_i (x - \hat{x})_j \,.$$

Note that the t-dependence of the terms on the right-hand side of the last equation has been suppressed. There are n^2 terms in the sum; we now obtain an upper bound on each term as a function of t. First, observe that for any n-vector z, it is true that $|z_i z_j|$ is bounded above by the sum of the squares of the z_i from 1 to n. Thus,

$$|a_{ij}(x-\hat{x})_i(x-\hat{x})_j| \leqslant \max_{ij}|a_{ij}(t)| \sum_{k=1}^{n} [x(t)-\hat{x}(t)]_k{}^2$$
$$= \max_{i,j}|a_{ij}(t)|z(t).$$

Thus (summing up):

$$\dot{z} \leqslant 2n^2 \max_{i,j}|a_{ij}(t)|z(t), \quad t \in \mathbf{R},$$

and

$$z(t_o) = 0.$$

Now let $a : \mathbf{R} \to \mathbf{R}$ be defined by

$$a(t) = 2n^2 \max_{i,j}|a_{ij}(t)|, \quad t \in \mathbf{R}.$$

Multiplying both sides by the *positive* function ρ given by

$$\rho(t) = \exp[-\int_{t_o}^{t} a(\tau)d\tau], \quad t \in \mathbf{R},$$

we get (since $\dot{z} \leqslant az$, $\dot{\rho} = a\rho$):

$$\frac{d}{dt}[\rho(t)z(t)] = -a(t)\rho(t)z(t) + \rho(t)\dot{z}(t)$$
$$\leqslant -a\rho z + a\rho z = 0,$$

or

$$\frac{d}{dt}[\rho(t)z(t)] \leqslant 0, \quad t \in [s_1, t_1],$$

$$\rho(t_o)z(t_o) = 0.$$

Thus (since $z(t) \geqslant 0$ by definition), $\rho(t)z(t)$ is a nonnegative function, with value zero at t_o, which has a non-positive derivative; hence it may not be increasing with t. We conclude that, for all t between t_o and t_1,

$$\rho(t)z(t) = 0.$$

A symmetric argument yields the result for $s_1 \leqslant t \leqslant t_o$. □

2.6 Exercise: Complete the proof of Theorem 2.5; i.e., prove $x(t) = \hat{x}(t)$ for all $t \in [s_1, t_o]$. □

Thus we have for vector differential equations a result analogous to Theorem 2.1. To prove that a solution x exists (and get a formula for x similar to that in Theorem 2.3), we follow a strategy similar to our earlier one. First, we solve the equations

$$\begin{aligned} \frac{d}{dt} M(t) &= A(t)M(t) \\ M(t_o) &= I_n \end{aligned} \qquad (**)$$

for an *invertible* $(n \times n)$ matrix function $t \to \Phi(t, t_o)$. (Invertibility corresponds to non-zeroness of $\phi(t, t_o)$ in the scalar case.) Then we examine

$$\begin{aligned} \dot{x}(t) &= A(t)x(t) + f(t), \qquad t \in \mathrm{R} \\ x(t_o) &= x_o . \end{aligned}$$

Just as in the scalar case, note that

$$\frac{d}{dt}[\Phi^{-1}(t, t_o)x(t)] = \frac{d}{dt}[\Phi^{-1}(t, t_o)]x(t) + \Phi^{-1}(t, t_o)\dot{x}(t).$$

Now, since

$$\frac{d}{dt}[\Phi^{-1}(t, t_o)\Phi(t, t_o)] = \frac{d}{dt}[I_n] = 0,$$

we conclude that

$$\begin{aligned} \frac{d}{dt}\Phi^{-1}(t, t_o) &= -\Phi^{-1}(t, t_o)\frac{d}{dt}[\Phi(t, t_o)]\Phi^{-1}(t, t_o) \\ &= -\Phi^{-1}(t, t_o)A(t)\Phi(t, t_o)\Phi^{-1}(t, t_o). \end{aligned}$$

Thus

$$\begin{aligned} \frac{d}{dt}[\Phi^{-1}(t, t_o)x(t)] &= -\Phi^{-1}(t, t_o)A(t)\Phi(t, t_o)\Phi^{-1}(t, t_o)x(t) \\ &\qquad + \Phi^{-1}(t, t_o)[A(t)x(t) + f(t)] \\ &= \Phi^{-1}(t, t_o)f(t). \end{aligned}$$

Hence, we are reduced to an equation which may be solved for the bracketed quantity on the left by an integration:

$$\begin{aligned} \Phi^{-1}(t, t_o)x(t) &= \Phi^{-1}(t_o, t_o)x(t_o) + \int_{t_o}^{t} \Phi^{-1}(\tau, t_o)f(\tau)d\tau \\ &= x_o + \int_{t_o}^{t} \Phi^{-1}(\tau, t_o)f(\tau)d\tau, \end{aligned}$$

or:

$$x(t) = \Phi(t, t_o)x_o + \int_{t_o}^{t} \Phi(t, t_o)\Phi^{-1}(\tau, t_o)f(\tau)d\tau \qquad (***)$$

We turn now to the solution of equations (**) above; performing Picard Iteration, as before, we obtain an infinite series of matrix functions

$$I_n + \int_{t_o}^{t} A(\tau_1)d\tau_1 + \int_{t_o}^{t}\int_{t_o}^{\tau_1} A(\tau_1)A(\tau_2)d\tau_2 d\tau_1 + \ldots, \qquad t \in \mathbf{R},$$

where the k th integral is

$$M_k(t) = \int_{t_o}^{t}\int_{t_o}^{\tau_1}\cdots\int_{t_o}^{\tau_{k-1}} A(\tau_1)A(\tau_2)\ldots A(\tau_k)d\tau_k \ldots d\tau_1, \qquad t \in \mathbf{R}.$$

To show that the series converges, we'll show that, for every t in any given interval $[s_1, t_1]$ with $s_1 < t_o < t_1$, $M_k(t)$ satisfies

$$\max_{i,j} |[M_k(t)]_{ij}| \leqslant c_k$$

for a sequence of constants c_k for which

$$\sum_{k=0}^{\infty} c_k < \infty .$$

The Weierstrass m-test will then imply that for each i, j, $1 \leqslant i, j \leqslant n$,the series

$$\sum_{k=0}^{\infty} [M_k(t)]_{ij}$$

converges uniformly on $[s_1, t_1]$ to a function $t \to [M(t)]_{ij}$. It is clear that for every $\tau_1, \ldots, \tau_k \in [s_1, t_1]$,

$$\max_{i,j} |[A(\tau_1)A(\tau_2)]_{ij}| \leqslant n \max_{i,j} |[A(\tau_1)]_{ij}| \max_{i,j} |[A(\tau_2)]_{ij}| ,$$

and similarly that (by induction)

$$\max_{i,j} |[A(\tau_1)\ldots A(\tau_k)]_{ij}|$$

is bounded above by

$$n^{k-1} \max_{i,j} |[A(\tau_1)]_{ij}| \ldots \max_{i,j} |[A(\tau_k)]_{ij}|$$

Define a new function $t \to a(t)$, $t \in \mathbf{R}$, via

$$a(t) = \max_{i,j} |[A(t)]_{ij}| , \qquad t \in \mathbf{R}.$$

Then

$$\begin{aligned}
\max_{i,j} |[M_k(t)]_{ij}| &= \max_{i,j} \left| \int_{t_o}^{t} \ldots \int_{t_o}^{\tau_{k-1}} A(\tau_1)\ldots A(\tau_k) d\tau_k \ldots d\tau_1 \right| \\
&\leqslant \max_{i,j} \int_{s_1}^{t} \ldots \int_{s_1}^{\tau_{k-1}} |[A \ldots A]_{ij}| d\tau_k \ldots d\tau_1 \\
&\leqslant \int_{s_1}^{t_1} \ldots \int_{s_1}^{\tau_{k-1}} n^{k-1} a(\tau_1)\ldots a(\tau_k) d\tau_k \ldots d\tau_1 \\
&= \frac{n^{k-1}}{k!} \left[\int_{s_1}^{t_1} a(\tau) d\tau\right]^k = c_k .
\end{aligned}$$

The lower limit t_o metamorphosed to s_1 because the first inequality would not have held otherwise for $t < t_o$. Now we need to prove that $\sum_{k=0}^{\infty} c_k < \infty$. To see this, note that

$$\sum_{k=0}^{\infty} c_k = \frac{1}{n} \sum_{k=0}^{\infty} \frac{n^k}{k!} \left[\int_{s_1}^{t_1} a(\tau) d\tau\right]^k = \frac{1}{n} \exp\left[n \int_{s_1}^{t_1} a(\tau) d\tau\right] < \infty .$$

The Weierstrass m-test then implies that, for each i, j with $1 \leqslant i, j \leqslant n$, the infinite series

$$\sum_{k=0}^{\infty} [M_k(t)]_{ij} , \qquad t \in [s_1, t_1],$$

converges uniformly to a function $M_{ij} : [s_1, t_1] \to \mathbf{R}$. To see this, observe that for every i, j and $t \in [s_1, t_1]$

$$|[M_k(t)]_{ij}| \leqslant \max_{i,} j \; |[M_k(t)]_{ij}| \leqslant c_k .$$

The series defining $t \to M(t)$ has a name; it is called the *Peano-Baker series for A.*

2.7 Theorem: *Let A be a continuous $(n \times n)$ matrix function defined for all $t \in \mathbf{R}$; let $s \in \mathbf{R}$ be given. Then there exists a unique $(n \times n)$ matrix function $t \to \Phi(t, s)$ defined for every $t \in \mathbf{R}$ which satisfies*

$$\begin{aligned}
\dot{M}(t) &= A(t) M(t) \qquad t \in \mathbf{R} \\
M(s) &= I_n
\end{aligned}$$

The solution is given for all t by the Peano-Baker series

$$\Phi(t, s) = \sum_{k=0}^{\infty} M_k(t), \qquad t \in \mathbf{R},$$

where

$$M_k(t) = \int_s^t \int_s^{\tau_1} \cdots \int_s^{\tau_{k-1}} A(\tau_1)A(\tau_2)\ldots A(\tau_k)d\tau_k \ldots d\tau_1 .$$

Proof: Our earlier work shows, since $s_1 < s$, $t_1 > s$ were arbitrary, that the Peano-Baker series does indeed sum to a an $(n \times n)$ matrix function $t \to \Phi(t, s)$ defined for every $t \in \mathbf{R}$. It remains to show that $t \to \Phi(t, s)$ satisfies the differential equation and initial condition, and is unique in that regard.

Observe that, by continuity of A:

$$\begin{aligned}\dot{M}_k(t) &= A(t) \int_s^{\tau_1} \cdots \int_s^{\tau_{k-1}} A(\tau_2)\ldots A(\tau_k)d\tau_k \ldots d\tau_2 \\ &= A(t)M_{k-1}(t).\end{aligned}$$

Thus

$$\sum_{k=0}^{N} \dot{M}_k(t) = A(t)\sum_{k=0}^{N-1} M_k(t), \qquad t \in \mathbf{R},$$

and so, since A is continuous and since the M_k-series converges uniformly on any interval $[s_1, t_1]$ containing $[s, t]$, the series

$$\sum_{k=1}^{\infty} \dot{M}_k(t), \qquad t \in [s_1, t_1],$$

converges uniformly to an $(n \times n)$ matrix function $t \to \dot{M}(t)$ defined on $[s_1, t_1]$. Since the convergence is uniform, the function $\dot{M}(t)$ is the derivative of the function $t \to \Phi(t, s)$ for $t \in [s_1, t_1]$. Observe that

$$\dot{M}(t) = \sum_{k=0}^{\infty} \dot{M}_k(t) = A(t)M(t).$$

(A term-by-term analysis of the series makes the last equation apparent.) Since s_1 and t_1 were arbitrary, we conclude that $t \to \Phi(t, s)$ is a solution to the differential equation of (**); a glance at the Peano-Baker series shows that $\Phi(s, s) = I_n$, so that $t \to \Phi(t, s)$ satisfies the initial condition.

Uniqueness of $t \to \Phi(t, s)$ is an elementary consequence of Theorem 2.5; the differential equation and initial condition in the statement of Theorem 2.7 may be regarded as specifying separate n-vector equations (for the columns of M) the solutions to which we know are unique. □

In order to apply Theorem 2.7 to the solution of our original problem (DE) and (IC), we must show that $\Phi(t, t_o)$ is invertible for every $t \in \mathbf{R}$.

2.8 Theorem: *Let the* $(n \times n)$ *matrix function* $t \rightarrow \Phi(t, t_o)$, $t \in \mathbf{R}$, *satisfy*

$$\frac{d}{dt}\Phi(t, t_o) = A(t)\Phi(t, t_o); \quad \Phi(t_o, t_o) = I_n .$$

Then $\Phi(t, t_o)$ *is invertible for all* $t \in \mathbf{R}$*; in fact,*

$$\det\Phi(t, t_o) = \exp\left[\int_{t_o}^{t} \mathrm{Tr}(A(\tau))d\tau\right] > 0, \quad t \in \mathbf{R},$$

where $\mathrm{Tr}(A(t)) = \sum_{i=1}^{n} [A(t)]_{ii}$.

Proof:

We show first that, for every $t, s, \tau \in \mathbf{R}$,

$$\Phi(t, s)\Phi(s, \tau) = \Phi(t, \tau) \qquad \text{(SG)}$$

by proving that the two sides of the identity satisfy the same differential equation and initial condition; it follows that they are the same by *uniqueness*. Regarding s and τ as fixed for the moment, denote by $L(t)$ the left-hand side and by $R(t)$ the right-hand side.

Taking derivatives with respect to t yields

$$\begin{aligned} \frac{d}{dt}L(t) &= \frac{d}{dt}[\Phi(t, s)\Phi(s, \tau)] \\ &= A(t)[\Phi(t, s)\Phi(s, \tau)] = A(t)L(t), \end{aligned}$$

and

$$\frac{d}{dt}R(t) = \frac{d}{dt}[\Phi(t, \tau)] = A(t)[\Phi(t, \tau)] = A(t)R(t).$$

Also, at $t = s$, we have

$$L(s) = \Phi(s, s)\Phi(s, \tau) = \Phi(s, \tau) = R(s).$$

Thus both sides of (SG), when regarded as functions of t, satisfy the same differential equation plus initial conditions, hence must be equal by uniqueness.

The very important identity (SG) - known as the *semi-group property* - also implies invertibility of $\Phi(t, t_o)$ for every (t, t_o). Just take $\tau = t$ in (SG):

$$\Phi(t,s)\Phi(s,t) = \Phi(t,t) = I_n, \qquad s,t \in \mathrm{R}.$$

Thus $\Phi^{-1}(t,s) = \Phi(s,t)$ for every $t,s \in \mathrm{R}$; in particular, $\Phi(t,t_o)$ is always invertible.

We now derive a formula for $\det\Phi(t,t_o)$. The formula is:

$$\det\Phi(t,t_o) = \exp\left[\int_{t_o}^{t} \mathrm{Tr}(A(\tau))d\tau\right],$$

where

$$\mathrm{Tr}(A(\tau)) = \sum_{i=1}^{n} [A(\tau)]_{ii}$$

is the sum of the diagonal elements in $A(\tau)$. (Tr stands for "trace.") To derive the formula, we'll show that $\det\Phi$ satisfies:

$$\frac{d}{dt}[\det[\Phi(t,t_o)]] = \mathrm{Tr}[A(t)]\det[\Phi(t,t_o)]$$

$$\det[\Phi(t_o,t_o)] = 1,$$

from which the equation above follows by way of the discussion after Theorem 2.3.

We employ some standard notation: the symbol $O(h)$ denotes any quantity which remains bounded as $h \to 0$ *after* being divided by h, while $o(h)$ denotes any quantity which *goes to zero* as $h \to 0$ after being divided by h. For example, h^2 is $o(h)$; h is $O(h)$ but not $o(h)$. Note that $hO(h)$ is $o(h)$.

Recall the definition of the derivative:

$$\frac{d}{dt}[\det\Phi(t,t_o)] = \lim_{h \to 0} \frac{\det\Phi(t+h,t_o) - \det\Phi(t,t_o)}{h}.$$

Using the semigroup property (SG) and Fact 1.31 on the determinant of a product, we get

$$\frac{d}{dt}[\det\Phi(t,t_o)] = \lim_{h \to 0}\left\{\frac{\det\Phi(t+h,t) - 1}{h}\right\}\det\Phi(t,t_o).$$

We'll show now that the quantity $\lim\{\ \}$ is equal to the trace of $A(t)$. Begin expanding $\Phi(t+h,t)$ in a Taylor series about $h = 0$ to get:

$$\Phi(t+h,t) = I_n + hA(t) + o(h).$$

It follows by an easy induction using Definition 1.30 of the determinant that for any $(n \times n)$ matrix B:

$$\det(I_n + hB) = 1 + h\,\mathrm{Tr}(B) + o(h).$$

Thus:

$$\begin{aligned}\det\Phi(t+h\,,t) &= \det[I_n + h(A(t) + O(h))] \\ &= 1 + h[\mathrm{Tr}(A(t)) + O(h)] + o(h) \\ &= 1 + h\,\mathrm{Tr}(A(t)) + o(h).\end{aligned}$$

Hence

$$\begin{aligned}\lim_{h\to 0}\left\{\frac{\det\Phi(t+h\,,t)-1}{h}\right\} &= \lim_{h\to 0}\left\{\frac{h\,\mathrm{Tr}(A(t)) + o(h)}{h}\right\} \\ &= \mathrm{Tr}(A(t)).\end{aligned}$$

We have arrived at

$$\frac{d}{dt}[\det\Phi(t\,,t_o)] = \mathrm{Tr}(A(t))\det\Phi(t\,,t_o)$$

$$\det\Phi(t_o\,,t_o) = 1,$$

which gives, for all $t \in \mathbf{R}$:

$$\det\Phi(t\,,t_o) = \exp\left[\int_{t_o}^{t} Tr(A(\tau))d\tau\right] > 0. \qquad \square$$

We may now apply Theorems 2.7 and 2.8 to the solution of (DE) and (IC). The entire discussion following Theorem 2.5 may be summarized as follows.

2.9 Theorem: *Let $t \to A(t)$ be a continuous $(n \times n)$, real-valued matrix function and let $t \to f(t)$ be a continuous n-vector function, both defined for $t \in \mathbf{R}$. Let $t_o \in \mathbf{R}$ and $x_o \in \mathbf{R}^n$ be given. Then there exists a unique n-vector function $t \to x(t)$, defined for every $t \in \mathbf{R}$, which satisfies*

$$\dot{x}(t) = A(t)x(t) + f(t), \quad t \in \mathbf{R} \qquad \text{(DE)}$$

$$x(t_o) = x_o \qquad \text{(IC)}$$

The solution is given by

$$x(t) = \Phi(t\,,t_o)x_o + \int_{t_o}^{t} \Phi(t\,,\tau)f(\tau)d\tau, \quad t \in \mathbf{R},$$

where $t \to \Phi(t\,,s)$ is the unique $(n \times n)$ matrix solution, given by Theorem 2.5, to the following matrix differential equation:

$$\dot{M}(t) = A(t)M(t), \qquad t \in \mathbf{R}$$
$$M(s) = I_n .$$

Proof: It suffices, by Theorem 2.5, to show that $x(t)$ given above, whose definition was prompted by the discussion following Exercise 2.6, satisfies (DE) and (IC). This, however, is a straightforward consequence of the definition of $\Phi(t, s)$ and continuity of f. Observe that the formula given here for $x(t)$ is the same as that in equation (***) because of (SG) in the proof of Theorem 2.8. □

2.10 Exercise: State and prove results analogous to Theorems 2.5, 2.7, 2.8, and 2.9 when the functions A and f are allowed to take on complex values. □

In general, the Peano-Baker series is the best one can do with regard to writing down a formula for $\Phi(t, t_o)$, which many authors call the *fundamental matrix* or *transition matrix* corresponding to (DE). When the matrix function $t \to A(t)$ is a *constant* function, however, the Peano-Baker series takes an especially simple form. Since most of our attention later on will be directed at system theoretic applications of Theorem 2.9, and especially at the case where A is constant, the ensuing discussion is crucial.

First, let us say a few words about the "transition" matrix function $t \to \Phi(t, t_o)$. Given (DE), $\Phi(t, t_o)$ describes the way in which the initial condition (IC), which is specified at "time" t_o, determines the value at time t of the solution $t \to x(t)$ to (DE) and (IC). Suppose for the moment that $f(t) = 0$ for all t. Then the value at any time t_1 of solution to (DE) and (IC) is

$$x_1 = x(t_1) = \Phi(t_1, t_o)x_o .$$

Suppose that x_1 is used as initial condition for

$$\dot{x}(t) = A(t)x(t), \qquad t \in \mathbf{R}$$
$$x(t_1) = x_1 .$$

Solving these equations may be regarded as *extending* the solution to (DE) and (IC) beyond t_1, but there is a better system-theoretic interpretation. Observe that the value at any time t_2 of the solution to the last pair of equations is given by

$$\begin{aligned} x_2 &= x(t_2) \\ &= \Phi(t_2, t_1)x_1 \\ &= \Phi(t_2, t_1)\Phi(t_1, t_o)x_o \,. \end{aligned}$$

The semigroup property (SG) in the proof of Theorem 2.8 yields

$$\begin{aligned} x_2 &= \Phi(t_2, t_1)\Phi(t_1, t_o)x_o \\ &= \Phi(t_2, t_o)x_o \,. \end{aligned}$$

In other words, as one might expect, if we "follow" the solution to (DE) and (IC) for "time" (t_2-t_o), we "get to the same place" in $\mathbf{R}^n$ as if we had followed it for time (t_1-t_o), rested for a moment, and subsequently followed it for time (t_2-t_1). The last assertion remains true when $f(t)$ is nonzero. Such a *geometric* interpretation of the semigroup property will be of great importance later on.

Finally, we consider the situation where the matrix function $t \to A(t)$ is constant; i.e., $A(t) \equiv A$, a constant $(n \times n)$ matrix, for every $t \in \mathbf{R}$. The kth term in the Peano-Baker series defining $\Phi(t, s)$ becomes

$$\int_s^t \int_s^{\tau_1} \cdots \int_s^{\tau_{k-1}} A^k \, d\tau_1 d\tau_2 \dots d\tau_k = \frac{(t-s)^k A^k}{k!} \,.$$

Thus

$$\Phi(t, s) = \sum_{k=0}^{\infty} \frac{[(t-s)A]^k}{k!} \,.$$

We denote the infinite sum in this case by $\exp[(t-s)A]$ or $e^{(t-s)A}$ in analogy with the power series expansion for the scalar exponential function. Observe that for every real t, τ, and s, we have

$$\Phi(t+\tau, s+\tau) = \Phi(t, s) = \exp[(t-s)A] \,.$$

Thus, when A is constant, the transition matrix is invariant under an equal shift of initial and final times. In later sections, we shall present a detailed discussion of the *matrix exponential function* $t \to \exp[tA]$ and its central importance in linear system theory.

3. *Linear Difference Equations*

This brief section treats linear difference equations, which play a role in the theory of discrete-time linear systems similar to that played by linear differential equations in the theory of continuous-time linear systems. The existence and uniqueness theorems for difference equations are much easier than those for differential equations; as we shall see, there is essentially "nothing to prove." On the other hand, there is a certain temporal asymmetry associated with difference equations which will manifest itself later on as a stumbling block that prevents a fully parallel discussion of continuous- and discrete-time linear systems.

First, we establish some notation. Denote by Z the set of integers $\{\ldots, -2, -1, 0, 1, 2, 3, \ldots\}$. Integer variables will be denoted by letters such as k and l; a (real or complex) *discrete-time n-vector function* $k \to x(k)$ is a mapping which assigns to each integer k a (real or complex) n-vector $x(k)$. Another common notation is $x : \mathrm{Z} \to \mathbf{F}^n$, where $\mathbf{F}$ is either $\mathbf{R}$ or $\mathbf{C}$. Discrete-time matrix functions $k \to A(k)$ are defined similarly. We shall be concerned with the following equations:

$$x(k+1) = A(k)x(k) + f(k), \quad k \geqslant k_o \qquad \text{(DfcE)}$$

$$x(k_o) = x_o \qquad \text{(IC)}$$

where $k \to f(k)$ is a known discrete-time n-vector function, $k \to A(k)$ is a known discrete-time $(n \times n)$ matrix function, and $x_o \in \mathbf{F}^n$ is given along with $k_o \in \mathrm{Z}$. The unknown n-vector discrete-time function $k \to x(k)$ is to be solved for.

Equations (DfcE) and (IC) may be interpreted as specifying an *algorithm* for computing $x(k)$, $k \geqslant k_o$, given $x(k_o) = x_o$. Such an interpretation enables one to see that existence and uniqueness of solutions to (DfcE) and (IC) are indeed trivialities. We do, however, present the results in a manner which should reveal parallels with the development in §2.

To begin with, observe that (DfcE) and (IC) yield immediately that

$$x(k_o+1) = A(k_o)x_o + f(k_o).$$

This last equation may now be inserted in (DfcE) to give

$$x(k_o+2) = A(k_o+1)A(k_o)x_o + A(k_o+1)f(k_o) + f(k_o+1).$$

Continuing in this fashion, we obtain

$$x(k) = \Phi(k, k_o)x_o + \sum_{l=k_o}^{k-1} \Phi(k, l+1)f(l), \quad k \geqslant k_o, (*)$$

where $\Phi(k, k_o)$ is defined as

$$\Phi(k, l) = \begin{cases} A(k-1)A(k-2)\ldots A(l) & k > l \\ I_n & k = l. \end{cases}$$

Observe that $(k, l) \rightarrow \Phi(k, l)$ is an $(n \times n)$ matrix function of two integer variables which is defined only for pairs (k, l) satisfying $k \geqslant l$. The domain of Φ could be extended to contain all integer pairs by putting

$$\Phi(k, l) = 0, \quad k < l,$$

but such an extended definition doesn't prove to be especially useful. Observe that $k \rightarrow \Phi(k, k_o)$, $k \geqslant k_o$, is the unique $(n \times n)$ matrix solution to the equations

$$M(k+1) = A(k)M(k), \quad k \geqslant k_o$$

$$M(k_o) = I_n.$$

Note the similarity between this last pair of equations and Equations (**) in §2; note also the resemblance which the formula (*) for $x(k)$, $k \geqslant k_o$, bears to its continuous-time counterpart in Theorem 2.9.

The fundamental difference between the results of the present section and those of §2 is that *the solution to a difference equation is defined only for times greater than or equal to the initial time*. Consider the following example:

$$x(k+1) = \begin{bmatrix} 0 & 1 \\ 0 & 0 \end{bmatrix} x(k) + \begin{bmatrix} k \\ 0 \end{bmatrix}, \quad k \geqslant k_o$$

$$x(k_o) = \begin{bmatrix} 0 \\ 1 \end{bmatrix}.$$

Clearly, the difference equation and initial condition are not satisfied for any choice of $x(k_o-1)$. Algebraically, the key to being able to define the solution to a differential equation with initial condition *for all* $t \in \mathbf{R}$ is the invertibility for all t of the transition matrix $\Phi(t, t_o)$. The "transition matrix" $\Phi(k, k_o)$ for a difference equation need not, on the other hand, be invertible for *any* value of k.

3.1 Exercise: Show that if $A(k)$ is invertible for every $k \in \mathbf{Z}$, then there exists a discrete-time n-vector function $k \to x(k)$ defined for *all* $k \in \mathbf{Z}$ and satisfying (DfcE) and (IC).□

As in the case of differential equations, if the matrix function $k \to A(k)$ is a constant function, then the formulas for $\Phi(k, l)$ and for the solution $k \to x(k)$, $k \geqslant k_o$, for (DfcE) and (IC), are simpler. Specifically, if for all $k \in \mathbf{Z}$ we have $A(k) \equiv A$, a constant $(n \times n)$ matrix, then

$$x(k) = A^{k-k_o} x_o + \sum_{l=k_o}^{k-1} A^{k-l-1} f(l), \qquad k \geqslant k_o ,$$

where A^0 is understood to be the $(n \times n)$ identity matrix I_n. In the constant-A case, the transition matrix function is given by

$$\Phi(k, l) = \begin{cases} A^{k-l} & k > l \\ I_n & k = l . \end{cases}$$

4. *Some More Linear Algebra*

It is often the case in science and engineering that a bit of abstract thinking helps one understand the "real world" somewhat better; the present section and the one which follows are written with that thought in mind. In this section, we discuss some fundamental techniques and results from "abstract" linear algebra. Although *full* generality is not our objective (for example, we deal only with real and complex vector spaces), most results are proven in fairly general contexts. The presentation is slanted in the direction of finite-dimensional vector spaces, since these will be our primary objects of interest in what follows. An outstanding reference for the topics to be discussed is [Halmos]. The reader should always bear in mind the definitions and constructions outlined in §1, as these are concrete examples of some of the abstract notions described here.

As always, by $\mathbf{F}$ we mean either $\mathbf{R}$ or $\mathbf{C}$; elements of $\mathbf{F}$ are called *scalars*.

4.1 Definition: A *vector space V over* $\mathbf{F}$ is a set V of objects (called vectors, naturally) on which are defined two operations:

(a) *addition*: given v^1 and v^2 in V, there is defined a vector $v^1 + v^2 \in V$ called the *sum* of v^1 and v^2.

(b) *scalar multiplication*: given $v \in V$ and $\alpha \in \mathbf{F}$, there is defined a vector $\alpha v \in V$, called the *scalar multiple of v by α*. It is assumed, moreover, that addition is commutative and associative, and that there exists an identity element in V (denoted by 0) for addition. It is also assumed that scalar multiplication distributes over addition (i.e., $\alpha(v^1 + v^2) = \alpha v^1 + \alpha v^2$), and that $\alpha(\beta v) = (\alpha\beta)v$ for all $\alpha, \beta \in \mathbf{F}$, and $v \in V$. □

Before giving examples of vector spaces, a word on terminology: a vector space V over **R** is often referred to as a *real vector space*, and one over **C** is often referred to as a *complex vector space*.

4.2 Examples:

(a) $\mathbf{F}^n$, for any n, is a vector space over **F** with addition and scalar multiplication defined in the usual (componentwise) fashion. Observe that $\mathbf{C}^n$ is a vector space over **R** as well as over **C**, but that $\mathbf{R}^n$ is not a vector space over **C** with respect to the usual operations.

(b) The *Euclidean Plane* E^2 is defined as follows: consider a plane with a distinguished point, called 0, but *no "coordinate axes."* Define E^2 (as a set) to be the family consisting of arrows emanating from 0. Each arrow is specified by giving its magnitude and direction. Addition and scalar multiplication are defined as in the following pictures:

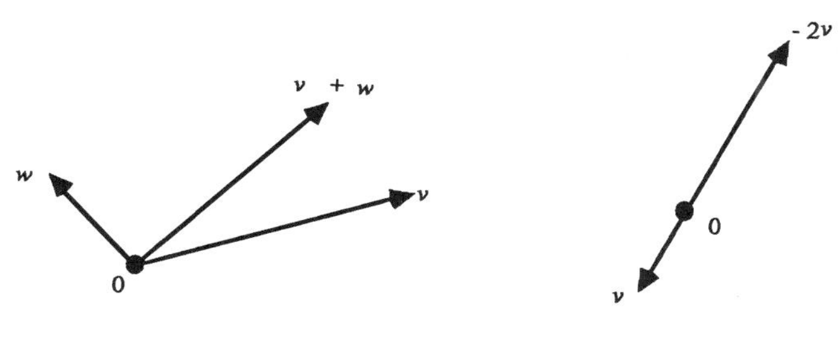

Figure 4.1 Vector operations in the Euclidean plane

With these operations, E^2 becomes a real vector space. Note that E^2 (by definition) is *not* "the same set" as $\mathbf{R}^2$. Still, E^2 and $\mathbf{R}^2$ bear a strong resemblance; we'll have more to say about this point later on.

(c) Denote by $C([0,1])$ the set of all real-valued continuous functions defined on the unit interval $[0, 1] \subset \mathbf{R}$. We claim that $C([0,1])$ has a natural real vector space structure. Define addition of f^1, $f^2 \in C([0, 1])$ by

$$(f^1 + f^2)(t) = f^1(t) + f^2(t), \qquad t \in [0, 1],$$

and scalar multiplication, for $f \in C([0, 1])$, $\alpha \in \mathbf{R}$ by

$$(\alpha f)(t) = \alpha(f(t)), \qquad t \in [0, 1].$$

Since these operations clearly take (scalars and) elements of $C([0,1])$ and yield new elements of $C([0,1])$, they define appropriate vector space operations on $C([0,1])$, making $C([0,1])$ into a real vector space.

(d) Denote by $\mathbf{U}$ the set of all infinite sequences $c = \{\ldots, c_{-2}, c_{-1}, c_0, c_1, c_2, c_3, \ldots\}$ of real numbers. Then $\mathbf{U}$ is a real vector space with addition and scalar multiplication defined by

$$(c + d)_k = c_k + d_k, \qquad c, d \in \mathbf{U}$$

$$(\alpha c)_k = \alpha c_k, \qquad \alpha \in \mathbf{R}, \; c \in \mathbf{U}.$$

(e) If $p > 1$, denote by L^p the set of functions $f : \mathbf{R} \to \mathbf{R}$ which satisfy $\int_{-\infty}^{\infty} |f(t)|^p \, dt < \infty$. Then L^p is a real vector space with addition and scalar multiplication defined by

$$(f^1 + f^2)(t) = f^1(t) + f^2(t), \qquad t \in \mathbf{R}; \; f^1, f^2 \in L^p$$

$$(\alpha f)(t) = \alpha(f(t)), \qquad t \in \mathbf{R}; \; \alpha \in \mathbf{R}; \; f \in L^p.$$

Observe that if f^1 and f^2 are in L^p, then $f^1 + f^2$ is also in L^p since

$$|f^1(t) + f^2(t)|^p \leqslant 2^{p-1}[|f^1(t)|^p + |f^2(t)|^p].$$

(f) For p > 1, the set l^p of real sequences $(\ldots, c_{-2}, c_{-1}, c_0, c_1, \ldots)$ which satisfy $\sum_{k=-\infty}^{\infty} |c_k|^p < \infty$ is a real vector space with addition and scalar multiplication defined as in example 4.2(d) above. □

4.3 Definition: A subset W of a vector space V is said to be a *subspace of* V if W is *closed* under the operations 4.1 (a), (b) defined on V. □

Evidently, not every *subset* of V is a sub*space* of V. Some examples (and non-examples) of subspaces of vector spaces follow.

4.4 Examples:

(a) $\{0\}$ and V are (somewhat trivial) subspaces of any vector space V.

(b) Consider the subset W of F^n defined by $W = \{x \in F^n : x_n = 0\}$. W is clearly closed under addition and scalar multiplication in F^n, and is therefore a subspace of F^n.

(c) If A is any $(m \times n)$ matrix, then $N(A) = \{x \in F^n : Ax = 0\}$, as defined in §1, is a subspace of F^n.

(d) $W = \{x \in F^n : x_n = 3\}$ is *not* a subspace of F^n. Why?

(e) $W = \{f \in C([0,1]) : f(1/2) = 0\}$ is a subspace of $C([0,1])$. To see this, observe that functions $t \to f(t)$ on [0,1] which vanish at t = 1/2, when added and multiplied by real numbers, yield functions on [0,1] which vanish at 1/2.

(f) The vector space l^p in Example 4.2 (f) is a subspace of the vector space U in Example 4.2 (d). □

An alternate definition of a subspace W of a vector space V is "a subset of V which is itself a vector space under the same operations which make V into a vector space;" as W is a subset of V, it makes sense to talk about performing operations on vectors in W which are defined for vectors in V. A somewhat universal example of how subspaces may be "built" is given by the following definition.

4.5 Definition: Let V be a vector space over F. Let $S \subset V$ be any subset. Define $W = \text{span}(S)$ to be the set of all finite "linear combinations" of vectors in S; i.e.,

$$W = \{\alpha_1 v^1 + \ldots + \alpha_n v^n : \alpha_i \in F,\ v^i \in S,\ 0 < n < \infty\}.$$

Then W is called *the subspace of V spanned by S;* any S' (including S) for which $W = \text{span}(S')$ is called a *spanning set* *for the subspace W.* □

Observe that Definition 4.5 is loaded; we have not shown that $W = \text{span}(S)$ is indeed a *subspace* of V. This easy task is left as an exercise for the reader. Some examples of spanned subspaces and spanning sets follow.

4.6 Examples:

(a) any vector space V is a (somewhat trivial) spanning set for itself.

(b) $\left\{ \begin{bmatrix} 1 \\ 0 \end{bmatrix}, \begin{bmatrix} 0 \\ 1 \end{bmatrix} \right\}$ is a spanning set for $\mathbf{R}^2$.

(c) $\left\{ \begin{bmatrix} 1 \\ 0 \end{bmatrix}, \begin{bmatrix} 0 \\ 1 \end{bmatrix}, \begin{bmatrix} 2 \\ 2 \end{bmatrix} \right\}$ is also a spanning set for $\mathbf{R}^2$.

(d) $\left\{ \begin{bmatrix} 2 \\ 0 \\ 0 \end{bmatrix}, \begin{bmatrix} 0 \\ 1 \\ 0 \end{bmatrix} \right\}$ is a spanning set for the subspace $W = \{x \in \mathbf{R}^3 : x_3 = 0\}$ of $\mathbf{R}^3$, as is $\left\{ \begin{bmatrix} 0 \\ 2 \\ 0 \end{bmatrix}, \begin{bmatrix} 1 \\ 0 \\ 0 \end{bmatrix} \right\}$. □

Examples 4.6 show that spanning sets for vector spaces are by no means unique; they may have different sizes, and, even if two spanning sets are of the same size, the sets may well be different. Writing down a spanning set for a vector space V may be regarded as completely specifying V in a more compact fashion than by "listing all the vectors in V." It behooves us, then, to seek spanning sets for V which are *as small as possible*.

4.7 Definition: A vector space V is said to be *finite-dimensional* if and only if V possesses a *finite* spanning set; that is, V is finite-dimensional if and only if there exists a finite subset $S = \{v^1, \ldots, v^k\} \subset V$ such that $V = \text{span}(S)$. □

Observe that "finite-dimensionality" has been defined while "dimension" hasn't; this apparent backhandedness should cause the reader no alarm. Examples 4.6 all exhibit finite spanning sets for certain vector spaces; consequently, each vector space in Example 4.6 is finite-dimensional. Observe, in fact, that $\mathbf{F}^n$, for any n, is finite-dimensional; one finite spanning set for $\mathbf{R}^n$ is the family $\mathbf{e} = \{e^1, \ldots, e^n\}$ of n-vectors satisfying $[e^i]_j = \delta_{ij}$, $1 \leqslant i, j \leqslant n$. Observe also that the Euclidean plane E^2, defined in Example 4.2(b), is finite-dimensional; any finite set of two or more noncollinear "arrows" is a finite spanning set for E^2. As it happens, the only vector spaces mentioned in the foregoing which are *not* finite-dimensional are $C([0,1])$, U, L^p, and l^p; this fact is a consequence of the proof of Theorem 4.11 below.

4.8 Definition: A set of vectors $\{v^1, \ldots, v^k\}$ in a vector space V is said to be a *linearly independent set* if and only if the only scalars $\alpha_1, \ldots, \alpha_n$ satisfying

$$\alpha_1 v^1 + \ldots + \alpha_n v^n = 0$$

are $\alpha_1 = \alpha_2 = \ldots = \alpha_n = 0$. □

It is easy to see that the finite sets of vectors in Examples 4.6(b), (d) are linearly independent, while that in Example 4.6(c) is not.

4.9 Theorem: *Every finite-dimensional vector space $V \neq \{0\}$ possesses a spanning set $\mathbf{v} = \{v^1, \ldots, v^n\}$ which is also a linearly independent set.*

Proof: Let $S_k = \{v^1, \ldots, v^k\}$ be a spanning set for V. If S_k is linearly independent, we're done; if not, by renumbering the v^i if necessary, there exist α_i, $1 \leqslant i \leqslant k$, with $\alpha_1 \neq 0$, such that

$$\alpha_1 v^1 + \alpha_2 v^2 + \ldots + \alpha_k v^k = 0 .$$

Since v^1 may be expressed as a linear combination of $v^2, \ldots, v^k$, it is seen that $S_{k-1} = \{v^2, \ldots, v^k\}$ is also a spanning set for V. Continue eliminating vectors in this fashion until a linearly independent set $S_n = \{v^1, \ldots, v^n\}$ is obtained. The assumption $V \neq \{0\}$ guarantees that the process terminates for some $n \geqslant 1$. □

4.10 Definition: Let V be a finite dimensional vector space. A *basis for* V is a spanning set $\mathbf{v} = \{v^1, \ldots, v^n\}$ which is also a linearly independent set. □

Theorem 4.9 ensures that every finite dimensional vector space possesses a basis. Observe that the finite sets of vectors in Examples 4.6 (b), (d) are bases for their spans, while that in Example 4.6 (c) is not. Example 4.6 (d) shows that a given vector space may possess more than one basis; as it happens, any real or complex vector space possesses an uncountable infinity of bases. The number of vectors in any basis for a given finite-dimensional V is, however, an invariant; we call it the *dimension* of V.

4.11 Theorem/Definition: *Every basis* $\mathbf{v}$ *for a given finite-dimensional vector space* V *contains the same number of vectors; this number is called the* dimension *of* V.

Proof: Let $\mathbf{v} = \{v^1, \ldots, v^m\}$ and $\mathbf{w} = \{w^1, \ldots, w^n\}$ be two bases for V with $m < n$. Proceeding exactly as in the proof of Theorem 1.21, we find scalars a_{ij}, $1 \leqslant i \leqslant n$ and $1 \leqslant j \leqslant m$, such that

$$w^i = \sum_{j=1}^{m} a_{ij} v^j , \qquad 1 \leqslant i \leqslant n ,$$

and an n-vector c such that

$$\sum_{i=1}^{n} c_i a_{ij} = 0 .$$

We conclude that $\sum_{i=1}^{n} c_i w^i = 0$, contradicting linear independence of $\mathbf{w}$. □

Theorem 4.11 and its proof indicate other ways in which one may define "basis for V." A direct consequence of the proof of Theorem 4.11 is the following statement: *there are at most* n *vectors in any linearly independent subset of an* n*-dimensional vector space* V. It may also be concluded that *any spanning set for an* n*-dimensional vector space* V *contains at least* n *vectors*. Along with the Exercise 4.12 below, these assertions yield two alternative *definitions* of basis:

(*i*) a basis for V is a maximal linearly independent set of vectors in V, and

(*ii*) a basis for V is a minimal spanning set for V.

4.12 Exercise: Show that any set $\mathbf{v}$ of n linearly independent vectors in an n-dimensional vector space V is a basis for

V. [*Suggestion:* If $\mathbf{v}$ does not span V, there exists w not in $\mathbf{v}$ such that $\mathbf{v} \cup \{w\}$ is a linearly independent set.] □

4.13 Exercise: By imitating the proof of Theorem 1.20, show that any subspace W of an n-dimensional vector space V is finite-dimensional with dimension at most n. □

4.14 Exercise: Show that the dimension of $\mathbf{F}^n$ is n for every n. [Generalize Example 4.6 (b) to arbitrary n from n = 2.] □

We turn now to the study of linear mappings between vector spaces. Most results on linear mappings between finite-dimensional vector spaces and their properties bear a strong resemblance to (and may be proved using the same techniques as are used in proving) properties of $(m \times n)$ matrices such as those discussed in §1. We begin with a definition.

4.15 Definition: Let V and W be vector spaces over $\mathbf{F} = \mathbf{R}$ or $\mathbf{C}$. A *linear mapping* $T : V \to W$ (also written $v \to T(v)$) is an ordinary mapping from V to W which in addition satisfies, for every $v^1, v^2 \in V$ and $\alpha_1, \alpha_2 \in \mathbf{F}$:

$$T(\alpha_1 v^1 + \alpha_2 v^2) = \alpha_1 T(v^1) + \alpha_2 T(v^2) .$$ □

4.16 Examples: (a) For any V and W, the mapping $T : V \to W$ defined for every $v \in V$ by $T(v) = 0 \in W$ is a (somewhat trivial) linear mapping; it is called the *zero mapping* from V to W.

(b) For any V, the mapping $T : V \to V$ defined by $T(v) = v$, $v \in V$, is a linear mapping from V to V; it is called the *identity mapping* on V.

(c) Any $(m \times n)$ matrix with entries in $\mathbf{F}$ defines by way of

$$x \to Ax , \quad x \in \mathbf{F}^n ,$$

a linear mapping from $\mathbf{F}^n$ into $\mathbf{F}^m$. □

We often call a linear mapping $T : V \to V$ from a vector space to itself a *linear transformation* on V; note, however, that many authors describe more general linear mappings as "transformations." Linear mappings may be manipulated in the same ways that arbitrary mappings are manipulated; linearity, however, provides additional structure. For example, if $T : V \to W$ and $S : W \to X$ are linear mappings between vector spaces, then the ordinary composition $S \circ T : V \to X$ is a *linear* mapping from V to X. The proof is straightforward, and is left to the reader. We summarize in the following definition some terminology which will be useful later on.

4.17 Definition: Let $T : V \to W$ be a linear mapping from a vector space V to a vector space W. Then T is said to be

(a) *injective* if $v^1 \neq v^2$ implies $T(v^1) \neq T(v^2)$;

(b) *surjective* if, for every $w \in W$, there exists $v \in V$ such that $T(v) = w$;

(c) *bijective* if it is both injective and surjective; and

(d) *invertible*, or an *isomorphism*, if there exists a *linear* mapping $S : W \to V$ such that $S \circ T$ and $T \circ S$ are the identity mappings (c.f. Example 4.16 (b)) respectively of V and of W. V and W are said to be *isomorphic* if there exists an isomorphism $T : V \to W$. □

4.18 Exercise: (a) Show that $T : V \to W$ is injective if and only if the only $v \in V$ satisfying $T(v) = 0$ is $v = 0$.

(b) Show that $T : V \to W$ is invertible if and only if it is bijective. [*Hint*: the subtlety is that T may not be "linearly invertible;" show that, under the bijectivity assumption, the "set theoretic" inverse for T is indeed linear.] □

It is time once again to specialize to the finite-dimensional situation. We shall see how matrices arise from linear mappings between finite-dimensional vector spaces, and how studying an n-dimensional vector space over F is in some sense "the same" as studying $\mathbf{F}^n$. Accordingly, let V be an n-dimensional vector space over F and let $\mathbf{v} = \{v^1, \ldots, v^n\}$ be a basis for V. Given any $x \in V$, we may write x uniquely in terms of $\mathbf{v}$; that is,

$$x = x_1 v^1 + x_2 v^2 + \ldots + x_n v^n$$

where x_i, $1 \leqslant i \leqslant n$, are scalars. The reader should check that the x_i are indeed uniquely determined by x given $\mathbf{v}$, since $\mathbf{v}$ is a linearly independent set. Denote by $x_{\mathbf{v}}$ the n-vector whose i th component is x_i; we call $x_{\mathbf{v}}$ the *coordinate vector* of x with respect to the basis $\mathbf{v}$. Observe that every n-vector is $x_{\mathbf{v}}$ for some $x \in V$.

The mapping $x \to x_{\mathbf{v}}$ of V to $\mathbf{F}^n$ is evidently linear; moreover, by the arguments in the preceding paragraph, it is bijective, and hence, by Exercise 4.18(b), it is *an isomorphism between* V *and* $\mathbf{F}^n$. Thus every n-dimensional vector space over $\mathbf{F}$ is isomorphic to $\mathbf{F}^n$ (c.f. Definition 4.17 (d)); a different isomorphism between V and $\mathbf{F}^n$ is defined by each choice of basis for V.

4.19 Definition: If $T : V \to W$ is a linear mapping from an n-dimensional vector space V into an m-dimensional vector space W, and if $\mathbf{v} = \{v^1, \ldots, v^n\}$ and $\mathbf{w} = \{w^1, \ldots, w^m\}$ are bases for V and W, respectively, then the *matrix of* T *with respect to the basis pair* $(\mathbf{v}, \mathbf{w})$, which we write $T_{\mathbf{v},\mathbf{w}}$, is defined as follows:

$$[T_{\mathbf{v},\mathbf{w}}]_{ij} = [T(v^j)_{\mathbf{w}}]_i \,, \qquad \begin{matrix} 1 \leqslant i \leqslant m \\ 1 \leqslant j \leqslant n \,. \end{matrix}$$ □

That is: the (i,j)-element of the $(m \times n)$ matrix $T_{\mathbf{v},\mathbf{w}}$ is the i th component in the coordinate vector, with respect to the basis $\mathbf{w}$, of the vector $T(v^j) \in W$. If we write $[T_{\mathbf{v},\mathbf{w}}]_{ij}$ as a_{ij}, then we get

$$T(v^j) = \sum_{i=1}^{m} a_{ij} w^i \,, \qquad 1 \leqslant j \leqslant n \,.$$

Each choice of basis pair $(\mathbf{v}, \mathbf{w})$, then, gives rise to a "matrix representation" for the linear mapping T. In the special case when T is a linear *transformation* on an n-dimensional vector space V, we speak of the *matrix of* T *with respect to the basis* $\mathbf{v}$ *for* V; this matrix, which is $(n \times n)$ and is denoted by $T_{\mathbf{v}}$, is given by

$$[T_{\mathbf{v}}]_{ij} = [T(v^j)_{\mathbf{v}}]_i \,.$$

That is, for $1 \leqslant j \leqslant n$, we have

$$T(v^j) = \sum_{i=1}^{n} [T_{\mathbf{v}}]_{ij} v^i \,.$$

Slightly abusing the notation of Definition 4.19, we may write $T_{\mathbf{v}} = T_{\mathbf{v},\mathbf{v}}$.

The profusion of new notation, along with ubiquitous summations, sometimes over several indices, will no doubt confuse the reader who is unfamiliar with the constructions described in the foregoing. One item which should be emphasized is *the distinction between a linear mapping T and its matrix with respect to choices of basis for its domain and codomain vector spaces.* A glance at Example 4.21 below might provide at least some measure of clarification.

4.20 Observations: *Let V, W, and Y be finite-dimensional vector spaces with respective bases given by $\mathbf{v} = \{v^1, \ldots, v^n\}$, $\mathbf{w} = \{w^1, \ldots, w^m\}$, and $\mathbf{y} = \{y^1, \ldots, y^r\}$. Then*

(a) if $T : V \to W$ is a linear mapping, then for any $x \in V$ the coordinate vector of $T(x)$ with respect to $\mathbf{w}$ is the matrix $T_{\mathbf{v},\mathbf{w}}$ times the coordinate vector of x with respect to $\mathbf{v}$; i.e.,

$$T(x)_{\mathbf{w}} = (T_{\mathbf{v},\mathbf{w}})(x_{\mathbf{v}}).$$

(b) If $S : W \to Y$ is another linear mapping, then the $(r \times n)$ matrix of $S \circ T$ with respect to the basis pair $(\mathbf{v}, \mathbf{y})$ is given by the matrix product

$$S \circ T_{\mathbf{v},\mathbf{y}} = S_{\mathbf{w},\mathbf{y}} T_{\mathbf{v},\mathbf{w}}.$$

Proof: (a) Let x_j be $[x_{\mathbf{v}}]_j$ and z_i be $[T(x)_{\mathbf{w}}]_i$. Then:

$$\begin{aligned}
\sum_{i=1}^{m} z_i w^i &= T(x) \\
&= T\Big(\sum_{k=1}^{n} x_k v^k\Big) \\
&= \sum_{k=1}^{n} x_k T(v^k) \\
&= \sum_{i=1}^{m} \Big[\sum_{k=1}^{n} [T_{\mathbf{v},\mathbf{w}}]_{ik} x_k\Big] w^i .
\end{aligned}$$

By linear independence of $\mathbf{w}$, the bracketed quantity, which is the ith component of the m-vector $T_{\mathbf{v},\mathbf{w}} x_{\mathbf{v}}$, equals z_i, the ith component of $T(x)_{\mathbf{w}}$, and the result follows.

(b) We have

$$S \circ T(v^j) = S(T(v^j)) = S(\sum_{k=1}^{m} [T_{\mathbf{v},\mathbf{w}}]_{kj} w^k);$$

and, by linearity of S,

$$S \circ T(v^j) = \sum_{k=1}^{m} [T_{\mathbf{v},\mathbf{w}}]_{kj} S(w^k) = \sum_{i=1}^{r} \{ \sum_{k=1}^{m} [S_{\mathbf{w},\mathbf{y}}]_{ik} [T_{\mathbf{v},\mathbf{w}}]_{kj} \} y^i .$$

The quantity in braces, by definition, is the (i, j) element of $S_{\mathbf{w},\mathbf{y}} \cdot T_{\mathbf{v},\mathbf{w}}$; by definition of the matrix $S \circ T_{\mathbf{v},\mathbf{y}}$, the result follows. □

Note that the matrix of the zero linear mapping $T : V \to W$ (c.f. Example 4.16 (a)) with respect to *any* basis pair is the zero matrix. Observe also that the identity transformation on an n-dimensional vector space V (Example 4.16 (b)) has matrix I_n with respect to any basis. The next example is somewhat less elementary.

4.21 Example: Consider the Euclidean plane E^2 (Example 4.2 (b)). Choose as a basis e for E^2 any two unit length perpendicular vectors (arrows) e^1 and e^2; let $T : E^2 \to E^2$ be defined as the (evidently linear) mapping which rotates every vector in E^2 through an angle θ counterclockwise.

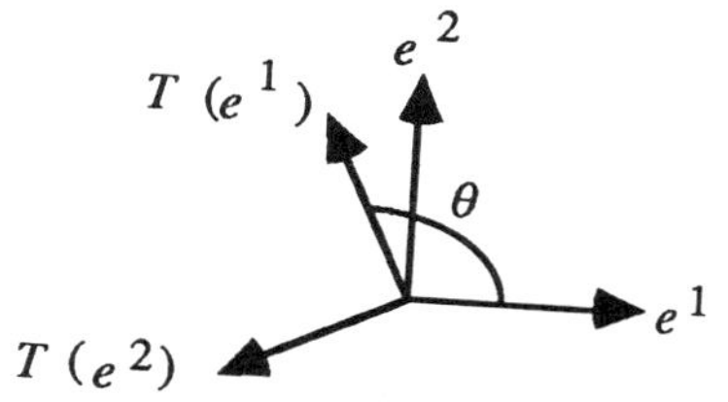

Figure 4.2 Rotation in the Euclidean plane

The laws of plane geometry show that

$$T(e^1) = \cos\theta \, e^1 + \sin\theta \, e^2$$
$$T(e^2) = -\sin\theta \, e^1 + \cos\theta \, e^2 .$$

Thus the matrix of T with respect to e is given by

$$T_{\mathbf{e}} = \begin{bmatrix} \cos\theta & -\sin\theta \\ \sin\theta & \cos\theta \end{bmatrix}.$$

[N.b.: the equations "transpose" to give the matrix.] □

The interplay between linear mappings and their matrices with respect to various bases is sometimes confusing. As we have seen, for example, every $(m \times n)$ matrix A with entries in $\mathbf{F}$ defines a linear mapping - call it ${}_AT$ - from $\mathbf{F}^n$ to $\mathbf{F}^m$. Now, $\mathbf{F}^n$ and $\mathbf{F}^m$ are vector spaces, and as a result each possesses many bases. An important basis for any $\mathbf{F}^k$ is the *standard basis* $\mathbf{e}^k$, to which we have alluded many times without naming; it is the set of k-vectors $e^1, \ldots, e^k$ satisfying

$$[e^i]_j = \delta_{ij} = \begin{cases} 1 & i = j \\ 0 & i \neq j \, . \end{cases}$$

Observe that *the matrix of* ${}_AT$ *with respect to the basis pair* (e^n, e^m) *for* $\mathbf{F}^n$, $\mathbf{F}^m$ *is* A *itself!* Thus A defines a mapping, and the mapping in turn has matrix A with respect to a "canonical" basis pair — quite a mind-bending sequence of events.

It has been mentioned that a crucial idea which underlies the study of finite-dimensional vector spaces and linear mappings between them is that the theory of n-dimensional vector spaces V over $\mathbf{F}$, m-dimensional vector spaces W over $\mathbf{F}$, and mappings $T : V \to W$ is "the same" as the theory of $\mathbf{F}^n$, $\mathbf{F}^m$, and $(m \times n)$ matrices with entries in $\mathbf{F}$. This "sameness," however, is not *natural*, in the sense that one must choose bases for V and W in order to draw exact correspondences $\mathbf{F}^n \to V$, $\mathbf{F}^m \to W$, and $A \to {}_AT$. The correspondences arise immediately, via Observations 4.20, from the isomorphisms $x \to [x]_{\mathbf{v}}$ and $z \to [z]_{\mathbf{w}}$ between V and $\mathbf{F}^n$, W and $\mathbf{F}^m$, respectively.

It is worth making a precise determination of how a change in choice of bases for V and W affects the matrix of a linear mapping $T : V \to W$. Suppose that $\mathbf{v} = \{v^1, \ldots, v^n\}$ and $\hat{\mathbf{v}} = \{\hat{v}^1, \ldots, \hat{v}^n\}$ are two bases for the n-dimensional vector space V. Since $\mathbf{v}$ is a basis for V, we may find scalars p_{ij}, $1 \leqslant i, j \leqslant n$, such that

$$\hat{v}^j = \sum_{i=1}^{n} p_{ij} v^i \, , \qquad 1 \leqslant j \leqslant n \, .$$

We claim that the $(n \times n)$ matrix P with (i, j) element p_{ij} is invertible; otherwise, there would exist by Theorem 1.5 a nonzero n-vector x such that for every i, $1 \leqslant i \leqslant n$:

$$0 = \sum_{j=1}^{n} p_{ij} x_j ,$$

implying that

$$\sum_{j=1}^{n} x_j \hat{v}^j = 0 ,$$

and contradicting linear independence of $\hat{\mathbf{v}}$. We leave it as an exercise for the reader to show that for every $v \in V$, it is true that the coordinate vectors of v with respect to the bases $\mathbf{v}$ and $\hat{\mathbf{v}}$ are related by

$$v_{\hat{\mathbf{v}}} = P^{-1} v_{\mathbf{v}} .$$

Similarly, suppose $\mathbf{w}$ and $\hat{\mathbf{w}}$ are bases for an m-dimensional vector space W; then there exists an invertible $(m \times m)$ matrix Q such that (where $q_{ij} = [Q]_{ij}$)

$$\hat{w}^j = \sum_{i=1}^{m} q_{ij} w^i , \quad 1 \leqslant j \leqslant m .$$

A simple manipulation shows that

$$w^k = \sum_{i=1}^{m} [Q^{-1}]_{ik} \hat{w}^i , \quad 1 \leqslant k \leqslant m .$$

(To derive the last formula, multiply both sides of the preceding equation by $[Q^{-1}]_{jk}$ and sum over j from 1 to m.)

Given a linear map $T : V \to W$, we wish to relate the two matrices $T_{\mathbf{v},\mathbf{w}}$ and $T_{\hat{\mathbf{v}},\hat{\mathbf{w}}}$ of T with respect to the two basis pairs $(\mathbf{v}, \mathbf{w})$ and $(\hat{\mathbf{v}}, \hat{\mathbf{w}})$. The formulas seem rather messy at first; let $A = T_{\mathbf{v},\mathbf{w}}$ to economize on subscripts. Now,

$$\begin{aligned} T(\hat{v}^j) &= T(\sum_{l=1}^{n} p_{lj} v^l) \\ &= \sum_{l=1}^{n} p_{lj} T(v^l) \\ &= \sum_{l=1}^{n} p_{lj} (\sum_{k=1}^{m} [A]_{kl} w^k), \end{aligned}$$

so that

$$T(\hat{v}^j) = \sum_{l=1}^{n} p_{lj}\{\sum_{k=1}^{m} [A]_{kl}(\sum_{i=1}^{m} [Q^{-1}]_{ik}\hat{w}^i)\}$$

$$= \sum_{i=1}^{m}\{\sum_{k=1}^{m} [Q^{-1}]_{ik}(\sum_{l=1}^{n} [A]_{kl}\, p_{lj})\}\hat{w}^i$$

$$= \sum_{i=1}^{m} [Q^{-1}AP]_{ij}\,\hat{w}^i \; .$$

We conclude, by definition of matrices with respect to basis pairs, that

$$T_{\hat{\mathbf{v}},\hat{\mathbf{w}}} = Q^{-1}\, T_{\mathbf{v},\mathbf{w}}P \; . \qquad (*)$$

In particular, if $T : V \rightarrow V$ is a transformation, and $\mathbf{v}, \hat{\mathbf{v}}$ are two bases for V related as above, then

$$T_{\hat{\mathbf{v}}} = P^{-1}T_{\mathbf{v}}P \; . \qquad (**)$$

The two formulas (*) and (**) will be important later on when we discuss realization theory for linear systems.

We now introduce "abstract" analogues of some of the constructions made in §1.

4.22 Definition: Let $T : V \rightarrow W$ be a linear mapping between two vector spaces. The *nullspace*, or *kernel* of T, written $N(T)$, is the subspace of V consisting of all vectors $v \in V$ such that $T(v) = 0$. The *range* of T, written $R(T)$, is the subspace of W consisting of those w for which $w = T(v)$ for some $v \in V$. □

4.23 Exercise: (a) Show that $N(T)$ and $R(T)$ for $T : V \rightarrow W$ are indeed subspaces of V and W, respectively.

(b) Show that T is surjective if and only if $R(T) = W$.

(c) Show that T is injective if and only if $N(T) = \{0\}$.

(d) Show that if V and W are finite-dimensional, and $\dim(W) < \dim(V)$, then T cannot be injective. [*Hint*: use Lemma 1.12 after choosing bases for V and W.]

(e) Show that isomorphic finite-dimensional vector spaces must have the same dimension. □

The next two results should serve to draw correspondences between some of the "abstract" material just presented and some of the "concrete" material in §1. First, we prove an assertion which parallels Theorem 1.5.

4.24 Theorem: *Let V be an n-dimensional vector space over $\mathbf{F} = \mathbf{R}$ or $\mathbf{C}$; let $T : V \to V$ be a linear transformation. The following four statements are equivalent:*

(a) T is invertible.

(b) $\mathrm{N}(T) = \{0\}$; i.e., T is injective.

(c) $\mathrm{R}(T) = V$; i.e., T is surjective.

(d) The matrix $T_{\mathbf{v}}$ of T with respect to any basis $\mathbf{v}$ for V is an invertible $(n \times n)$ matrix.

Proof: Suppose T is invertible; let $\mathbf{v}$ be a basis for V and let S be the inverse of T. Then by observation 4.20(b) and the discussion which followed:

$$(S \circ T)_{\mathbf{v}} = S_{\mathbf{v}} T_{\mathbf{v}} = I_n = T_{\mathbf{v}} S_{\mathbf{v}} = (T \circ S)_{\mathbf{v}} .$$

Thus $S_{\mathbf{v}}$ is the inverse of $T_{\mathbf{v}}$, and $T_{\mathbf{v}}$ is invertible. We conclude that (a) implies (d).

By Theorem 1.5, we know that the invertibility of the matrix $T_{\mathbf{v}}$ is equivalent to the statement that the only $x \in \mathbf{F}^n$ satisfying $T_{\mathbf{v}} x = 0$ is $x = 0$. Observation 4.20 (a) therefore says that (d) is equivalent to (b), since $T(v) = 0$ precisely $T_{\mathbf{v}} x = 0$, where x is the coordinate vector of v with respect to $\mathbf{v}$. Similarly, Theorem 1.5 and Observation 4.20 (a) enable us to conclude that (d) and (c) are equivalent; indeed, given $w \in V$, let y be the coordinate vector of w with respect to $\mathbf{v}$. Then (d) holds if and only if for every such y so constructed, there exists an n-vector x such that $T_{\mathbf{v}} x = y$. But such an x, by Observation 4.20 (a), is the coordinate vector of some $v \in V$ such that $T(v) = w$.

Thus (a) implies (d), which is equivalent to both (b) and (c). To complete the proof, recall the discussion following Definition 4.17, which asserts that T is invertible if and only if it is bijective. Thus (b) and (c) together imply (a), and the proof is complete. □

It is important to observe that Theorem 4.24 is *not* true for infinite-dimensional vector spaces. For example, the mapping T from $C([0,1])$ into itself defined by

$$(Tf)(t) = t^2 f(t), \quad t \in [0, 1]$$

is linear and injective but is not surjective, since every function in $R(T)$ must vanish at $t = 0$.

The following result relates the range and "rank" of a linear mapping to those of its matrix with respect to bases.

4.25 Theorem: *Let V and W be vector spaces over $F = R$ or C having respective dimensions n and m. Let $T : V \to W$ be a linear mapping; let $\mathbf{v}$ and $\mathbf{w}$ be bases for V and W, respectively. Then the rank of the $(m \times n)$ matrix $T_{\mathbf{v},\mathbf{w}}$ is the same as the dimension of $R(T)$.*

Proof: We have $w \in R(T)$ if and only if $w = T(v)$ for some $v \in V$. If $w = T(v)$, then by Observation 4.20 (a), the coordinate vectors x and y of v and w with respect to the given bases satisfy

$$y = T_{\mathbf{v},\mathbf{w}} x .$$

Hence if $w \in R(T)$ then its coordinate vector y is in $R(T_{\mathbf{v},\mathbf{w}})$, the range of the $(m \times n)$ matrix $T_{\mathbf{v},\mathbf{w}}$. Similarly, if $y \in R(T_{\mathbf{v},\mathbf{w}})$, then $w = T(v)$, where $v = \sum_{i=1}^{n} x_i v^i$ and $x \in F^n$ is such that $y = T_{\mathbf{v},\mathbf{w}} x$.

The mapping which takes each $w \in R(T)$ to its coordinate vector with respect to $\mathbf{w}$ induces a linear mapping from $R(T) \subset W$ into $R(T_{\mathbf{v},\mathbf{w}}) \subset F^m$; it is clearly a bijective mapping between those two vector spaces, and thus the two vector spaces must have the same dimension. To see this, consult Exercises 4.18(b) and 4.23(e). We conclude that the rank of $T_{\mathbf{v},\mathbf{w}}$, which by definition equals the dimension of $R(T_{\mathbf{v},\mathbf{w}})$, is equal to the dimension of $R(T)$. □

We conclude this section by discussing constructions which enable us to "put together" several given vector spaces so as to make a single vector space which, at least in some sense, "contains" the given vector spaces as subspaces. These constructions will be of paramount importance later on; they are easy to define abstractly, but efforts to apply them often lead to a considerable amount of confusion. We begin with

4.26 Definition: Let $V^1, \ldots, V^k$ be subspaces of a vector space X over F. The *vector sum of* $V^1, \ldots, V^k$, written $V^1 + \ldots + V^k$, is the subspace of X consisting of all sums of vectors $v^1 + v^2 + \ldots + v^k$ with $v^i \in V^i$, $1 \leqslant i \leqslant k$; i.e., $V^1 + \ldots + V^k$ is given by

$$\{v^1 + v^2 + \ldots + v^k : v^i \in V^i, \quad 1 \leqslant i \leqslant k\}. \qquad \square$$

4.27 Definition: Let V and W be subspaces of a vector space X. The subspaces V and W are said to be *disjoint* if and only if

$$V \cap W = \{0\}.$$

More generally, k subspaces $V^1, \ldots, V^k$ of X are said to be *mutually disjoint* if and only if, for each i, $1 \leqslant i \leqslant k$

$$V^i \cap (V^1 + \ldots + V^{i-1} + V^{i+1} + \ldots + V^k) = \{0\},$$

where + in the last equation denotes vector sum (c.f. Definition 4.26). $\square$

The vector sum of a family of subspaces of X may be viewed as the subspace of X which is *spanned* by the union of the given subspaces; observe that the union of a family of subspaces of X is rarely itself a subspace of X. Note that the disjointness of two subspaces of X is *not* the same as "set-theoretic" disjointness; indeed, every subspace of X must contain the zero element of X.

4.28 Observations: *Let $V^1, \ldots, V^k$ be subspaces of a vector space X over* $\mathbf{F}$. *Then*

(*a*) *$V^1, \ldots, V^k$ are mutually disjoint if and only if the relation*

$$v^1 + v^2 + \ldots + v^k = 0,$$

with $v^i \in V^i$, $1 \leqslant i \leqslant k$, implies $v^1 = v^2 = \ldots = v^k = 0$.

(*b*) *If X is finite-dimensional, then the dimension of the vector sum of $V^1, \ldots, V^k$ is less than or equal to the sum $dim(V^1) + dim(V^2) + \ldots + dim(V^k)$ of the dimensions of the V^i.*

(*c*) *If X is finite-dimensional, then $dim(V^1 + \ldots + V^k)$ is equal to $dim(V^1) + \ldots + dim(V^k)$ if and only if $V^1, \ldots, V^k$ are mutually disjoint.*

(*d*) *If $V^1, \ldots, V^k$ are mutually disjoint subspaces of X, X is finite-dimensional, and $\mathbf{v}_1, \mathbf{v}_2, \ldots, \mathbf{v}_k$ are bases respec-*

tively for $V^1, \ldots, V^k$*, then* $\mathbf{v}_1 \cup \mathbf{v}_2 \cup \ldots \cup \mathbf{v}_k$ *is a basis for the vector sum* $V^1 + V^2 + \ldots + V^k$.

Proof: (a) Suppose $V^1, \ldots, V^k$ are mutually disjoint, and $v^1 + \ldots + v^k = 0$, with $v^i \in V^i$, $1 \leqslant i \leqslant k$; then for each i, $1 \leqslant i \leqslant k$:

$$-v^i = v^1 + \ldots + v^{i-1} + v^{i+1} + \ldots + v^k;$$

since the right-hand side is an element of $V^1 + \ldots + V^{i-1} + V^{i+1} + \ldots + V^k$, and since the left-hand side is an element of V^i, we conclude that $v^i = 0$, by mutual disjointness. Since i was arbitrary, $v^1 = v^2 = \ldots = v^k = 0$.

Conversely, if $V^1, \ldots, V^k$ are not mutually disjoint, there exists some i, $1 \leqslant i \leqslant k$, and some nonzero $v^i \in V^i$ such that

$$v^i = v^1 + v^2 + \ldots + v^{i-1} + v^{i+1} + \ldots + v^k$$

with $v^l \in V^l$, $1 \leqslant l \leqslant k$, $l \neq i$. Then

$$v^1 + v^2 + \ldots + v^{i-1} + (-v^i) + v^{i+1} + \ldots + v^k = 0$$

with at least one term nonzero, contradicting the (true) statement that $\sum_{j=1}^{k} v^i = 0$ with each $v^i \in V^i$ can happen only when $v^i = 0$, $1 \leqslant j \leqslant k$.

(b) Let $n_i = dim(V^i)$, and let $\mathbf{v}_i$ be a basis for V^i, $1 \leqslant i \leqslant k$. Since every element of $V^1 + \ldots + V^k$ is a sum of elements v^i of the V^i, and since each v^i is a linear combination of vectors in $\mathbf{v}_i$, we conclude that any $v \in V^1 + \ldots + V^k$ may be expressed as a linear combination of vectors in $\mathbf{v}_1 \cup \mathbf{v}_2 \cup \ldots \cup \mathbf{v}_k$, which means that $\mathbf{v}_1 \cup \ldots \cup \mathbf{v}_k$ is a *spanning set* for $V^1 + \ldots + V^k$. Thus the dimension of $V^1 + \ldots + V^k$ is at most equal to the size of $\mathbf{v}_1 \cup \mathbf{v}_2 \cup \ldots \cup \mathbf{v}_k$, which is $n_1 + n_2 + \ldots + n_k$.

(c) We saw in the proof of (b) that $\mathbf{v} = \mathbf{v}_1 \cup \mathbf{v}_2 \cup \ldots \cup \mathbf{v}_k$ is a spanning set for the vector sum $V^1 + \ldots + V^k$; it remains to show that $\mathbf{v}$ (which contains $\sum_{i=1}^{k} dim(V^i) = \sum_{i=1}^{k} n_i$ vectors) is *also a linearly independent set* if and only if $V^1, \ldots, V^k$ are mutually disjoint.

Suppose first that the V^i are mutually disjoint; if $\mathbf{v}$ were not a linearly independent set, we could find constants α_{ij}, $1 \leqslant i \leqslant k$, $1 \leqslant j \leqslant n_i$, such that

$$\sum_{i=1}^{k} \left(\sum_{j=1}^{n_i} \alpha_{ij} v^{ij} \right) = \sum_{i=1}^{k} v^i = 0$$

where in an obvious notation $\mathbf{v}_i = \{v^{i1}, \ldots, v^{in_i}\}$ and $v^i = \sum_{j=1}^{n_i} \alpha_{ij} v^{ij}$. Since the V^i are disjoint, by (a) above $v^i = 0$, $1 \leqslant i \leqslant k$. But since $\mathbf{v}_i$ is a basis for V^i (hence a linearly independent set), we conclude that for each i, $\alpha_{ij} = 0$, $1 \leqslant j \leqslant n_i$, implying that *all* the α_{ij} are zero. Hence $\mathbf{v}$ is a linearly independent set.

Conversely, if $V^1, \ldots, V^k$ are not mutually disjoint, there exists a relation $v^1 + v^2 + \ldots + v^k = 0$ with $v^i \in V^i$, $1 \leqslant i \leqslant k$, not all zero; this assertion follows from (a) above. Expanding each v^i as a combination of the vectors in $\mathbf{v}_i$, we obtain a linear combination

$$\sum_{i=1}^{n} \sum_{j=1}^{n_i} \alpha_{ij} v^{ij} = 0$$

with not every α_{ij} equal to zero, which implies that $\mathbf{v}$ is not a linearly independent set of vectors.

The foregoing argument proves (d), as well. □

4.29 Definition: Let V and W be vector spaces over $\mathbf{F}$. The *direct sum* of V and W, written $V \oplus W$, is the vector space over $\mathbf{F}$ defined as follows: as a set, $V \oplus W$ consists of all ordered pairs (v, w) with $v \in V$, $w \in W$; i.e.,

$$V \oplus W = \{(v, w) : v \in V,\ w \in W\}.$$

The vector space operations on $V \oplus W$ are defined as follows:

$$(v^1, w^1) + (v^2, w^2) = (v^1 + v^2, w^1 + w^2)$$
$$\alpha(v, w) = (\alpha v, \alpha w),$$

where v and v^i are in V, w and w^i are in W, and $\alpha \in \mathbf{F}$. The "zero element" in $V \oplus W$ is (clearly) $(0, 0)$. □

Thus $V \oplus W$ is a vector space obtained by "putting together" the two vector spaces V and W in a prescribed fashion. Definition 4.29 is easily generalized by defining the direct sum of k vector spaces $V^1, \ldots, V^k$ as the set of all k-tuples $(v^1, \ldots, v^k)$ of vectors, with $v^i \in V^i$, $1 \leqslant i \leqslant k$, and vector space operations defined componentwise.

4.30 Exercise: Show that if V and W are finite-dimensional vector spaces, with respective dimensions n and m, then $V \oplus W$ is finite-dimensional with dimension $(n+m)$. [*Suggestion:* Let $\{v^1, \ldots, v^n\}$, $\{w^1, \ldots, w^m\}$ be bases for V and W. Show that the set of all $(v^i, 0)$ such that $1 \leqslant i \leqslant n$ taken together with the set of all $(0, w^j)$ such that $1 \leqslant j \leqslant m$ forms a basis for $V \oplus W$.] □

In a sense, $V \oplus W$ is obtained from V and W in the same way that $\mathbf{R}^2$ is obtained from two copies of $\mathbf{R}^1$; one may view the direct sum of V and W as a vector space having a "V-axis" and a "W-axis." $V \oplus W$ contains "copies" of V and W; the copy of V is $\{(v, 0): v \in V\}$, and the copy of W is $\{(0, w): w \in W\}$.

As *defined*, the direct sum of two vector spaces is a set of *ordered pairs of vectors*. In what follows, however, we shall be more interested in analyzing a given "large" vector space X by exhibiting vector spaces V and W for which X is *isomorphic* to $V \oplus W$. The reader should think of "isomorphic" vector spaces as being "almost the same;" we have exhibited isomorphisms between $\mathbf{F}^n$ and any n-dimensional vector space V over $\mathbf{F}$, and thus may regard any such V as being "almost the same" as $\mathbf{F}^n$.

The isomorphisms between X and $V \oplus W$ which we are about to describe reflect our desire to regard $\mathbf{R}^3$ as being "almost the same" as the direct sum of $\mathbf{R}^2$ and $\mathbf{R}^1$. The confusion arises because $\mathbf{R}^2 \oplus \mathbf{R}^1$ is "really" a set of ordered pairs (x, α) where x is a real 2-vector and α is a real scalar, whereas $\mathbf{R}^3$ is a set of real 3-vectors. Still, the mapping from $\mathbf{R}^2 \oplus \mathbf{R}$ to $\mathbf{R}^3$ given by

$$(x, \alpha) \to \begin{bmatrix} x_1 \\ x_2 \\ \alpha \end{bmatrix}$$

is evidently an isomorphism, so we like to think of $\mathbf{R}^3$ as $\mathbf{R}^2 \oplus \mathbf{R}^1$ in disguise. We may draw the correspondence between $\mathbf{R}^3$ and $\mathbf{R}^2 \oplus \mathbf{R}^1$ as follows: $\mathbf{R}^3$ is, evidently, the (disjoint) *vector sum* of the "x^1, x^2 plane" and the "x^3 axis," each regarded as a subspace of $\mathbf{R}^3$; thus we require a way of relating direct sums to (disjoint) vector sums.

4.31 Theorem: *Let X be a vector space; let $V^1, \ldots, V^k$ be subspaces of X which are disjoint and whose vector sum is X. Then the mapping from the direct sum $V^1 \oplus \ldots \oplus V^k$ to X given by*

$$(v^1, \ldots, v^k) \to v^1 + \ldots + v^k, \quad v^i \in V^i, \ 1 \leqslant i \leqslant k,$$

is an isomorphism.

Proof: Denote by T the mapping defined above; T is obviously linear. In order to show that T is an isomorphism, we merely need show (Exercise 4.18) that it is bijective. Since $V^1 + \ldots + V^k = X$, we know that T is surjective; i.e., every $x \in X$ may be expressed as $v^1 + \ldots + v^k$ for some $v^i \in V^i$, $1 \leqslant i \leqslant k$. As for injectivity, suppose that $T((v^1, \ldots, v^k)) = 0$. By the disjointness assumption, along with Observation 4.28 (a), this implies that each v^i is zero. Thus, $N(T) = \{0\}$; hence T is injective by Exercise 4.18 (a). □

4.32 Observation: *If X and (hence) each of the V^i in Theorem 4.31 are finite dimensional, with $dim(V^i) = n_i$, then an easy generalization of Exercise 4.30 shows that*

$$dim(V^1 \oplus \ldots \oplus V^k) = dim(X) = \sum_{i=1}^{k} n_i .$$

(Compare Exercise 4.23(e).) □

5. *Dual Spaces, Norms, and Inner Products*

In §1, it was observed that two matrices of the same size $(m \times n)$ could be added and multiplied by scalars to form new $(m \times n)$ matrices. The way these operations are defined, in view of Definition 4.1, makes it clear that the family of all $(m \times n)$ matrices with entries in F forms a vector space over F. In fact, the dimension of this vector space is mn, since the mn matrices $E(k, l)$ (each is of size $(m \times n)$) whose (i, j) elements are given, for $1 \leqslant i \leqslant m$ and $1 \leqslant j \leqslant n$, by

$$[E(k, l)]_{ij} = \delta_{ik}\delta_{jl} = \begin{cases} 1 & i=k, j=l \\ 0 & otherwise \end{cases}$$

and which are defined for $1 \leqslant k \leqslant m$, $1 \leqslant l \leqslant n$, form a *basis* for the vector space of all $(m \times n)$ matrices. To see this, observe that if A is an arbitrary $(m \times n)$ matrix, then A may be expressed as

$$A = \sum_{i=1}^{m} \sum_{j=1}^{n} a_{ij} E(i,j),$$

and that the sum on the right is zero if and only if every a_{ij} is zero; these facts prove simultaneously that $\{E(i,j)\}$, where $1 \leqslant i \leqslant m$ and $1 \leqslant j \leqslant n\}$, is a linearly independent set and spans the vector space of all $(m \times n)$ matrices.

One might expect, again in the wake of §4, that a vector space structure could be given to the set of all linear mappings $T: V \to W$, where V and W are vector spaces over **F**. This assertion is indeed true; if we define, for each pair S and T of linear mappings from V to W, and for each $\alpha \in$ F,

$$\begin{aligned} (S+T)(v) &= S(v) + T(v), \quad v \in V \\ (\alpha T)(v) &= \alpha(T(v)), \quad v \in V, \end{aligned} \qquad (*)$$

then we see that the set of all linear mappings from V to W is endowed with natural vector space operations.

5.1 Definition: Let V and W be vector spaces over F; then

(a) Hom(V, W) is defined as the vector space of all linear mappings from V to W with vector space operations defined by (*) above.

(b) Hom(V, **F**), the vector space of all linear mappings from V into F, is called the *dual space* of V, and is denoted by V^*; elements of V^* are called *linear functionals* on V. □

We said a moment ago that the family of all $(m \times n)$ matrices had dimension mn as a vector space; a similar result holds for the vector space of all linear mappings from an n-dimensional vector space V into an m-dimensional vector space W. We need an auxiliary result in order to prove that assertion.

5.2 Lemma: *Let V be an n-dimensional vector space over* F; *let* $\mathbf{v} = \{v^1, \ldots, v^n\}$ *be a basis for V; let* W *be any other vector space over* F. *Then, given any $w^1, \ldots, w^n$ in* W, *there exists a unique linear mapping $T: V \to W$ which satisfies*

$$T(v^i) = w^i, \quad 1 \leqslant i \leqslant n.$$

Proof: Since **v** is a basis for V, we may express any $v \in V$ as a unique linear combination of the v^i. Define $T(v)$, when $v = \alpha_1 v^1 + \ldots + \alpha_n v^n$, by

$$T(v) = \sum_{i=1}^{n} \alpha_i w^i .$$

Then $T(v^i) = w^i$, $1 \leqslant i \leqslant n$; T is well-defined by uniqueness of the α_i given v; and T is evidently linear. T is uniquely determined since any other linear mapping T' satisfying $T'(v^i) = w^i$, $1 \leqslant i \leqslant n$, must also take $\sum \alpha_i v^i$ to $\sum \alpha_i w^i$. □

5.3 Theorem: *Let V be an n-dimensional vector space over* F; *let W be an m-dimensional vector space over* F. *Then*

(*a*) Hom(V, W) *has dimension mn.*

(*b*) *In particular, V^* has dimension n.*

Proof: Let **v** and **w** be bases for V and W, respectively. Define a mapping ϕ from Hom(V, W) to the vector space of all $(m \times n)$ matrices by

$$T \overset{\phi}{\rightarrow} T_{\mathbf{v},\mathbf{w}} .$$

Thus, $\phi(T)$ is the matrix of T with respect to the basis pair **v**, **w**.

It is a simple matter to show that ϕ is a *linear* mapping; ϕ is injective since if $\phi(T)$ is the zero matrix, then $T = 0$ (c.f. Exercise 4.18 (a)). It follows from Lemma 5.2 that ϕ is surjective, since given an $(m \times n)$ matrix A we may define a linear mapping $T : V \rightarrow W$ via

$$T(v^j) = \sum_{i=1}^{m} a_{ij} w^i , \quad 1 \leqslant j \leqslant n ,$$

and it is then evident that $A = \phi(T)$. Thus ϕ is a bijective mapping, hence is an isomorphism; it follows from Exercise 4.23 (e) that Hom(V, W) has the same dimension as the vector space of all $(m \times n)$ matrices, which is mn.

In particular, taking W = F and observing that W is then a *one*-dimensional vector space over F, we conclude that the dimension of V^* = Hom(V, F) is n. □

It is important to point out that associated with each basis $\mathbf{v}$ for a dimensional vector space V there is a special basis for V^*.

5.4 Definition: Let V be an n-dimensional vector space; let $\mathbf{v} = \{v^1, \ldots, v^n\}$ be a basis for V. The n linear functionals l_i, $1 \leqslant i \leqslant n$, defined by Lemma 5.1 and

$$l_i(v^j) = \delta_{ij} = \begin{cases} 1 & i = j \\ 0 & i \neq j \end{cases}$$

together form a basis for V^* called the *dual basis* associated with $\mathbf{v}$. □

5.5 Exercise: Prove that $l = \{l_1, \ldots, l_n\}$ given in Definition 5.4 is indeed a basis for V^*. [*Suggestion:* If $l \in V^*$ is an arbitrary linear functional on V, and $l(v^i) = c_i \in \mathbf{F}$, show that $l = \sum_{i=1}^{n} c_i l_i$.] □

We turn now to abstract analogues of two objects - the norm and inner product - which are no doubt familiar to the reader in the context of $\mathbf{F}^n$. If $x \in \mathbf{F}^n$ is an n-vector, the *Euclidean norm* of x, written $\|x\|_2$, is the nonnegative real number given by

$$\|x\|_2 = \left\{ \sum_{i=1}^{n} |x_i|^2 \right\}^{1/2}$$

where $|x_i|$ denotes the magnitude or absolute value of x_i according to whether $\mathbf{F} = \mathbf{C}$ or $\mathbf{F} = \mathbf{R}$. The number $\|x\|_2$ in some sense measures the "length" of x; in $\mathbf{R}^2$ or $\mathbf{R}^3$, this concept has obvious geometric significance. The Euclidean norm on $\mathbf{F}^n$ is but one example of a more general object.

5.6 Definition: A *norm* on a vector space V over $\mathbf{F}$ is a mapping $v \rightarrow \|v\|$ from V to the nonnegative real numbers which satisfies

(a) $\|\alpha v\| = |\alpha| \, \|v\|, \quad \alpha \in \mathbf{F}, v \in V.$

(b) $\|v + w\| \leqslant \|v\| + \|w\|, \quad v, w \in V.$

(c) $\|v\| = 0$ if and only if $v = 0$. □

Property (b) is known as the *triangle inequality*, and often plays a crucial role in convergence proofs (c.f. §2).

5.7 Exercise: Show that the following define norms on $\mathbf{F}^n$:

(a) $\| x \|_\infty = \max_{1 \leqslant i \leqslant n} \{ |x_i| \}$

(b) $\| x \|_1 = \sum_{i=1}^{n} |x_i|$ □

As is shown by Example 5.7, which is merely a drill in working with Definition 5.6, many norms may be defined on a given vector space V. Associated with each such norm is a notion of *convergence* of vectors in V.

5.8 Definition: Let V be a vector space over $\mathbf{F} = \mathbf{R}$ or $\mathbf{C}$; let $\| \ \|$ be a norm on V. We say that a sequence $\{ v^k : k = 0, 1, \ldots \}$ of vectors in V *converges* to the vector $\hat{v} \in V$ *with respect to the norm* $\| \ \|$ if and only if

$$\lim_{k \to \infty} \| v^k - \hat{v} \| = 0 .$$

In this case, we say that the vector $\hat{v}$ is the *limit with respect to the norm* $\| \ \|$ of the sequence $\{ v^k \}$. □

Naturally, different norms on a given vector space can be expected to give rise to different notions of convergence of vectors. We shall soon see that if V is any *finite-dimensional*, vector space, in particular $\mathbf{F}^n$, where $\mathbf{F} = \mathbf{R}$ or $\mathbf{C}$, then *any two norms* on V are *equivalent* in the sense that a sequence $\{ v^k \}$ of vectors in V converges with respect to one of the norms if and only if it converges with respect to the other. This result follows, as one might expect, from a result about $\mathbf{F}^n$. Throughout §2, we regarded the word *convergence* as meaning *componentwise convergence;* it is easily shown that this notion of convergence is the same as "convergence with respect to the Euclidean norm" on $\mathbf{R}^n$. This "standard" notion of convergence makes it possible to talk about *continuous functions* whose domains or codomains are $\mathbf{F}^n$.

5.9 Lemma: *Let* $\mathbf{F} = \mathbf{R}$ *or* $\mathbf{C}$. *If* $\| \ \|_2$ *is the standard Euclidean norm on* $\mathbf{F}^n$ *and* $\| \ \|$ *is any other norm on* $\mathbf{F}^n$, *then* *there exist positive real numbers* α and β such that for every $x \in \mathbf{F}^n$,

$$\alpha \| x \|_2 \leqslant \| x \| \leqslant \beta \| x \|_2 .$$

Proof: Consider the function $\eta: \mathbf{F}^n \to \mathbf{R}$ defined at $x \in \mathbf{F}^n$ by $\eta(x) = \| x \|$; we show first that η is continuous on $\mathbf{F}^n$. First note that if $x \in \mathbf{F}^n$, we have

$$\max_i |x_i| \leqslant [\sum_{i=1}^{n} |x_i|^2]^{1/2} = \| x \|_2 ;$$

to see this, square both sides. Now let $\{e^1, \ldots, e^n\}$ be the standard basis for $\mathbf{F}^n$, and let $M > 0$ be given by

$$M = \max_i \eta(e^i) .$$

By the triangle inequality 5.6(b) for $\| \ \|$, since any $x \in \mathbf{F}^n$ is given by $x = x_1 e^1 + \ldots + x_n e^n$, we have

$$\begin{aligned} \eta(x) &\leqslant \sum_{i=1}^{n} |x_i| \eta(e^i) \\ &\leqslant nM \max_i |x_i| \\ &\leqslant nM \| x \|_2 . \end{aligned}$$

(The final inequality proves one half of the lemma if we take $\beta = nM$.) If a sequence of n-vectors $\{x^k\}$ converges to $\hat{x} \in \mathbf{F}^n$, then $\| x^k - \hat{x} \|_2$ must converge to zero. By the last sequence of inequalities, it follows that $\eta(x^k - \hat{x})$ converges to zero; we conclude that $\eta(x^k) \to \eta(\hat{x})$ by the triangle inequality 5.6(b), since $\eta(x_k) \leqslant \eta(\hat{x}) + \eta(x_k - \hat{x})$.

Hence $\eta: \mathbf{F}^n \to \mathbf{R}$ is *continuous;* it follows that it assumes a minimum value $\alpha > 0$ on $\{x \in \mathbf{F}^n : \| x \|_2 = 1\}$. For any $x \in \mathbf{F}^n$, we conclude that

$$\alpha \leqslant \eta(\frac{x}{\| x \|_2}) ,$$

from which the lower bound on $\eta(x)$ follows. □

5.10 Corollary: *Let V be an n-dimensional vector space over* $\mathbf{F} = \mathbf{R}$ *or* $\mathbf{C}$. *If* $\| \ \|_a$ *and* $\| \ \|_b$ *be any two norms on* V, *then there exist positive constants* α *and* β *such that*

$$\| v \|_a \leqslant \alpha \| v \|_b$$

and

$$\|v\|_b \leqslant \beta \|v\|_a$$

for every $v \in V$. *It follows that a sequence of vectors* $\{v^k\} \subset V$ converges to $\hat{v} \in V$ with respect to $\| \ \|_a$ if and only if it converges to $\hat{v}$ with respect to $\| \ \|_b$.

Proof: Let $fv = \{v^1, \ldots, v^n\}$ be a basis for V; define $\eta_a, \eta_b : \mathbf{F}^n \to \mathbf{R}$ by

$$\eta_a(x) = \| \sum_{i=1}^{n} x_i v^i \|_a$$

and

$$\eta_b(x) = \| \sum_{i=1}^{n} x_i v^i \|_b .$$

It is easy to show that η_a and η_b are both norms on $\mathbf{F}^n$; by Lemma 5.9, there exist positive real numbers α_a, α_b, β_a, and β_b such that

$$\alpha_a \|x\|_2 \leqslant \eta_a(x) \leqslant \beta_a \|x\|_2$$

and

$$\alpha_b \|x\|_2 \leqslant \eta_b(x) \leqslant \beta_b \|x\|_2$$

for every $x \in \mathbf{F}^n$. If we take $\alpha = \beta_a \alpha_b^{-1}$ and $\beta = \beta_b \alpha_a^{-1}$, and remember that every $v \in V$ is of the form $x_1 v^1 + \ldots + x_n v^n$ for some $x \in \mathbf{F}^n$, then the result follows. □

We turn now to a discussion of the *norm of a linear mapping* $T : V \to W$ which is induced by norms on the vector spaces V and W. Let V and W be vector spaces over $\mathbf{F} = \mathbf{R}$ or $\mathbf{C}$; suppose that V is endowed with a norm $\| \ \|_V$ and W with a norm $\| \ \|_W$. Recall that $\mathrm{Hom}(V, W)$, the set of all linear mappings $T : V \to W$, is *also* (naturally) a vector space over $\mathbf{F}$. Consider the subset $\mathrm{Hom}_{\mathbf{B}}(V, W) \subset \mathrm{Hom}(V, W)$ consisting of those linear mappings $T : V \to W$ for which there exists an upper bound for the set of nonnegative real numbers given by

$$\{ \|T(v)\|_W : \|v\|_V = 1\} .$$

We say that $T : V \to W$ is a *bounded* linear mapping from V to W if and only if $T \in \mathrm{Hom}_{\mathbf{B}}(V, W)$.

It is evident that $\mathrm{Hom}_{\mathbf{B}}(V, W)$ is a *subspace* of $\mathrm{Hom}(V, W)$; to see this, observe that if T_1 and T_2 are in $\mathrm{Hom}(V, W)$, then

$$\{ \|T_1(v)+T_2(v)\|_W : \|v\|_V = 1\}$$

is *contained*, because of the triangle inequality 5.6(b) applied to $\| \ \|_W$, in the interval $[0, M_1+M_2]$, where M_i is an upper bound for

$$\{ \|T_i(v)\|_W : \|v\|_V = 1\} .$$

In case V is finite-dimensional, we can say even more.

5.11 Lemma: *If V is a finite-dimensional vector space over $\mathbf{F} = \mathbf{R}$ or $\mathbf{C}$, and W is an arbitrary vector space over* F, *then* $\mathrm{Hom_B}(V, W) = \mathrm{Hom}(V, W)$; *that is, if V is finite-dimensional, then every linear mapping $T : V \to W$ is bounded.*

Proof: Suppose that V has dimension n, and that $T \in \mathrm{Hom}(V, W)$. Let $\{v^1, \ldots, v^n\}$ be a basis for V; define the function $\eta : \mathbf{F}^n \to \mathbf{R}$ by $\eta(x) = \|x_1 v^1 + \ldots x_n v^n\|$. It's not hard to show that η defines a norm on $\mathbf{F}^n$; consequently, there exists by Lemma 5.9 some $\alpha > 0$ such that $\eta(x) \geqslant \alpha \|x\|_1$, where $\| \ \|_1$ is the "1-norm" on $\mathbf{F}^n$ (c.f. Example 5.7) given by

$$\|x\|_1 = \sum_{i=1}^{n} |x_i| , \qquad x \in \mathbf{F}^n .$$

It follows that if $x \in \mathbf{F}^n$ is such that $\|x_1 v^1 + \ldots + x_n v^n\|_V = 1$, then $\|x\|_1 \leqslant \alpha^{-1}$.

Now let $w^i = T(v^i)$ for each i, $1 \leqslant i \leqslant n$; let $M > 0$ be the maximum over i of $\|w^i\|_W$. If $v \in fV$ satisfies $\|v\|_V = 1$, write $v = x_1 v^1 + \ldots + x_n v^n$, and conclude that

$$\begin{aligned}\|T(v)\|_W &\leqslant \sum_{i=1}^{n} |x_i| \, \|w^i\|_W \\ &\leqslant \alpha^{-1} M ,\end{aligned}$$

from which the boundedness of T follows. □

We shall now demonstrate that norms $\| \ \|_V$ and $\| \ \|_W$ on arbitrary vector spaces V and W give rise to a norm $\| \ \|_{V,W}$ on the vector space $\mathrm{Hom_B}(V, W)$.

5.12 Definition: Let V and W be vector spaces equipped with respective norms $\| \ \|_V$ and $\| \ \|_W$. Given $T \in \mathrm{Hom_B}(V, W)$, define $\|T\|_{V,W}$, the norm of the linear mapping T *induced by* $\| \ \|_V$ *and* $\| \ \|_W$, as follows:

$$\|T\|_{V,W} = \sup\{\|T(v)\|_W : \|v\|_V = 1\} . \qquad \square$$

The supremum exists by boundedness of the linear mapping T ; we leave it as an exercise for the reader to show that $\| \ \|_{V,W}$ satisfies properties (a) through (c) in Definition 5.6. The following exercises should enhance the reader's understanding of the *operator norm* $\| \ \|_{V,W}$.

5.13 Exercises: (a) Show that in the above context, one may also define $\| \ \|_{V,W}$ for each $T \in \mathrm{Hom}_{\mathbf{B}}(V, W)$ by way of

$$\|T\|_{V,W} = \sup\{\|T(v)\|_W : \|v\|_V \leqslant 1\} ,$$

or

$$\|T\|_{V,W} = \sup\{\frac{\|T(v)\|_W}{\|v\|_V} : \|v\|_V \neq 0\} .$$

(b) Show that $\|T(v)\|_W \leqslant \|T\|_{V,W} \|v\|_V$. whenever $T \in \mathrm{Hom}_{\mathbf{B}}(V, W)$ and $v \in V$. $\square$

The intuition behind the definition of $\| \ \|_{V,W}$ should be apparent; the idea is that $T : V \to W$ should, in some sense, be considered *large* if it "stretches" vectors in V by a large amount. Although there might be no apparent relationship between $\| \ \|_V$ and $\| \ \|_W$ which renders the two norms commensurate, the statement that $T \in \mathrm{Hom}_{\mathbf{B}}(V, W)$ is the same as the statement that T maps the *unit sphere* in V into a *bounded set* in W. The "largest factor" by which T stretches the length of vectors in V's unit sphere is then defined to be the norm of T induced by the norms on V and W.

From Lemma 5.11, we may conclude that *every* linear mapping $T : \mathbf{F}^n \to \mathbf{F}^m$ is bounded, no matter *what* norms are used on the domain and codomain vector spaces. We have seen (c.f. Exercise 5.7) three examples of norms on $\mathbf{F}^n$: the *Euclidean norm* $\| \ \|_2$, the so-called *1-norm* $\| \ \|_1$, and the so-called *infinity norm* $\| \ \|_\infty$. Suppose we are given a linear *transformation* $T : \mathbf{F}^n \to \mathbf{F}^n$; we may define $\|T\|_2$, $\|T\|_1$, and $\|T\|_\infty$ as the norms of T induced by using, respectively, the Euclidean norm, the 1-norm, and the infinity norm on $\mathbf{F}^n$. Let ${}_AT$ be the matrix of the linear transformation T with respect to the standard basis for $\mathbf{F}^n$. It is not hard to show that

$$\|T\|_1 = \max_j \sum_{i=1}^{n} |[{}_AT]_{ij}| ,$$

and that

$$\|T\|_\infty = \max_i \sum_{j=1}^{n} |[{}_AT]_{ij}| .$$

The proofs of these formulas are left as exercises for the reader. It is easy to show (by means of Exercise 5.13(b)) that both equations hold with $\leqslant$ replacing $=$; it is then a simple matter to construct, in each case, some $x \in \mathbf{F}^n$ having norm one for which the inequality in Exercise 5.13(b) holds with equality. It turns out that $\|T\|_2$ is more difficult to specify; the interested reader is referred to Exercise 12.13.

The notions of *convergence* arising from norms on vector spaces, as we have noted, give rise to a corresponding notions of *continuity* of mappings between the vector spaces. Suppose that V and W are vector spaces over F, each equipped with its own norm, and that $T : V \to W$ is a linear mapping. It turns out that the continuity of T and the boundedness of T are equivalent.

5.14 Lemma: *Suppose that V and W are vector spaces over* F *which are equipped with norms. Then $T \in \mathrm{Hom}(V, W)$ is continuous with respect to the given norms if and only if $T \in \mathrm{Hom}_{\mathrm{B}}(V, W)$.*

Proof: Suppose first that T is bounded; denote its induced norm by $\|T\|$. If $\{v^k\}$ converges to $\hat{v}$ as $k \to \infty$, then $\{v^k - \hat{v}\}$ converges to $0 \in V$. Let $w^k = T(v^k)$ and $\hat{w} = T(\hat{v})$. For every $k \geqslant 0$, we have

$$\|w^k - \hat{w}\|_W \leqslant \|T\| \, \|v^k - \hat{v}\|_V ;$$

hence $\|w^k - \hat{w}\|_W$ must go to zero as $k \to \infty$, proving continuity of T.

Conversely, if T is continuous, then T must be bounded. To see this, suppose temporarily that T is *not* bounded. By definition, there exists a sequence $\{v^k\}$ of vectors in V, each having norm 1, such that

$$\lim_{k \to \infty} \|T(v^k)\|_W = \infty .$$

Let q^k be defined by

$$q^k = \frac{v^k}{\|T(v^k)\|_W}, \qquad 0 \leq k < \infty;$$

then the sequence $\{q^k\}$ converges to $0 \in V$, but the sequence $\{T(q^k)\}$ does *not* converge to $T(0) = 0 \in W$; indeed, each $T(q^k)$ has W-norm 1. We conclude that T is not continuous, which contradicts our initial assumption. □

We turn now to the abstract generalization of the well-known *scalar product*, or *dot product*, of vectors in $\mathbf{R}^2$ or $\mathbf{R}^3$. Given two vectors x^1 and x^2 in $\mathbf{R}^2$ or $\mathbf{R}^3$, we define their scalar product by means of the formula

$$<x^1, x^2> = \|x^1\|_2 \|x^2\|_2 \cos\theta,$$

where θ is the angle between x^1 and x^2.

The n-dimensional generalization of the scalar product is the *Euclidean inner product* on $\mathbf{F}^n$, where, as usual, F denotes **R** or **C**. Even more generally, we have

5.15 Definition: Let $\mathbf{F} = \mathbf{R}$ or $\mathbf{C}$. An *inner product* on a vector space V over F is a mapping $(v, w) \to <v, w> \in \mathbf{F}$ which takes pairs of vectors in V to scalars in **F** and which obeys the following rules: (Here, a "star superscript" means complex conjugate; observe that $\alpha^* = \alpha$ if and only if $\alpha \in \mathbf{R}$.)

(a) For all $v, w, v', w' \in V$:

$$\begin{aligned} <v + v', w> &= <v, w> + <v', w> \\ <v, w + w'> &= <v, w> + <v, w'> \\ <w, v> &= <v, w>^* . \end{aligned}$$

(b) For all $v, w \in V$ and $\alpha \in \mathbf{F}$:

$$<\alpha v, w> = \alpha <v, w> = <v, \alpha^* w> .$$

(c) For every $v \in V$, $<v, v> \geq 0$; moreover, $<v, v> = 0$ if and only if $v = 0$. □

5.16 Theorem: *Every inner product* $< \, , \, >$ *on a real or complex vector space V over induces a norm on V by way of*

$$\|v\| = (<v, v>)^{1/2}$$

Proof: Properties 5.6 (a) and (c) are direct consequences of 5.8 (a) and (c). As for the triangle inequality 5.6 (b), we must resort to a trick. Let α be any real scalar; by 5.8 (c), we have

$$\begin{aligned} 0 &\leqslant <v+\alpha w, v+\alpha w> \\ &\leqslant <v,v> + \alpha <v,w> + \alpha <w,v> + \alpha^2 <w,w>. \end{aligned}$$

Now, the middle two terms sum to $2\alpha \mathrm{Re}\{<v,w>\}$, where Re{ } means real part. Thus we have a quadratic inequality in α of the form $a\alpha^2 + b\alpha + c \geqslant 0$; by elementary analytic geometry, this equation may not have two distinct real roots, which implies that $b^2 - 4ac \leqslant 0$. In the present context, we obtain

$$Re\{<v,w>\} \leqslant (<v,v><w,w>)^{1/2}.$$

Taking $\alpha = 1$, substitution of the last inequality into the previous equation yields

$$<v+w, v+w> \leqslant [<v,v>^{1/2} + <w,w>^{1/2}]^2,$$

whereby

$$\| v+w \| = <v+w, v+w> \leqslant \| v \| + \| w \|. \qquad \square$$

In the proof of Theorem 5.16, we nearly proved another important result, the *Schwarz inequality*, which says that if we are given an inner product $< , >$ on a vector space V, then

$$|<v,w>| \leqslant \| v \| \, \| w \|,$$

where the norms on the right-hand side are induced via Theorem 5.16 by $< , >$. To prove the Schwarz inequality, pick α to be given by

$$\alpha = \frac{-|<v,w>|}{<w,w>} \exp(i \; arg <v,w>),$$

and the first equation in the proof becomes

$$\begin{aligned} 0 &\leqslant <v+\alpha w, v+\alpha w> \\ &= <v,v> + \alpha <w,v> + \alpha^* <v,w> + |\alpha|^2 <w,w> \\ &= <v,v> - \frac{|<v,w>|^2}{<w,w>}, \end{aligned}$$

from which we obtain

$$|<v,w>| \leqslant <v,v>^{1/2} <w,w>^{1/2} = \| v \| \, \| w \|,$$

which is the Schwarz inequality. (Note: α is real if V is a real vector space.)

5.17 Definition: Let v and w be vectors in a vector space V; let $< , >$ be an inner product on V. Then v and w are said to be *orthogonal* if and only if

$$<v, w> = 0 .$$

More generally, if $v^1, \ldots, v^k \in V$, then the v^i are said to be *orthogonal with respect to* $< , >$ if and only if

$$<v^i, v^j> = 0, \quad i \neq j, \quad 1 \leqslant i, j \leqslant k,$$

and *orthonormal with respect to* $< , >$ if in addition

$$<v^i, v^i> = 1, \quad 1 \leqslant i \leqslant k . \qquad \square$$

As it happens, it is always possible (and often desirable) to obtain a basis $\mathbf{w}$ for a finite dimensional vector space V in such a way that $\mathbf{w}$ is also an orthonormal set with respect to a given inner product on V. Our final result presents a systematic means for obtaining an orthonormal basis for V from an arbitrary basis.

First, we consider *orthogonal projections* in a finite-dimensional vector space V which is endowed with an inner product $< , >$. Let $\{w^1, \ldots, w^k\}$ be an orthonormal set of vectors in V; let $v \in V$ be arbitrary. Consider the subspace W of V spanned by the w^i. We seek a vector $w^o \in W$ which is "closest to v" in the sense that

$$\| v - w^o \| \leqslant \| v - w \|$$

for every $w \in W$, where the indicated norm is that induced by $< , >$ as in Theorem 5.16.

Our intuition dictates that w^o should be obtained by "dropping a perpendicular" from v to the subspace W:

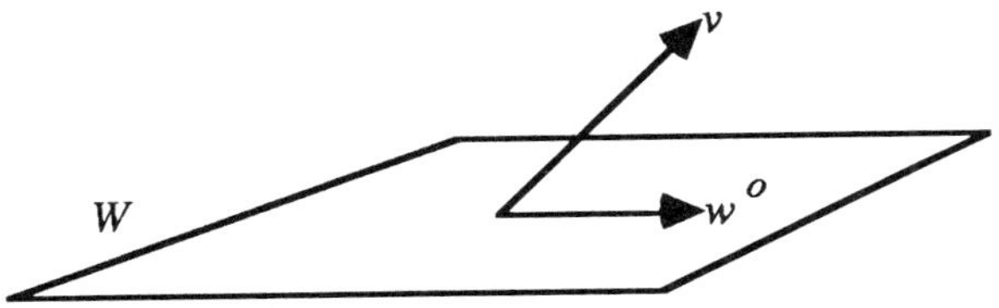

Figure 5.1 Orthogonal projection

If this may be accomplished, we have $w^o \in W$ such that $v - w^o$

is orthogonal to every $w \in W$. If $w \in W$ is arbitrary, then

$$\begin{aligned} \|v-w\|^2 &= \|v-w^o + w^o-w\|^2 \\ &= <(v-w^o)+(w^o-w),(v-w^o)+(w^o-w)> \\ &= \|v-w^o\|^2 + \|w^o-w\|^2 \geqslant \|v-w^o\|^2, \end{aligned}$$

where the last equality follows from the fact that $v-w^o$ is orthogonal to w^o-w. Hence w^o so chosen is the vector in W closest to v; w^o is called *the orthogonal projection of v onto the subspace W*.

A formula for w^o may be obtained as follows: since $w^o \in W$, there exist α_i such that

$$w_o = \sum_{i=1}^{k} \alpha_i w^i ;$$

moreover, since $v-w^o$ must in particular be orthogonal to each w^j, it must be true, for $1 \leqslant j \leqslant k$, that

$$<v-w^o, w^j> = <v-\sum_{i=1}^{k} \alpha_i w^i, w^j> = 0 .$$

Expanding the inner product, we obtain

$$<v, w^j> = \alpha_j <w^j, w^j> = \alpha_j, \qquad 1 \leqslant j \leqslant k,$$

since $\{w^i\}$ is an orthonormal set of vectors. Thus

$$w^o = \sum_{i=1}^{k} <v, w^i> w^i .$$

Now, suppose that $\mathbf{v} = \{v^1, \ldots, v^n\}$ is an arbitrary basis for V and $< , >$ is an inner product on V. We obtain an orthonormal basis $\mathbf{w}$ for V by the *Gram-Schmidt procedure*, which may be described as follows.

5.18 Gram-Schmidt Procedure: Given that the n-tuple $\mathbf{v} = \{v^1, \ldots, v^n\}$, is a basis for an n-dimensional vector space V, and $< , >$ is an inner product on V, define an orthonormal basis $\mathbf{w} = \{w^1, \ldots, w^n\}$ for V inductively as follows: (Here, $\|v\| = <v, v>^{1/2}$.)

$$\begin{aligned} w^1 &= \frac{v^1}{\|v^1\|} \\ w^{i+1} &= \frac{z^{i+1}}{\|z^{i+1}\|}, \qquad 1 \leqslant i \leqslant n-1, \end{aligned}$$

where $z^{i+1} = v^{i+1}$ − projection of v^{i+1} on span$\{w^1, ..., w^i\}$; that is,

$$z^{i+1} = v^{i+1} - \sum_{j=1}^{i} < v^{i+1}, w^j > w^j \ . \qquad \square$$

Note that $z^{i+1} \neq 0$ for each i since $\{w^1, w^2, \ldots, w^i, v^{i+1}\}$ is a linearly independent set; this last assertion follows from the fact that, by construction, we have

$$\text{span}\{w^1, \ldots, w^i\} = \text{span}\{v^1, \ldots, v^i\}, \quad 1 \leqslant i \leqslant n .$$

5.19 Exercise: Prove the last identity in the preceding paragraph. [*Hint* : Since w^i is a linear combination of the v^j for $j \leqslant i$, the left-hand side is included in the right-hand side. The reverse inclusion follows since v^i is a linear combination of the w^j for $j \leqslant i$.] □

We now discuss another important construction which hinges on the notion of inner product, and which should serve to sharpen the reader's geometric intuition concerning orthogonality of vectors.

5.20 Definition: Let V be a vector space; let $< , >$ be an inner product on V. Let W be a subspace of V; the *orthogonal complement of* W, written $W^\perp$, is the subspace of V consisting of all vectors in V which are orthogonal to *every* vector in W; i.e.,

$$W^\perp = \{v \in V : <v, w> = 0 \text{ for } all \ w \in W\}. \qquad \square$$

5.21 Exercise: Show that $W^\perp$ is, indeed, a subspace of V. □

5.22 Theorem: *Let V be an n-dimensional vector space with inner product $< , >$; let W be a k-dimensional subspace of V. Then*

(a) W and $W^\perp$ are disjoint subspaces of V; i.e., $W \cap W^\perp = \{0\}$; and

(b) $V = W + W^{\perp}$, *where* + *denotes vector sum, implying* (by Observation 4.28(c)) that $dim(W^{\perp}) = n-k$.

Proof: (a) Suppose $w_o \in W \cap W^{\perp}$; since $w \in W^{\perp}$, w_o is orthogonal to every $w \in W$, and, in particular, since $w_o \in W$, $<w_o, w_o> = 0$, which implies (by 5.16 (c)) that $w_o = 0$.

(b) Given $v \in V$, let w_o denote the orthogonal projection of v on W; then

$$v = w_o + (v - w_o),$$

where $(v - w_o) \in W^{\perp}$ by construction. Hence every $v \in V$ may be written as the sum of a vector in W and a vector in $W^{\perp}$, implying that $V = W + W^{\perp}$. □

5.23 Exercise: Let A be a real $(m \times n)$ matrix.

(a) Show that the range of A^T is the orthogonal complement (taken with respect to the Euclidean inner product on F^n) of the nullspace of A. (The range of A^T, for obvious reasons, is often called the *rowspace of* A.)

(b) Conclude from (a) that the rank of A^T is the same as the rank of A.

(c) Is part (a) true if A is complex and A^T is replaced by $A^\dagger$, the conjugate transpose of A (c.f. 1.1(f))? Try to generalize part (a) to complex matrices. □

PART II
STATE SPACE LINEAR SYSTEMS

6. *State Space Linear Systems: Formal Definitions And General Properties*

In this section, we present formal definitions of continuous- and discrete-time state space linear systems and discuss some general properties which such systems exhibit. The treatment may be viewed as an effort to make precise some of the notions described heuristically in the Introduction. Specifically, *linear systems* should serve as *mathematical models* for "real-world" processes which take input "time functions" to output "time functions" in some linear fashion. The real-world processes which we seek to model with, say, continuous-time state space linear systems are those which take m real-valued suitably well-behaved input time-functions $t \to u_i(t)$, $1 \leqslant i \leqslant m$, and "process" them through a box in such a way as to yield p real-valued output time functions $t \to y_j(t)$, $1 \leqslant j \leqslant p$.

The box is assumed to have associated with it some family of possible *internal situations;* if one knows the internal situation at some time, and one is given the input for all later times, one should be able to figure out precisely the future time-evolution both of the internal situation and of the output. Moreover, all of this should be happening "linearly," in some sense.

The inputs and outputs in such real-world processes are usually ordinary time-functions that are handed to us by the conditions under which the processes operate. Inputs and outputs in a mechanical process may be, respectively, forcing functions and measurements of positions and velocities; in an electrical process, inputs might come from voltage and current sources, and outputs from voltmeters and current meters attached strategically at various points.

In what sense, though, may the time-evolution of the "internal situation" in a real-world process be regarded as a *time-function?* This "function" must take values in the "set of all possible internal situations;" this set must have some structure in order to render the time-function concept meaningful. In practical situations, the "internal situation set" is usually easily modeled as some space of n-tuples of real numbers; the sets of possible instantaneous values of capacitor voltages and inductor currents, of particles' positions and velocities, and even of bank balances, all spring immediately to mind as appropriate models for sets of internal situations in various real-world processes. The time-evolution of the "state" of such a process, then, has an evident interpretation as a bona fide, real n-vector-valued time function.

Nonetheless, as is often the case in applied mathematics, it is conceptually useful to extract the fundamental components and properties of such specific mathematical models and construct a general framework for analyzing processes which may be modeled similarly. By so doing, one is often able to draw significant parallels between disparate processes and their behaviors; moreover, the generality of a broad mathematical framework can, in many cases, facilitate the proofs of conclusions which hold in a variety of very different contexts. It is with these thoughts in mind that we present in this section a very general (and perhaps, to some, annoyingly abstract) definition of "state space linear system." The reader who is interested only in applications might skip ahead to §7, or even to §8, since essentially all of the results about linear-systems-in-the-abstract have concrete manifestations which may be (and are, in this book) presented in a form more palatable to those who are averse to formalism.

We sum up the above philosophical digression in the following minimal list of items which our definition of "state-space system" should contain:

(1) a mathematical specification as to what input and output time-functions are allowed;

(2) a set X serving as a *state space* in the sense that elements of X ("states") should correspond to possible internal situations;

(3) a mathematical specification of how the current "state" and future inputs determine the future "state;" and

(4) a mathematical specification of how the current "state" and present input value determine the current output.

The following definitions accomplish these objectives. Similar definitions are given in [Kalman, Falb, and Arbib] and in [Padulo and Arbib].

6.1 Definition: A *real m-input, p-output finite-dimensional continuous-time state space linear system* consists of the following:

(a) A vector space U of *input functions* $u : \mathbf{R} \to \mathbf{R}^m$;

(b) For each $t_o \in \mathbf{R}$, a vector space Y_{t_o} of *output functions* $y : [t_o, \infty) \to \mathbf{R}^p$;

(c) An n-dimensional real vector space X, the *state space* of the system;

(d) For each pair of (real) times (t_1, t_o) with $t_1 \geqslant t_o$, a *state transition mapping* from $X \oplus \mathrm{U}$ to X defined for each $x_o \in X$, $u \in \mathrm{U}$ as

$$(x^\circ, u) \to \phi(t_1, t_o, x^\circ, u),$$

which satisfies the following conditions:

(*i*) *linearity:* The above mapping is linear from $X \oplus \mathrm{U}$ to X for every $t_1, t_o \in \mathbf{R}$ with $t_1 \geqslant t_o$;

(*ii*) *consistency:* $\phi(t_o, t_o, x^\circ, u) = x^\circ$ for all $x^\circ \in X$, $u \in \mathrm{U}$, $t_o \in \mathbf{R}$;

(*iii*) *causality:* If $t_1 \geqslant t_o$, and if u, $\hat{u} \in \mathrm{U}$ are such that $u(\tau) = \hat{u}(\tau)$ for every $\tau \in [t_o, t_1)$, then

$$\phi(t, t_o, x^\circ, u) = \phi(t, t_o, x^\circ, \hat{u})$$

for every $t \in [t_o, t_1]$, $x^\circ \in X$;

(*iv*) *semigroup property:* If $t_2 \geqslant t_1 \geqslant t_o$, then for every $x^\circ \in X$ and $u \in \mathrm{U}$:

$$\phi(t_2, t_o, x^\circ, u) = \phi(t_2, t_1, \phi(t_1, t_o, x^\circ, u), u);$$

(*e*) For each $t \in \mathbf{R}$, a linear *readout mapping* from $X \oplus \mathbf{R}^m$ to $\mathbf{R}^p$ defined, for each $x_o \in X$, $w \in \mathbf{R}^m$, as

$$(x, w) \to \rho(t, x, w). \qquad \square$$

6.2 Definition: A *real, m-input, p-output finite-dimensional discrete-time state space linear system* consists of the following:

(a) A vector space U of *input functions* $u : Z \to R^m$;

(b) For each $k_o \in Z$, a vector space Y_{k_o} of *output functions* $y : [k_o, \infty) \to R^p$ defined on the semi-infinite sequence $\{k_o, k_o + 1, \ldots\}$;

(c) An n-dimensional real vector space X, the *state space* of the system;

(d) For each pair of (integer) times (k_1, k_o) with $k_1 \geqslant k_o$ a *state-transition mapping* from $X \oplus U$ to X defined for each $x^\circ \in X$, $u \in U$ as

$$(x^\circ, u) \to \phi(k_1, k_o, x^\circ, u),$$

which satisfies the following conditions:

(*i*) *linearity:* The above mapping is linear from $X \oplus U$ to X for every $k_1 \geqslant k_o$;

(*ii*) *consistency:* $\phi(k_o, k_o x^\circ, u) = x^\circ$ for every $k_o \in Z$, $x^\circ \in X$, $u \in U$;

(*iii*) *causality:* If $u, \hat{u} \in U$ are such that $u(l) = \hat{u}(l)$ for every l with $k_o \leqslant l < k_1$, then

$$\phi(k, k_o, x^\circ, u) = \phi(k, k_o, x^\circ, \hat{u})$$

for every k with $k_o \leqslant k \leqslant k_1$ and for every $x^\circ \in X$;

(*iv*) *semigroup property:* If $k_2 \geqslant k_1 \geqslant k_o$, then for every $x^\circ \in X$ and $u \in U$:

$$\phi(k_2, k_o, x^\circ, u) = \phi(k_2, k_1, \phi(k_1, k_o, x^\circ, u), u);$$

(e) For each $k \in Z$, a linear *readout mapping* from $X \oplus R^m$ to R^p defined, for each $x \in X$, $w \in R^m$, as

$$(x, w) \to \rho(k, x, w) \quad x \in X, w \in R^m. \qquad \square$$

A few words should be said about how these definitions provide us with a class of mathematical models which contain the "desired" components (1) to (4) above. Certainly, items (a) and (b) of Definitions 6.1 and 6.2 are self-explanatory; specification of U, along with the Y_{t_o} and Y_{k_o}, identifies those inputs which are admissible and those outputs which may arise when the system "operates" on these inputs. Item (c) in Definitions 6.1 and 6.2 answers requirement (2) above. We assume (as we must) that the vector spaces Y_{t_o} in Definition 6.1 are "large enough" to include every output function $t \rightarrow y(t)$, defined on $[t_o, \infty)$, which might arise from the application of an input function in U to the system when the state at t_o is an arbitrary $x^\circ \in X$. A similar assumption is made about the spaces Y_{k_o} in Definition 6.2.

The state-transition mapping, as described under items (d) in the definitions, is perhaps the most complicated component of a linear system. The idea (in the context of Definition 6.1) is that $\phi(t_1, t_o, x^\circ, u)$ *is the state of the system at time* t_1 *given that the state was* x° *at time* t_o *and input* u *was applied between* t_o *and* t_1. Thus, item 6.1 (d) satisfies requirement (3). Similarly, item (e), the readout mapping, is the answer to (4) above. In words, $\rho(t, x, w)$ is *the output at time t given that the state at time t is x and the input value at time t, i.e.* $u(t)$, *is* w. Similar verbal descriptions may be given for the corresponding components of discrete-time systems.

The conditions (d) *(i)*, *(ii)*, *(iii)*, *and* *(iv)* *on the* state transition mappings in Definitions 6.1 and 6.2 are natural, as well. Property *(i)* is the embodiment of linearity: the state at some later time should be "jointly linear" in the state "now" and the input function which is applied "between now and then." Property *(ii)* says that the state "now" determines the state "now" regardless of what is happening at the input. Property *(iii)* serves a dual purpose; it asserts that, given the state at the beginning of some time interval, the state evolution during that time interval depends only on the input applied *during that time interval*. That is, inputs which were applied *before* time t_o are irrelevant to the state evolution during the interval $[t_o, t_1]$ *once we know the state at time* t_o; simultaneously, inputs to be applied in the *future* (i.e. *after* time t_1) are irrelevant to the state evolution *before* time t_1. The second clause in the preceding sentence is what one might ordinarily think of as a causality condition, but we shall use the term "causality" to cover both statements. The state of the system at a given time, then, summarizes the entire history of the system up to that time, and constitutes "everything one needs to know" in order to determine future states of the system given future inputs.

Conditions (d) *(iv)*, the semigroup properties, appear, at least initially, to be somewhat less fathomable. In essence, they attest to the equivalence of two different "experiments" which one may perform on the same system. In one case, assuming once again that we're in the continuous-time context, an input u is applied to the system between t_o and t_2; the system begins in state x° at time t_o and ends up in state $\phi(t_2, t_o, x^\circ, u)$ at time t_2. In the other "experiment," the same input is applied from t_o to t_1, again starting from state x° at t_o, and a state $x^1 = \phi(t_1, t_o, x^\circ, u)$ is reached at time t_1; subsequently, x^1 is used as initial condition at time t_1, and the input u continues to be applied until t_2, at which point the system has reached state

$$\phi(t_2, t_1, x^1, u) = \phi(t_2, t_1, \phi(t_1, t_o, x^\circ, u), u).$$

The semigroup property says that these two procedures lead to the same "final" system state at time t_2, as we might expect they should.

Implicit in Definitions 6.1 and 6.2 is that the state of a linear system might not be directly observable; the spaces Y_{t_o} and Y_{k_o} play the role of "experimental result spaces." An experimenter may, in principle, attempt to determine by observing input-output behavior whether the system at time t_o was in state x° or $\hat{x}^\circ$; this measurement of external (i.e., input-output) behavior, however, is the only *a priori* means by which such a determination may be made.

We make two additional definitions before giving some examples of linear systems. These definitions are actually statements of terminology more than anything else.

6.3 Definition: The *overall response function* **S** of an m-input, p-output continuous-time finite-dimensional state space linear system is defined as the mapping which takes values in $\mathbf{R}^p$ and has domain $\mathbf{R}_+^2 \times X \oplus \mathbf{U}$, where $\mathbf{R}_+^2$ is the set of all $(t, t_o) \in \mathbf{R}^2$ such that $t \geqslant t_o$, and which is given at points in that domain by

$$\mathbf{S}(t, t_o, x^\circ, u) = \rho(t, \phi(t, t_o, x^\circ, u), u(t)). \qquad \square$$

6.4 Definition: The *overall response function* **S** of an m-input, p-output discrete-time finite-dimensional state space linear system is defined as the mapping which takes values in $\mathbf{R}^p$ and has domain $\mathbf{Z}_+^2 \times X \oplus \mathbf{U}$, where $\mathbf{Z}_+^2$ is the set of all $(k, k_o) \in \mathbf{Z}^2$ such that $k \geqslant k_o$, and which is given at points in its domain by

$$S(k, k_o, x^\circ, u) = \rho(k, \phi(k, k_o, x^\circ, u), u(k)). \qquad \square$$

In words, the overall response function of a (continuous-time) linear system may be described as follows: $S(t, t_o, x^\circ, u)$ is the value of the output at time $t \geqslant t_o$ given that the state of the system at time t_o was $x^\circ \in X$ and input $u \in \mathbf{U}$ was applied between t_o and t. Thus S describes the way in which initial times, initial states, and input functions lead to future values of the output. Observe once again that the output value at a given time depends only on the state *at that time* and the input value *at that time*. As far as the output is concerned, in other words, the present state summarizes the entire history of the system, and only the present value of the input function has additional influence on the present output value.

It is important at this point to give two examples of how linear systems —that is, composite mathematical objects satisfying Definitions 6.1 and 6.2 —may be constructed. As we shall see in the next two sections, these examples are universal in a very fundamental sense. The problem of analyzing arbitrary finite dimensional state space linear systems may, in a manner of speaking, be reduced to the problem of analyzing systems such as the ones in the examples.

6.5 Example: *A continuous-time linear system with state space* $\mathbf{R}^n$.

To define the system, we'll need to define the various components which make up Definition 6.1. We let $\mathbf{U}$ be the vector space of all continuous real m-vector-valued functions $u : \mathbf{R} \to \mathbf{R}^m$; for each $t_o \in \mathbf{R}$, let $\mathbf{Y}_{t_o}$ be the set of all continuous real p-vector-valued functions $y : [t_o, \infty) \to \mathbf{R}^p$. Let $t \to A(t)$, $t \to B(t)$, $t \to C(t)$, and $t \to D(t)$ be continuous matrix functions, defined on $\mathbf{R}$, where the matrices are of respective sizes $(n \times n)$, $(n \times m)$, $(p \times n)$, and $(p \times m)$. Take X to be $\mathbf{R}^n$. For each real t, s with $t \geqslant s$, let $t \to \Phi(t, s)$ be the $(n \times n)$ matrix solution to

$$\dot{M}(t) = A(t)M(t) \quad M(s) = I_n.$$

(See §2.) Define, for each $t_1 \geqslant t_o$, $x^\circ \in \mathbf{R}^n$, $u \in \mathbf{U}$:

$$\phi(t_1, t_o, x^\circ, u) = \Phi(t_1, t_o)x^\circ + \int_{t_o}^{t_1} \Phi(t_1, \tau)B(\tau)u(\tau)d\tau,$$

and for each $t \in \mathbf{R}$, $x \in \mathbf{R}^n$, $w \in \mathbf{R}^m$

$$\rho(t, x, w) = C(t)x + D(t)w .$$

We claim that these assignments comprise an m-input, p-output continuous-time finite-dimensional state space linear system with state space $X = \mathbf{R}^n$. We need only make certain that the various components meet the prerequisites given in Definition 6.1. Let's check, for example, the state transition mapping. For each t_1, t_o, $(x^\circ, u) \to \phi(t_1, t_o, x^\circ, u)$ is obviously jointly linear in x° and u. Conditions (d) *(ii)* and *(iii)* in Definition 6.1 are also clearly satisfied; observe that $\phi(t, t_o, x^\circ, u)$, for $t \in [t_o, t_1]$, depends only on $u(\tau)$ for $\tau \in [t_o, t_1]$. As for the semigroup property, recall from the proof of Theorem 2.8 that the matrix function Φ satisfies a "semigroup property" of its own; namely, for all t, s, τ,

$$\Phi(t, s)\Phi(s, \tau) = \Phi(t, \tau) . \tag{SG}$$

Thus, in the present example, if $t_2 \geqslant t_1 \geqslant t_o$,

$$\begin{aligned}
\phi(t_2, t_o, x^\circ, u) &= \Phi(t_2, t_o)x^\circ + \int_{t_o}^{t_2} \Phi(t_2, \tau)B(\tau)u(\tau)d\tau \\
&= \Phi(t_2, t_1)[\Phi(t_1, t_o)x^\circ + \int_{t_o}^{t_1} \Phi(t_1, \tau)B(\tau)u(\tau)d\tau] \\
&\quad + \int_{t_1}^{t_2} \Phi(t_2, \tau)B(\tau)u(\tau)d\tau \\
&= \phi(t_2, t_1, \phi(t_1, t_o, x^\circ, u), u) .
\end{aligned}$$

The remaining properties in Definition 6.1 are easily proven for this example. Notice that, for every $t \geqslant t_o$, $x^\circ \in \mathbf{R}^n$, $u \in \mathbf{U}$, the n-vector function

$$t \to \phi(t, t_o, x^\circ, u) , \qquad t \geqslant t_o ,$$

is the unique solution, for $t \geqslant t_o$, to the differential equation and initial condition

$$\dot{x}(t) = A(t)x(t) + B(t)u(t) , \qquad t \in \mathbf{R} \tag{DE}$$

$$x(t_o) = x^\circ \tag{IC}$$

If we add to this pair of equations the additional "state-output" relation

$$\begin{aligned} y(t) &= C(t)x(t) + D(t)u(t) \\ &= \rho(t, x(t), u(t)), \end{aligned} \qquad \text{(SO)}$$

then we see why the equations (DE), (IC), (SO) are often said to "define a linear system with state space $\mathbf{R}^n$."

Note that the overall response function $\mathbf{S}$ of the system in this example is given by

$$\begin{aligned} \mathbf{S}(t, t_o, x^\circ, u) &= \rho(t, \phi(t, t_o, x^\circ, u), u(t)) \\ &= C(t)\phi(t, t_o, x^\circ, u) + D(t)u(t) \\ &= C(t)\Phi(t, t_o)x^\circ + \int_{t_o}^{t} C(t)\Phi(t, \tau)B(\tau)u(\tau)d\tau \\ &\quad + D(t)u(t). \end{aligned}$$

□

6.6 Example: *A discrete-time system with state space* $\mathbf{R}^n$.

This example parallels Example 6.5; let $\mathbf{U}$ be the vector space of *all* real discrete-time m-vector functions $u : \mathbf{Z} \to \mathbf{R}^m$ (c.f. Example 4.2(d)); for $k_o \in \mathbf{Z}$, let $\mathbf{Y}_{k_o}$ be the set of all real p-vector functions $y : \{k_o, k_o+1, k_o+2, \ldots\} \to \mathbf{R}^p$. Let $X = \mathbf{R}^n$; let $k \to A(k)$, $k \to B(k)$, $k \to C(k)$, and $k \to D(k)$ be discrete-time matrix functions of respective sizes $(n \times n)$, $(n \times m)$, $(p \times n)$, and $(p \times m)$. Define $\Phi(k, l)$, for $k \geqslant l$, by

$$\Phi(k, l) = \begin{cases} I_n & k = l \\ A(k-1)\ldots A(l) & k > l, \end{cases}$$

and, for $k_1 \geqslant k_o$, let the state-transition mapping be given for $x^\circ \in \mathbf{R}^n$ and $u \in \mathbf{U}$ by

$$\phi(k_1, k_o, x^\circ, u) = \Phi(k, k_o)x^\circ + \sum_{l=k_o}^{k_1-1} \Phi(k_1, l+1)B(l)u(l),$$

and define the readout mapping, for $k \in \mathbf{Z}$, $x \in \mathbf{R}^n$, $w \in \mathbf{R}^p$, by

$$\rho(k, x, w) = C(k)x + D(k)w.$$

Just as in Example 6.5, these specifications define an m-input, p-output discrete-time finite-dimensional state space linear system with state space $\mathbf{R}^n$. We leave it to the reader to verify, using techniques from §3, that the various components of the system so defined satisfy the requirements of Definition 6.2. Observe (c.f. Example 6.5) that the equations

$$x(k+1) = A(k)x(k) + B(k)u(k)\,, \qquad k \geqslant k_o$$

$$x(k_o) = x^\circ$$

$$y(k) = C(k)x(k) + D(k)u(k)\,, \qquad k \in \mathbf{Z}$$

may be viewed as "giving rise to" the system in the present example, since $k \to \phi(k, k_o, x^\circ, u)$, as defined, is the unique solution to the above difference equation and initial condition; the third equation, which is a "state-output relation," reveals the overall response function S of the system as being given by

$$S(k, k_o, x^\circ, u) = C(k)\Phi(k, k_o)x^\circ$$

$$\sum_{l=k_o}^{k-1} C(k)\Phi(k, l+1)B(l)u(l) + D(k)u(k)$$

for $k \geqslant k_o$, $x^\circ \in \mathbf{R}^n$, $u \in \mathbf{U}$. □

Because of the linearity hypotheses in Definitions 6.1 and 6.2, many special conclusions may be drawn about *linear* systems and their behavior. These conclusions are often referred to in the "classical" physics and engineering literature as *superposition principles;* as we shall see, they are direct consequences of linearity. First we introduce some more terminology.

6.7 Definition: Given an m-input, p-output continuous-time state space linear system as in Definition 6.1, and given $x^\circ \in X$, $t_o \in \mathbf{R}$, and $u \in \mathbf{U}$:

(a) The mapping $t \to \phi(t, t_o, x^\circ, u)$, defined for $t \in [t_o, \infty)$ and taking values in X, is called the *trajectory of the system starting from* x° *at time* t_o *under the input* u. The mappings $t \to \phi(t, t_o, x^\circ, 0)$ and $t \to \phi(t, t_o, 0, u)$ are called, respectively, the *zero-input trajectory starting from* x° *at time* t_o and the *zero-state trajectory starting at* t_o *under the input* u.

(b) The function $t \to S(t, t_o, x^\circ, 0)$, defined for $t \in [t_o, \infty)$ and taking values in $\mathbf{R}^p$, is called the *zero-input response starting from* x° *at time* t_o; the mapping $t \to S(t, t_o, 0, u)$ is called the *zero-state response starting at time* t_o *under the input* u. □

The trajectory of a linear system starting from a given initial state x° at a given time t_o under an input u may be viewed as a *curve* or *path* in the state space X which emanates from x° and is parametrized by time. The state of the system "follows" this path as time goes on. The zero-input and zero-state responses of a system (given x°, t_o, u) are output *functions* defined for $t \geqslant t_o$, and arise as follows. The zero input response starting from x° at time t_o reflects what happens at the output as time evolves given that zero input is applied. The zero-state response starting at t_o under the input u reflects the behavior of the output if the system is "forced" with input u for $t \geqslant t_o$ and begins in a "state of rest" at time t_o.

We introduce now the three superposition principles which follow directly from the definition of linearity.

6.8 Superposition Principles:

(a) *Superposition of zero-input and zero-state trajectories:* The trajectory of an m-input, p-output continuous-time state space linear system starting from x° at t_o under the input $u \in \mathbf{U}$ is the "linear superposition" of the zero-input trajectory starting from x° at t_o and the zero-state trajectory starting at t_o under u; i.e., for $t \geqslant t_o$,

$$\phi(t, t_o, x^\circ, u) = \phi(t, t_o, x^\circ, 0) + \phi(t, t_o, 0, u)$$

(b) *Superposition of zero-input and zero-state responses:* The output function, defined for $t \geqslant t_o$, of an m-input, p-output continuous-time state space linear system starting from x° at time t_o under input $u \in \mathbf{U}$, is the "linear superposition" of the zero-input and zero-state responses; i.e., for $t \geqslant t_o$:

$$\mathbf{S}(t, t_o, x^\circ, u) = \mathbf{S}(t, t_o, x^\circ, 0) + \mathbf{S}(t, t_o, 0, u)$$

(c) *Superposition of zero-state responses to two inputs:* Given an m-input, p-output continuous-time state space linear system, and given two input functions $u, \hat{u} \in \mathbf{U}$, the zero-state response to the sum $(u + \hat{u})$ starting from any $t_o \in \mathbf{R}$ is the "linear superposition" of the zero state responses starting from t_o to u and $\hat{u}$; i.e., for every $t \geqslant t_o$,

$$\mathbf{S}(t, t_o, 0, u + u') = \mathbf{S}(t, t_o, 0, u) + \mathbf{S}(t, t_o, 0, \hat{u}). \qquad \square$$

The proofs of these properties are extremely simple; they follow from the fact that, for every $t \geqslant t_o \in \mathbf{R}$, the mappings

$$(x^\circ, u) \rightarrow \phi(t, t_o, x^\circ, u)$$

and

$$(x, w) \rightarrow \rho(t, x, w)$$

are linear on their domains. Take 6.8(b), for example:

$$\begin{aligned} \mathbf{S}(t, t_o, x^\circ, u) &= \rho(t, \phi(t, t_o, x^\circ, u), u(t)) \\ &= \rho(t, \phi(t, t_o, x^\circ, 0) + \phi(t, t_o, 0, u), u(t)) \\ &= \rho(t, \phi(t, t_o, x^\circ, 0), 0) + \rho(t, \phi(t, t_o, 0, u), u(t)) \\ &= \mathbf{S}(t, t_o, x^\circ, 0) + \mathbf{S}(t, t_o, 0, u), \end{aligned}$$

where the first and last lines hold by definition, the second line follows from Superposition Principle 6.8(a), and the third line follows from the linearity of ρ.

Similar superposition principles hold for discrete-time systems; simply replace t and t_o with k and k_o everywhere in 6.8.

6.9 Exercise: State and prove a discrete-time version of 6.8(b). □

There also exists a notion of *trajectory* for a discrete-time system; it is a sequence of points in state space X which begins at $x^\circ \in X$ and is indexed by k, for $k \geqslant k_o$, where k_o is the initial time.

We conclude this section with definitions of time-invariance for continuous- and discrete-time state space systems. Since the next several sections and most of Part III deal almost exclusively with time-invariant systems, these concepts are of great importance. First, we introduce some notation. Given $u : \mathbf{R} \rightarrow \mathbf{R}^m$, and given $s \in \mathbf{R}$, define ${}_s u : \mathbf{R} \rightarrow \mathbf{R}^m$, a new m-vector function, by

$${}_s u(t) = u(t - s), \quad t \in \mathbf{R}.$$

That is, ${}_s u$ is a "shifted version of u." Similarly, given an m-vector discrete-time function $u : \mathbf{Z} \rightarrow \mathbf{R}^m$, and given $j \in \mathbf{Z}$, let us define ${}_j u : \mathbf{Z} \rightarrow \mathbf{R}^m$ by

$${}_j u(k) = u(k - j), \quad k \in \mathbf{Z}.$$

6.10 Definition: An m-input, p-output continuous-time state space system is said to be *time-invariant* if and only if

(a) For every $s \in \mathbf{R}$, we have

(*i*) If $u \in \mathbf{U}$, then ${}_s u \in \mathbf{U}$, and

(*ii*) If $y \in \mathbf{Y}_{t_0}$, then the function $t \to y(t+s)$, defined on $[t_0 - s, \infty)$, is in $\mathbf{Y}_{t_0 - s}$;

(b) For every $t, s, t_0 \in \mathbf{R}$, with $t \geqslant t_0$, every $x^\circ \in X$, and every $u \in \mathbf{U}$:

$$\phi(t+s, t_0+s, x^\circ, {}_s u) = \phi(t, t_0, x^\circ, u);$$

and

(c) For every $t, s \in \mathbf{R}$, every $x \in X$ and every $w \in \mathbf{R}^m$:

$$\rho(t, x, w) = \rho(s, x, w). \qquad \square$$

6.11 Definition: An m-input, p-output discrete-time state space linear system is said to be *time-invariant* if and only if

(a) For every $j \in \mathbf{Z}$, we have

(*i*) If $u \in \mathbf{U}$, then ${}_j u \in \mathbf{U}$, and

(*ii*) If $y \in \mathbf{Y}_{k_0}$, then the function $k \to y(k+j)$, defined on $[k_0 - j, \infty)$, is in $\mathbf{Y}_{k_0 - j}$;

(b) For every $k, l, k_0 \in \mathbf{Z}$, with $k \geqslant k_0$, every $x^\circ \in X$, and every $u \in \mathbf{U}$:

$$\phi(k+l, k_0+l, x^\circ, {}_l u) = \phi(k, k_0, x^\circ, u);$$

and

(c) For every $k, l \in \mathbf{Z}$, every $x \in X$, and every $w \in \mathbf{R}^m$:

$$\rho(k, x, w) = \rho(l, x, w). \qquad \square$$

Properties 6.10(a) and 6.11(a) ensure that both $\mathbf{U}$ and the possible output function spaces are invariant under time-shifting; 6.10(b) and 6.11(b) say that if the system is started up at a certain time and is forced for a certain length of time with some input function, then the same thing happens to the state of the system as would happen if the same procedure were followed starting at some other time. Properties 6.10(c) and 6.11(c) assert that the way in which the present state and the present input value determine the present output value does not change over time.

Observe that if the matrix functions (A, B, C, D) in Example 6.5 are *constant*, then the system defined in that example is time-invariant. To see this, first note that 6.10(a) holds, since continuous functions shift to give continuous functions. Secondly, observe that 6.10(c) obviously holds by definition of ρ. To see that 6.10(b) is true, recall from §2 that, when A is constant, we have, for every $t, \tau \in \mathbf{R}$,

$$\Phi(t, \tau) = \exp[(t-\tau)A].$$

In the notation of the continuous-time Example 6.5, we obtain

$$\begin{aligned}\phi(t, t_o, x^\circ, u) &= \exp[(t-t_o)A]x^\circ + \int_{t_o}^{t} \exp[(t-\tau)A]Bu(\tau)d\tau \\ &= e^{[(t-t_o)A]}x^\circ + \int_{t_o+s}^{t+s} \exp[(t+s-\sigma)A]Bu(\sigma-s)d\sigma;\end{aligned}$$

To obtain the last equation, we changed variables in the integral from τ to $\sigma = \tau + s$, and used the fact (c.f. the discussion at the end of §2) that $\exp[(t-t_o)A] = \exp[(t+s-t_o-s)A]$. We conclude that

$$\phi(t, t_o, x^\circ, u) = \phi(t+s, t_o+s, x^\circ, {}_s u).$$

6.12 Exercise: Show that if the discrete-time matrix functions A, B, C, and D in Example 6.6 are constant, then the discrete-time system in that example is time-invariant. □

It is worthwhile to say a few words about what happens when the hypothesis of finite-dimensionality of the state space X is removed from Definitions 6.1 and 6.2. By making this adjustment, one obtains appropriate definitions of *infinite-dimensional state space linear systems;* such systems serve as models for a variety of real-world processes, some of which may be described quite simply.

Consider, for example, how one might model a "pure delay" process as an infinite-dimensional state space linear system. In such a process, a continuous input function $t \to u(t) \in \mathbf{R}^m$ is fed into a box; for each $t_o \in \mathbf{R}$, the output $y(t)$ of the box at time $t \geqslant t_o$ is $u(t-T)$, where $T > 0$ is a fixed "delay." What is an appropriate notion of "internal situation" for such a process? The internal situation at time t_o should be all one needs to know in order to determine $y(t)$ for

$t \geqslant t_o$ given $u(t)$ for $t \geqslant t_o$. Evidently, knowledge of $u(\tau)$ for every $\tau \in [t_o - T, t_o]$ is necessary in order to make that determination. The following infinite-dimensional state space linear system would seem to be a suitable model for such a delay.

6.13 Example: Let U be the vector space of all continuous functions $u : \mathbf{R} \to \mathbf{R}^m$; for each $t_o \in \mathbf{R}$, let $\mathbf{Y}_{t_o}$ be the vector space of all continuous functions $y : [t_o, \infty) \to \mathbf{R}^m$. Let X be the vector space of all continuous functions which map $[0, T]$ into $\mathbf{R}^m$.

Given $t_1 \geqslant t_o$ and $u \in \mathbf{U}$, and given some $x^\circ \in X$, we define $\phi(t_1, t_o, x^\circ, u)$ as follows. If $t_1 - t_o \geqslant T$, then $\phi(t_1, t_o, x^\circ, u)$ is the function $f : [0, T] \to \mathbf{R}^m$ given by

$$f(\tau) = u(t_1 - \tau), \qquad \tau \in [0, T].$$

If $t_1 - t_o < T$, then $\phi(t_1, t, x^\circ, u)$ is the function $f : [0, T] \to \mathbf{R}^m$ given by

$$f(\tau) = \begin{cases} x^\circ(\tau - [t_1 - t_o]) & \tau \in (t_1 - t_o, T] \\ u(t_1 - \tau) & \tau \in [0, t_1 - t_o). \end{cases}$$

Finally, define the readout mapping $\rho : R \times X \times R^m \to R^m$ by

$$\rho(t, x, w) = x(T).$$

With a little effort, the reader may verify that the state-transition mapping ϕ satisfies properties (d) in Definition 6.1; intuitively, the "delay box" may be visualized as a window of length T which is being slid to the *left* along a graph of the time-reversed input function $t \to u(-t)$. At time t_o, the window reveals the portion of this graph which lies between t_o and $t_o - T$; at time $t_1 \geqslant t_o$, the window's right-hand edge hits the graph of $t \to u(-t)$ at time point $t_1 - T$. The state of the system at any time is the function with the domain $[0, T]$ whose graph looks like the piece of the graph of $t \to u(-t)$ which lies in the window.

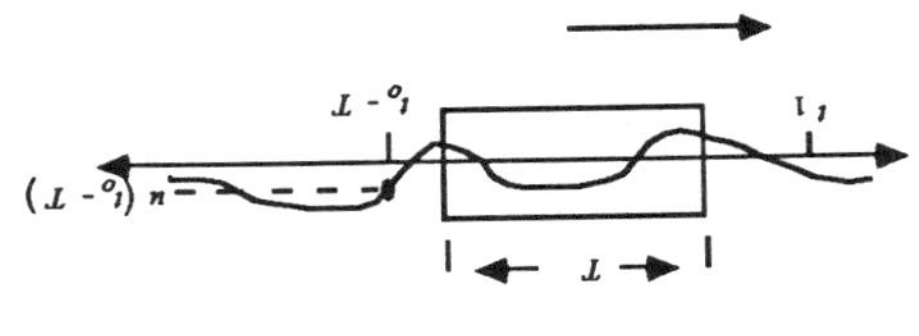

Figure 6.1 A delay system

The output, given the current input value and the current state, is the state's value (as a function) at T, namely, the value assumed by $u(-t)$ at the right-hand end of the window. □

Infinite-dimensional state space linear systems also serve as appropriate models for many *distributed-parameter systems*, such as those which arise in continuum mechanics and in problems involving heat diffusion. For an interesting discussion of potential applications of infinite-dimensional linear system theory, see [Brockett], [Baras, Brockett, Fuhrmann], and the references therein.

We close this section by mentioning that if all the word "real" in Definitions 6.1 and 6.2 is replaced by "complex," one arrives at definitions of *complex state space linear systems*. For some processes, complex systems seem to serve as more natural models than do real systems. Moreover, the algebraic structure of complex systems makes *parametrized families* of such systems easier to deal with than parametrized families of real systems; see, for example, [Byrnes].

7. *Realizations*

Mathematical models for "real-world" processes often appear to be as complicated and cumbersome as the processes themselves. Indeed, an adequate mathematical model must reflect at least a majority of the behavioral subtleties of the modeled process. If a model is to be *useful*, however, it must be *computationally approachable*. The abstract Definitions 6.1 and 6.2 provide a coherent, mathematically attractive class of models for a broad range of real-world processes; still, unless they can be connected firmly with "honest" equations involving *numbers* and *time functions*, they are nothing but arcane, somewhat creaky pieces of theoretical machinery. It is the goal of the present section to forge such a connection.

The central result is that "most" state space linear systems arise in the same fashion as the systems described in Examples 6.5 and 6.6. That is, in most cases, one may find matrix functions (A, B, C, D) so that the systems with state space $\mathbb{R}^n$ defined via Examples 6.5 and 6.6 through the equations

$$\dot{x}(t) = A(t)x(t) + B(t)u(t)$$

$$x(t_o) = x^\circ \tag{I}$$

$$y(t) = C(t)x(t) + D(t)u(t)$$

and

$$x(k+1) = A(k)x(k) + B(k)u(k)$$

$$x(k_o) = x^\circ \tag{II}$$

$$y(k) = C(k)x(k) + D(k)u(k)$$

are "exact replicas" of the abstract systems under consideration. The theory of real finite-dimensional state space linear systems may then be reduced, essentially, to the study of state space linear systems with state space $\mathbf{R}^n$. The reduction procedure bears a strong resemblance to the way (c.f. §4) in which linear algebra involving finite-dimensional vector spaces and mappings between them may be reduced to the study of vectors and matrices. The crucial step, as the reader might have guessed already, is a choice of basis for the state space X.

We present the discrete-time result first, since no special assumptions are required and the proof is straightforward. As in §4, the notation $[v]_{\mathbf{x}}$, where $v \in X$ and $\mathbf{x}$ is a basis for X, means the coordinate vector of v with respect to the basis $\mathbf{x}$.

7.1 Theorem: *Let there be given real m-input, p-output discrete-time finite-dimensional state space linear system with components* $(\mathbf{U}, \mathbf{Y}_{k_o}, X, \phi, \rho)$ *as in Definition 6.2, with the dimension of* X *equal to* n. *Suppose that* $\mathbf{U}$ *contains all discrete-time functions* $u : \mathbf{Z} \to \mathbf{R}^m$. *For each choice of basis* $\mathbf{x}$ *for* X, *there exist uniquely defined discrete-time matrix functions* $k \to A(k)$, $k \to B(k)$, $k \to C(k)$, *and* $k \to D(k)$ *of respective sizes* $(n \times n)$, $(n \times m)$, $(p \times n)$, *and* $(p \times m)$, *such that the system with state space* $\mathbf{R}^n$ *defined by (II) represents the given system in the following sense:*

(a) For every $x^\circ \in X$, $u \in \mathbf{U}$, $k_o \in \mathbf{Z}$, *the* n*-vector function* $k \to [\phi(k, k_o, x^\circ, u)]_{\mathbf{x}}$, $k \geqslant k_o$, *is the unique solution to*

$$x(k+1) = A(k)x(k) + B(k)u(k), \quad k \geqslant k_o$$
$$x(k_o) = [x^\circ]_{\mathbf{x}}$$

and

(*b*) *For every* $x^\circ \in X$, $u \in \mathbf{U}$, $k \geq k_o \in \mathbf{Z}$, $\rho(k, \phi(k, k_o, x^\circ, u), u(k))$ *is given by*

$$C(k)[\phi(k, k_o, x^\circ, u)]_{\mathbf{x}} + D(k)u(k).$$

Proof: First, we construct $k \to A(k), B(k)$ and show that (a) is satisfied. Let $k \in \mathbf{Z}$ be arbitrary. Let $A(k)$ be the matrix with respect to the basis $\mathbf{x}$ of the linear transformation on X defined by

$$x \to \phi(k+1, k, x, 0).$$

If $l \in \mathbf{Z}$ and $w \in \mathbf{R}^m$, define $u_{l,w} \in \mathbf{U}$ by

$$u_{l,w}(n) = \begin{cases} w & n = l \\ 0 & n \neq l \end{cases}$$

and let $B(k)$ be the matrix (with respect to the standard basis for $\mathbf{R}^m$ and the basis $\mathbf{x}$ for X) of the linear mapping from $\mathbf{R}^m$ into X defined by

$$w \to \phi(k+1, k, 0, u_{k,w}).$$

(Observe that $u_{l,w}$ *is* in $\mathbf{U}$, since $\mathbf{U}$ was assumed to contain all $u : \mathbf{Z} \to \mathbf{R}^m$.)

Thus, $A(k)$ and $B(k)$ are defined for all $k \in \mathbf{Z}$. Let $k_o \in \mathbf{Z}$, $x^\circ \in X$, and $u \in \mathbf{U}$ be arbitrary. By the *consistency property* (Definition 6.2(d)(*ii*)), we have

$$\phi(k_o, k_o, x^\circ, u) = x^\circ,$$

and hence

$$[\phi(k_o, k_o, x^\circ, u)]_{\mathbf{x}} = [x^\circ]_{\mathbf{x}}. \tag{1}$$

By the *semigroup property* (c.f. Definition 6.2(d)(iv)) and linearity, for $k \geq k_o$:

$$\phi(k+1, k_o, x^\circ, u) = \phi(k+1, k, \phi(k, k_o, x^\circ, u), u),$$

which in turn is equal to

$$\phi(k+1, k, \phi(k, k_o, x^\circ, u), 0) + \phi(k+1, k, 0, u). \tag{2}$$

If $u(k) = w$, by *causality* (c.f. Definition 6.2(d)(*iii*)), we must have

$$\phi(k+1, k, 0, u) = \phi(k+1, k, 0, u_{k,w}),$$

with notation as above. Taking coordinates with respect to $\mathbf{x}$ of both sides of (2), we obtain

$$[\phi(k+1, k_o, x^\circ, u)]_{\mathbf{x}} = A(k)[\phi(k, k_o, x^\circ, u)]_{\mathbf{x}} + B(k)u(k),$$

which, together with (1), shows that $k \to \phi(k, k_o, x^\circ, u)$, $k \geqslant k_o$, is the unique solution to the difference equation and initial condition in (II).

To prove (b), let $C(k)$ be the matrix with respect to the basis $\mathbf{x}$ for X and the standard basis for $\mathbf{R}^p$ of the linear mapping from X to $\mathbf{R}^p$ defined by

$$x \to \rho(k, x, 0),$$

and let $D(k)$ be the matrix with respect to standard bases of the linear mapping from $\mathbf{R}^m$ into $\mathbf{R}^p$ defined by

$$w \to \rho(k, 0, w).$$

Then, for arbitrary k_o, $k \geqslant k_o$, $x^\circ \in X$, and $u \in \mathrm{U}$, $\rho(k, \phi(k, k_o, x^\circ, u), u(k))$ is given by

$$C(k)[\phi(k, k_o, x^\circ, u)]_{\mathbf{x}} + D(k)u(k),$$

which proves (b). □

7.2 Definition: Given a discrete-time system satisfying the hypothesis of Theorem 7.1, a quadruple of matrix functions (A, B, C, D) for which the conclusion of Theorem 7.1 holds is called a *realization* for the given system. □

Roughly speaking, then, any discrete-time linear system whose state space has dimension n is in some sense "equivalent" to a discrete-time system with state space $\mathbf{R}^n$. This equivalence is not "natural," in the sense that we must first *choose a basis* for X in order to construct a realization for a system having state space X.

An important observation is the following: if the given discrete-time system is *time-invariant*, then the matrix functions A, B, C, and D constructed as in the proof of Theorem 7.1 will be *constant*. This assertion follows immediately from the manner in which $A(k)$, $B(k)$, $C(k)$, and $D(k)$ are defined: they are matrices of linear mappings between various vector spaces taken with respect to fixed basis pairs; these mappings, by Definition 6.11 of time-invariance, are independent of k. Note, however, that there do exist *time-varying* quadruples of matrix functions satisfying Definition 7.2 *even if the system in question is time-invariant*. We shall have more to say about this point in later sections.

We turn now to the continuous-time analogue of Theorem 7.1. Unfortunately, some additional assumptions are required in order to guarantee that such a result hold for a given continuous-time system. Throughout 7.3 below, we are speaking about a given real m-input, p-output continuous-time state space linear system (c.f. Definition 6.1) with state space X having dimension n.

7.3 Assumptions: (a) Given any basis $\mathbf{x}$ for X, any $x^\circ \in X$, and any $u \in \mathbf{U}$, the n-vector function

$$t \rightarrow [\phi(t, t_o, x^\circ, u)]_{\mathbf{x}}$$

is continuously differentiable on $[t_o, \infty)$.

(b) The input function space $\mathbf{U}$ contains all the continuous real m-vector functions and no others.

(c) Given any basis $\mathbf{x}$ for X, for each $t \geqslant t_o$ the linear mapping from $\mathbf{U}$ to $\mathbf{R}^n$ defined by

$$u \rightarrow [\phi(t, t_o, 0, u)]_{\mathbf{x}}$$

is given by an integral with "continuous kernel;" i.e., for each t_o and for every $t \geqslant t_o$, there exists an $(n \times m)$ matrix function

$$\tau \rightarrow \hat{G}_{t_o}(t, \tau)$$

which is continuous in τ at least on $[t_o, t]$, and satisfies

$$[\phi(t, t_o, 0, u)]_{\mathbf{x}} = \int_{-\infty}^{\infty} \hat{G}_{t_o}(t, \tau) u(\tau) d\tau. \qquad \square$$

Assumptions 7.3(a) and (b) are reasonable; moreover, 7.3(a) and (c) are true for all bases $\mathbf{x}$ for X if they are true for one basis. To see this, recall that during the discussion preceding Definition 4.22 we said that if P was the (invertible) $(n \times n)$ matrix relating two bases $\mathbf{x}$, $\tilde{\mathbf{x}}$ for X via

$$\tilde{x}^j = \sum_{i=1}^{n} p_{ij} x^i$$

and v was any vector in X, then

$$[v]_{\tilde{\mathbf{x}}} = P^{-1}[v]_{\mathbf{x}}.$$

Hence, in the present context

$$[\phi(t, t_o, x^\circ, u)]_{\tilde{\mathbf{x}}} = P^{-1}[\phi(t, t_o, x^\circ, u)]_{\mathbf{x}}$$

from which it follows immediately that 7.3(a) is true for any basis if it is true for one basis; the same conclusion follows for 7.3(c) if we take $x^\circ = 0$.

As for Assumption 7.3 (c), it is not as "loaded" as it might appear to be *prima facie*. As it happens, a very general class of real-valued linear functionals on vector spaces of functions are representable as integrals; a text on real analysis, such as [Royden] or [Rudin], is a good reference for such "representation theorems." In what follows, we make use of another elementary result from analysis, namely

7.4 Fact: *If $\tau \to H(\tau)$ and $\tau \to \tilde{H}(\tau)$ are continuous $(n \times m)$ matrix functions, and if*

$$\int_{-\infty}^{\infty} H(\tau)u(\tau)d\tau = \int_{-\infty}^{\infty} \tilde{H}(\tau)u(\tau)d\tau$$

for every continuous function $\underset{\sim}{u} : \mathbf{R} \to \mathbf{R}^m$ for which the integrals exist, then $H(\tau) = \tilde{H}(\tau)$ for every $\tau \in \mathbf{R}$. □

Fact 7.4 is quite believable if one considers a continuous function $\underset{\sim}{u}$ which is constant over some small interval on which H and $\tilde{H}$ disagree. We need three preliminary results before proving The Big Theorem.

7.5 Lemma: *Given a continuous-time state space linear system, as in Definition 6.1, which has an n-dimensional state space X, let $\mathbf{x}$ be any basis for X. For $t \geqslant s$, let $\Phi(t, s)$ be the $(n \times n)$ matrix (with respect to $\mathbf{x}$) of the linear transformation of X defined by*

$$x^\circ \to \phi(t, s, x^\circ, 0), \qquad x^\circ \in X.$$

Then, if $t \geqslant s \geqslant \tau$ are arbitrary,

$$\Phi(t, s)\Phi(s, \tau) = \Phi(t, \tau).$$

Proof: By the semigroup property in Definition 6.1(d)(iv), we have, for every $x^\circ \in X$,

$$\phi(t, \tau, x^\circ, 0) = \phi(t, s, \phi(s, \tau, x^\circ, 0), 0).$$

Take coordinates of both sides with respect to the basis $\mathbf{x}$ and use the definition of Φ to obtain

$$[\phi(t, \tau, x^\circ, 0)]_{\mathbf{x}} = \Phi(t, s)[\phi(s, \tau, x^\circ, 0)]_{\mathbf{x}}$$
$$\Phi(t, \tau)[x^\circ]_{\mathbf{x}} = \Phi(t, s)\Phi(s, \tau)[x^\circ]_{\mathbf{x}} .$$

Since the last equation holds for every $x^\circ \in X$, we conclude that

$$\Phi(t, \tau) = \Phi(t, s)\Phi(s, \tau). \qquad \square$$

7.6 Lemma: *For each* $t \geqslant t_o$, *let* $\tau \to \hat{G}_{t_o}(t, \tau)$ *be as in Assumption 7.3(c). Then*

(a) $\hat{G}_{t_o}(t, \tau) = 0$ *if* τ *lies outside of* $[t_o, t]$;

(b) $\hat{G}_{t_o}(t, \tau) = \hat{G}_s(t, \tau)$ *if* $s \in [t_o, t]$, $\tau \in [s, t]$.

Proof: (a) Suppose u is zero between t_o and t. Then we need (by causality):

$$\phi(t, t_o, 0, u) = 0 .$$

Thus

$$\begin{aligned} 0 &= [\phi(t, t_o, 0, u)]_{\mathbf{x}} \\ &= \int_{-\infty}^{t_o} \hat{G}_{t_o}(t, \tau)u(\tau)d\tau + \int_t^{\infty} \hat{G}_{t_o}(t, \tau)u(\tau)d\tau . \end{aligned}$$

Since u may be an arbitrary continuous function on $(-\infty, t_o]$ and $[t, \infty)$, we conclude from Fact 7.4 that

$$\hat{G}_{t_o}(t, \tau) = 0, \quad \tau < t_o ,$$

and

$$\hat{G}_{t_o}(t, \tau) = 0, \quad \tau > t .$$

(b) If $t \geqslant s \geqslant t_o$, we have (for $u \in \mathbf{U}$):

$$\phi(t, t_o, 0, u) = \phi(t, s, \phi(s, t_o, 0, u), u) \qquad (*)$$

by the semigroup property. If $u \in \mathbf{U}$ is zero on the interval $[t_o, s)$, we get

$$\phi(t, t_o, 0, u) = \phi(t, s, 0, u),$$

and taking coordinates gives (by (a))

$$\int_s^t \hat{G}_{t_o}(t, \tau)u(\tau)d\tau = \int_{t_o}^t \hat{G}_{t_o}(t, \tau)u(\tau)d\tau = \int_s^t \hat{G}_s(t, \tau)u(\tau)d\tau .$$

Since u is an arbitrary continuous function on $[s, t]$, Fact 7.4 implies that

$$\hat{G}_{t_0}(t, \tau) = \hat{G}_s(t, \tau)$$

for all $\tau \in [s, t]$ and $s \in [t_0, t]$. □

We may phrase the conclusion of Lemma 7.6 in another way. Observe that, for each $t \geq t_0$ and $s \in [t_0, t]$, the matrix functions $\tau \to \hat{G}_{t_0}(t, \tau)$ and $\tau \to \hat{G}_s(t, \tau)$ agree for all $\tau \geq s$; hence there must exist for each t an $(n \times m)$ matrix function $\tau \to G(t, \tau)$, continuous on $(-\infty, t]$, such that

$$\hat{G}_{t_0}(t, \tau) = G(t, \tau)1(\tau - t_0),$$

where $t \to 1(t)$ denotes the "unit step" function

$$1(t) = \begin{cases} 1 & t \geq 0 \\ 0 & t < 0. \end{cases}$$

We may therefore summarize Lemma 7.6 as follows: for each $t \in \mathbf{R}$, there exists an $(n \times m)$ matrix function $\tau \to G(t, \tau)$, continuous on $(-\infty, t]$, such that for every $u \in \mathbf{U}$

$$[\phi(t, t_0, 0, u)]_{\mathbf{x}} = \int_{t_0}^{t} G(t, \tau)u(\tau)d\tau.$$

Observe also that Assumption 7.3(a) implies that $t \to G(t, t)$ is continuous; this is clear if we choose $u \in \mathbf{U}$ constant on the interval $[t_0, \infty)$ and apply the Fundamental Theorem of Calculus to the last equation.

We now set out to prove an important identity, namely

7.7 Lemma: If $t \geq s \geq \tau$, then

$$G(t, \tau) = \Phi(t, s)G(s, \tau),$$

where $\Phi(t, s)$ is as in Lemma 7.5.

Proof: Using Superposition Principle 6.8(a) in Equation (*) above, we see that whenever $t \geq s \geq t_0$,

$$\phi(t, t_0, 0, u) = \phi(t, s, \phi(s, t_0, 0, u), 0) + \phi(t, s, 0, u).$$

Taking coordinates with respect to $\mathbf{x}$ an application of Assumption 7.3 (c) along with Lemma 7.6 yields

$$\int_{t_0}^{t} G(t, \tau)u(\tau)d\tau = \Phi(t, s)\int_{t_0}^{s} G(s, \tau)u(\tau)d\tau + \int_{s}^{t} G(t, \tau)u(\tau)d\tau.$$

If u is zero on $[s, t]$, we get

$$\int_{t_o}^{s} G(t,\tau)u(\tau)d\tau = \int_{t_o}^{s} \Phi(t,s)G(s,\tau)u(\tau)d\tau .$$

By Fact 7.4, since this holds for arbitrary continuous u on $[t_o, s)$, it follows that

$$G(t,\tau) = \Phi(t,s)G(s,\tau). \qquad \square$$

7.8 Theorem: *Let there be given a real m-input, p-output continuous-time finite-dimensional state space linear system with components* $(\mathrm{U}, \mathrm{Y}_{t_0}, X, \phi, \rho)$ *as in Definition 6.1, with the dimension of* X *equal to* n*, which satisfies Assumptions 7.3. For each choice of basis* $\mathbf{x}$ *for* X*, there exist uniquely defined matrix functions* $t \to A(t)$*,* $t \to B(t)$*,* $t \to C(t)$*, and* $t \to D(t)$ *of respective sizes* $(n \times n)$*,* $(n \times m)$*,* $(p \times n)$*, and* $(p \times m)$*, with* A *and* B *continuous, such that the system with state space* $\mathbf{R}^n$ *defined by (I) represents the given system in the following sense:*

(a) For every $x^\circ \in X$*,* $u \in \mathrm{U}$*, and* $t_o \in \mathbf{R}$*, the n-vector function* $t \to [\phi(t, t_o, x^\circ, u)]_{\mathbf{x}}$*,* $t \geqslant t_o$*, is the unique solution to*

$$\dot{x}(t) = A(t)x(t) + B(t)u(t), \quad t \geqslant t_o$$

$$x(t_o) = [x^\circ]_{\mathbf{x}} .$$

and

(b) For every $x^\circ \in X$*,* $u \in \mathrm{U}$*,* $t \geqslant t_o \in \mathbf{R}$*,* $\rho(t, \phi(t, t_o, x^\circ, u), u(t))$ *is given by*

$$C(t)[\phi(t, t_o, x^\circ, u)]_{\mathbf{x}} + D(t)u(t).$$

Proof: First, we define the matrix functions A, B, C, and D; subsequently, we'll show that (a) and (b) are satisfied. Let $\Phi(t,s)$ be as in Lemma 7.5; observe, by the consistency property (c.f. Definition 6.1(d)(ii)), that $\Phi(s,s) = I_n$ for all s. Assumption 7.3 (a) implies that, for every $x \in \mathbf{R}^n$, the n-vector function

$$t \to \Phi(t,s)x$$

is differentiable on $[s, \infty)$; since this holds for arbitrary x, we conclude that $t \to \Phi(t,s)$ is differentiable on $[s, \infty)$ for every $s \in \mathbf{R}$. By definition of the derivative:

$$\frac{d}{dt}\Phi(t,s) = \lim_{h\to 0}\left[\frac{\Phi(t+h,s)-\Phi(t,s)}{h}\right]$$
$$= \lim_{h\to 0}\left[\frac{\Phi(t+h,t)-\Phi(t,t)}{h}\right]\Phi(t,s)$$
$$= A(t)\Phi(t,s),$$

where $A(t)$ is defined as

$$A(t) = \lim_{h\to 0}\left[\frac{\Phi(t+h,t)-\Phi(t,t)}{h}\right].$$

Observe that Assumption 7.3(a) guarantees that $t \to A(t)$ is continuous. By Lemmas 7.6 and 7.7, we have

$$\int_{t_o}^{t} G(t,\tau)u(\tau)d\tau = \int_{t_o}^{t}\Phi(t,\tau)G(\tau,\tau)u(\tau)d\tau.$$

Now define

$$B(t) = G(t,t);$$

then B is a continuous $(n\times m)$ matrix function by the comment preceding Lemma 7.7. It is also seen that

$$\int_{t_o}^{t} G(t,\tau)u(\tau)d\tau = \int_{t_o}^{t}\Phi(t,\tau)B(\tau)u(\tau)d\tau.$$

Collecting information, we conclude that for every $x^\circ \in X$, $t \geqslant t_o \in \mathrm{R}$, and $u \in \mathrm{U}$,

$$[\phi(t,t_o,x^\circ,u)]_{\mathbf{x}} = [\phi(t,t_o,x^\circ,0)]_{\mathbf{x}} + [\phi(t,t_o,0,u]_{\mathbf{x}}$$
$$= \Phi(t,t_o)[x^\circ]_{\mathbf{x}} + \int_{t_o}^{t}\Phi(t,\tau)B(\tau)u(\tau)d\tau,$$

where $\Phi(t,t_o)$ is the unique solution to

$$\frac{d}{dt}\Phi(t,t_o) = A(t)\Phi(t,t_o)$$

$$\Phi(t_o,t_o) = I_n.$$

Hence, by Theorem 2.9, the n-vector function $t \to \phi(t,t_o,x^\circ,u)$ is the unique solution to

$$\dot{x}(t) = A(t)x(t) + B(t)u(t)$$

$$x(t_o) = [x^\circ]_{\mathbf{x}},$$

which proves Theorem 7.8 (a).

As for (b), let $C(t)$ be the matrix taken with respect to the basis $\mathbf{x}$ for X and the standard basis for R^p of the linear mapping from X into R^p defined by

$$x \to \rho(t, x, 0), \quad x \in X,$$

and let $D(t)$ be the matrix (with respect to standard bases) of the linear mapping from R^m into R^p defined by

$$w \to \rho(t, 0, w), \quad w \in \mathrm{R}^m.$$

Then, because of linearity, for arbitrary $x^\circ \in X$, $t \geqslant t_o$, and $u \in \mathrm{U}$, $\rho(t, \phi(t, t_o, x^\circ, u), u(t))$ is given by

$$C(t)[\phi(t, t_o, x^\circ, u)]_{\mathbf{x}} + D(t)u(t),$$

which is (b). □

7.9 Definition: Given a continuous-time system satisfying the hypothesis of Theorem 7.8, a quadruple of matrix functions (A, B, C, D) for which the conclusion of Theorem 7.8 holds is called a *realization* for the given system. □

We observe that if a continuous-time system is time-invariant, then the matrix functions (A, B, C, D) appearing in a realization as constructed in the proof of Theorem 7.8 are *constant;* nonetheless, there exist *non-constant* realizations for time-invariant systems in continuous time just as in discrete time.

7.10 Exercise: Prove the last two assertions using Definition 6.10 of time-invariance. □

8. *Eigenvectors, Eigenvalues, and Normal Modes*

In §7 we demonstrated that almost any real time-invariant m-input, p-output finite-dimensional state space linear system whose state space X has dimension n may be *realized;* that is to say, there exist for most such systems real constant matrices A, B, C, and D having respective sizes $(n \times n)$, $(n \times m)$, $(p \times n)$, and $(p \times m)$ such that the system's behavior may be analyzed in terms of the equations

$$\dot{x} = Ax(t) + Bu(t), \quad t \geq t_o$$
$$x(t_o) = x^\circ \tag{I}$$
$$y(t) = Cx(t) + Du(t)$$

for a continuous time system, or

$$x(k+1) = Ax(k) + Bu(k), \quad k \geq k_o$$
$$x(k_o) = x^\circ \tag{II}$$
$$y(k) = Cx(k) + Du(k),$$

in discrete time. In this section, we begin investigating (I) and (II) in detail. The development hinges on the material in §§1-5.

Recall from §2 that given a continuous $u : \mathbf{R} \to \mathbf{R}^m$, the unique $x : [t_o, \infty) \to \mathbf{R}^n$ satisfying (I) is given by

$$x(t) = e^{(t-t_o)A} x^\circ + \int_{t_o}^{t} e^{(t-\tau)A} Bu(\tau) d\tau, \quad t \geq t_o,$$

where

$$e^{(t-s)A} = \sum_{k=0}^{\infty} \frac{(t-s)^k A^k}{k!},$$

and that given $u : \mathbf{Z} \to \mathbf{R}^m$, the unique discrete-time real n-vector function $k \to x(k)$, $k \geq k_o$, which satisfies (I) is given by

$$x(k) = A^{(k-k_o)} x^\circ + \sum_{l=k_o}^{k-1} A^{k-l-1} Bu(l), \quad k \geq k_o.$$

The problem of analyzing (I) and (II), then, is in some sense reduced to the problem of determining $e^{(t-s)A}$ and $A^{(k-l)}$, given A. Since $e^{(t-s)A}$ is given by an infinite series, we should like to exhibit a technique which makes $e^{(t-s)A}$ computable, at least in principle, through a finite number of operations. As it happens, the approach we take is of independent system theoretic interest, and in addition provides a method for finding $A^{(k-l)}$ without performing $(k-l)$ matrix multiplications.

We begin with a formal definition which we shall subsequently tailor to suit our special needs.

8.1 Definition: *Let V be a vector space over* F, *where* F *is* R *or* C. *Let $T : V \to V$ be a linear transformation. An* *eigenvector of T is a nonzero vector $v \in V$ which satisfies, for some $\lambda \in \mathbf{F}$,*

$$T(v) = \lambda v \ .$$

In this case, λ is called an eigenvalue of T, and v is said to be an eigenvector of T corresponding to the eigenvalue λ. □

Our rationale for studying eigenvectors of linear transformations is based on many considerations. In the abstract context of Definition 8.1, an eigenvector v of a linear transformation T is a vector on which T "acts" in "the simplest possible fashion," namely by *scaling* with a scalar λ. If we are to understand what a complicated linear transformation T "does" to vectors in V, we should naturally like to isolate those vectors which T transforms in a straightforward manner. Very soon we shall discover other reasons for studying eigenvectors.

8.2 Examples: (a) Let V be any real or complex vector space; then every nonzero $v \in V$ is an eigenvector of the identity transformation T on V corresponding to the eigenvalue 1; i.e., for any $v \in V$:

$$T(v) = 1 \cdot v \ .$$

Similarly, every $v \neq 0$ is an eigenvector of the zero transformation corresponding to the eigenvalue 0.

(b) Consider E^2, the Euclidean plane; let $T : E^2 \to E^2$ be counterclockwise rotation through θ as in Example 4.21. Clearly, if $0 < \theta < \pi$, T has *no eigenvectors*, since there is no $v \in E^2$ for which $T(v)$ is a scalar multiple of v. □

We could proceed from here with an abstract development of the concepts of eigenvectors and eigenvalues in the realm of arbitrary linear transformations on arbitrary vector spaces; such an exposition is given (quite beautifully) in [Halmos]. Our purposes, however, are better served by the more concrete approach which we follow in the ensuing pages.

Recall from §4 that any real $(n \times n)$ matrix A defines a linear transformation ${}_AT : \mathbf{R}^n \to \mathbf{R}^n$ by means of the formula

$${}_AT(x) = Ax \ , \quad x \in \mathbf{R}^n \ .$$

The *same real matrix* may be used to define a complex linear transformation ${}_A\hat{T} : \mathbf{C}^n \to \mathbf{C}^n$ via precisely the same assignment:

$$_A\hat{T}(z) = Az \ , \quad z \in \mathbf{C}^n .$$

The transformation $_A\hat{T}$ is called the *complexification* of the transformation $_AT$. It should always be remembered that $_A\hat{T}$ and $_AT$ are *different objects*, even though both transformations arise from the same real $(n \times n)$ matrix A. A crucial observation (c.f. §4) is that the matrix of $_AT$ with respect to the standard basis for $\mathbf{R}^n$ is A; in fact, the matrix of $_A\hat{T}$ with respect to the standard basis for $\mathbf{C}^n$ is also A. Thus $z \in \mathbf{C}^n$ is an eigenvector of $_A\hat{T}$ corresponding to eigenvalue $\lambda \in \mathbf{C}$ if and only if $z \neq 0$ and $Az = \lambda z$.

Notice also that if $x \in \mathbf{R}^n$ is an eigenvector for $_AT$ corresponding to eigenvalue $\lambda \in \mathbf{R}$, then x (viewed as a vector in $\mathbf{C}^n$) is *also* an eigenvector for $_A\hat{T}$ corresponding to the same eigenvalue λ; the real scalar λ, of course, is an element of $\mathbf{C}$, too. An eigenvector of $_A\hat{T}$, however, is *not* necessarily an eigenvector of $_AT$. We are prompted to make the following definition.

8.3 Definition: Let A be a real $(n \times n)$ matrix. An *eigenvector* of A is an eigenvector of the transformation $_A\hat{T}$; i.e., $z \in \mathbf{C}^n$ is an eigenvector of A if and only if $_A\hat{T}(z) = \lambda z$ for some $\lambda \in \mathbf{C}$. In this case, λ is said to be an *eigenvalue* of A and z is said to be an *eigenvector of A corresponding to the eigenvalue* λ. □

The motivation for Definition 8.3 is best illustrated by an example.

8.4 Example: Define the matrix A by

$$A = \begin{bmatrix} 0 & -1 \\ 1 & 0 \end{bmatrix} .$$

Let us try to find eigenvectors for $_AT$ and $_A\hat{T}$. Since A is the matrix with respect to standard bases both for $_AT$ and for $_A\hat{T}$, finding a nonzero $x \in \mathbf{R}^2$ (or $z \in \mathbf{C}^2$) satisfying $_AT(x) = \lambda x$ (or $_A\hat{T}(z) = \lambda z$) for some real (or complex) λ is equivalent to finding a nonzero $x \in \mathbf{R}^2$ or $z \in \mathbf{C}^2$ satisfying one of the equations

$$Ax = \lambda x \ , \qquad \lambda \in \mathbf{R} \ ,$$

or

$$Az = \lambda z \ , \qquad \lambda \in \mathbf{C} \ .$$

These equations may be written in components as follows:

$$\begin{bmatrix} -x_2 \\ x_1 \end{bmatrix} = \begin{bmatrix} \lambda x_1 \\ \lambda x_2 \end{bmatrix} ; \qquad \begin{bmatrix} -z_2 \\ z_1 \end{bmatrix} = \begin{bmatrix} \lambda z_1 \\ \lambda z_2 \end{bmatrix} ,$$

from which we obtain

$$\begin{aligned} x_1 &= -\lambda^2 x_1 \\ x_2 &= -\lambda^2 x_2 \end{aligned} ; \qquad \begin{aligned} z_1 &= -\lambda^2 z_1 \\ z_2 &= -\lambda^2 z_2 \ . \end{aligned}$$

The constraints $x \neq 0$, $z \neq 0$ stipulate that $\lambda^2 = -1$; i.e., $\lambda = \pm i$. Hence, ${}_A T$ *has no eigenvectors*, but ${}_A \hat{T}$ does; it is a simple matter to check that

$$z^1 = \begin{bmatrix} i \\ 1 \end{bmatrix} \quad \text{and} \quad z^2 = \begin{bmatrix} -i \\ 1 \end{bmatrix}$$

are eigenvectors for ${}_A \hat{T}$ corresponding with eigenvalues i (for z^1) and $-i$ (for z^2). □

The central idea underlying Definition 8.3 is that we want *every real matrix* to have eigenvalues and eigenvectors, even if *no real eigenvalue or eigenvector exists*.

8.5 Observations: *Let A be a real $(n \times n)$ matrix.*

(a) If $z \in \mathbf{C}^n$ is an eigenvector of A corresponding to eigenvalue λ, then $z^* \in \mathbf{C}^n$ is an eigenvector of A corresponding to eigenvalue λ^*. (Here, a star superscript denotes complex conjugate.)

(b) If $z^1, \ldots, z^k \in \mathbf{C}^n$ are eigenvectors of A corresponding to eigenvalues $\lambda_1, \ldots, \lambda_k \in \mathbf{C}$, and the λ_i are all different, then $\{z^1, \ldots, z^k\}$ is a linearly independent set of vectors.

Proof: (a) Since z is an eigenvector of A corresponding to λ, $z \neq 0$ and $Az = \lambda z$. Taking complex conjugates and remembering that A is real, we get $Az^* = \lambda^* z^*$; since $z^* \neq 0$, z^* is an eigenvector of A corresponding to eigenvalue λ^*.

(b) We perform induction. If $k = 2$, we reorder the z's if necessary and assume there exist $c_1, c_2 \in \mathbf{C}$ with $c_1 \neq 0$ such that

$$c_1 z^1 + c_2 z^2 = 0 .$$

Multiply by A to obtain

$$c_1 \lambda_1 z^1 + c_2 \lambda_2 z^2 = 0 .$$

Using $z^1 = \dfrac{-c_2}{c_1} z^2$, we get

$$c_2 (\lambda_2 - \lambda_1) z^2 = 0$$

which can't occur unless $c_2 = 0$ since $(\lambda_2 - \lambda_1) z^2 \neq 0$. Thus $c_1 = 0$ must hold in the first equation, which contradicts our assumption.

Now suppose we've shown linear independence of any set of $(k-1)$ eigenvectors corresponding to $(k-1)$ different eigenvalues. Given the notation in (b), reorder if necessary and assume a relation

$$c_1 z^1 + \ldots + c_k z^k = 0$$

with $c_1 \neq 0$. Solve for z^1 in terms of z^j, $j > 1$; then multiply by A and subtract equations, as before, to get

$$d_2 z^2 + \ldots + d_k z^k = 0$$

where $d_j = \dfrac{c_j}{c_1}(\lambda_j - \lambda_1)$. By induction, the assumption $\lambda_j \neq \lambda_1$ implies $d_j = 0$, $j > 1$; finally, deduce that $c_1 = 0$ from the first equation and the result follows. □

8.6 Corollary: *Let A be a real $(n \times n)$ matrix. Then*

(a) If $\lambda \in \mathbf{C}$ is an eigenvalue of A, then so is λ^*; consequently, if n is odd, then A has at least one real eigenvalue.

(b) There exist at most n different eigenvalues for A.

Proof: (a) is obvious given Observation 8.5(a). As for (b), Observation 8.5(b) would yield k linearly independent eigenvectors for any k different eigenvalues; thus $k \leqslant n$ by Lemma 1.18, since the dimension of $\mathbf{C}^n$ is n. (For the last fact, the reader is referred to the remarks following Theorem 1.21.) □

A reader well-versed in linear algebra might feel that our approach of eigenvectors and eigenvalues is in some sense circuitous. The view taken here, however, is that eigen-*vectors* are the original objects of interest; we seek vectors on which ${}_A\hat{T}$ acts in a "simple" way, namely by scaling. In searching for those vectors, we discover (and are somewhat surprised at the fact) that a given ${}_A\hat{T}$ may only scale vectors by certain *special* scale factors; indeed, we find that (since A is real) complex eigenvalues come in conjugate pairs; moreover, there may be only n different eigenvalues for a given $(n \times n)$ matrix.

It is worthwhile at this point to take a look back at the differential equation in (I) and get a preliminary idea about why eigenvectors and eigenvalues will play a central role in our analysis of time-invariant state space linear systems. Assume for the moment that $t_o = 0$ and the input function u is taken to be zero; thus the differential equation solution, for a given initial condition $x°$, is

$$x(t) = e^{tA}x°, \quad t \geqslant 0 .$$

Suppose now that $x°$ is a *real* eigenvector of the $(n \times n)$ matrix A corresponding to the (necessarily) real eigenvalue $\lambda_o \in \mathbf{R}$. Since $A^k x° = \lambda^k x°$ for every $k \geqslant 0$, we conclude that, for $t \geqslant 0$,

$$\begin{aligned} e^{tA}x° &= [\sum_{k=0}^{\infty} \frac{t^k A^k}{k!}]x° \\ &= \sum_{k=0}^{\infty} (\frac{t^k}{k!})\lambda_o^k x° \\ &= (\sum_{k=0}^{\infty} \frac{(\lambda_o t)^k}{k!})x° = e^{\lambda_o t}x° . \end{aligned}$$

Thus, for every $t \geqslant 0$, $x(t) = e^{tA}x°$ is a scalar multiple of $x°$. If equations (I) are viewed as defining a state space linear system with state space $\mathbf{R}^n$ (compare Example 6.5), the foregoing discussion is seen to imply that if the initial state at time t_o is on the "line" through $0 \in \mathbf{R}^n$ spanned by a real eigenvector $x°$ of A, then the zero-input trajectory, for all $t \geqslant t_o$, *stays on the line*. The "line," of course, is the one-dimensional subspace of $\mathbf{R}^n$ *spanned* by $\{x°\}$. According to whether $\lambda_o > 0$ or $\lambda_o < 0$, the trajectory either goes off toward infinity or converges toward the origin — all along the line spanned by $\{x°\}$. Since the trajectory we've been talking about is a zero-input trajectory, we see that the particular *line* along a real eigenvector of A is an *intrinsic feature of the system*, independent of the input. For this reason, we often call the line, or $x°$ itself, or the n-vector function $e^{\lambda_o t}x°$, a *normal mode* or *natural mode* of the system defined by (I).

A similar but somewhat more intricate analysis may be done if $z° \in \mathbf{C}^n$ is nonreal eigenvector of A corresponding to a nonreal eigenvalue λ_o. Now, since $z°$ is not an element of $\mathbf{R}^n$, $z°$ may *not* be used to initialize the differential equation in (I). Still, if we recall from Observation 8.5(a) that $z°^*$ is another eigenvector of A corresponding to λ_o^*, we may maneuver our way back into $\mathbf{R}^n$ as follows. Define

$$x° = \mathrm{Re}\{z°\} = \frac{1}{2}(z° + z°^*)$$

$$y° = \mathrm{Im}\{z°\} = \frac{1}{2i}(z° - z°^*) .$$

While $x°$ and $y°$ are *not* eigenvectors of A, they span a 2-dimensional subspace of $\mathbf{R}^n$ which plays a role similar to the real normal mode just discussed. First, we check that $x°$ and $y°$ are linearly independent; if they weren't, there would exist $c_1, c_2 \in \mathbf{R}$ not both zero such that $c_1x° + c_2y° = 0$, and that would imply that

$$\left(\frac{c_1 - ic_2}{2}\right)z° + \left(\frac{c_1 + ic_2}{2}\right)z°^* = 0 ,$$

which would contradict linear independence of $\{z°, z°^*\}$, which is true by Observation 8.5(b).

Now let us see what happens if the differential equation $\dot{x} = Ax$ is initialized at $t_o = 0$ with a vector in the 2-dimensional subspace of $\mathbf{R}^n$ spanned by $\{x°, y°\}$. Observe that, since A and consequently e^{tA} are real, we have

$$\begin{aligned} e^{tA}x° &= \mathrm{Re}\{e^{tA}z°\} \\ &= \mathrm{Re}\{\sum_{k=0}^{\infty} \frac{(\lambda_o t)^k}{k!} z°\} \\ &= \mathrm{Re}\{e^{\lambda_o t} z°\} , \end{aligned}$$

where the second and third equalities follow precisely as in the case where λ_o is real. If $\lambda_o = \mu + i\omega$, then

$$\begin{aligned} Re\{e^{\lambda_o t} z°\} &= Re\{e^{\mu t}(\cos\omega t + i\sin\omega t)(x° + iy°)\} \\ &= (e^{\mu t}\cos\omega t)x° - (e^{\mu t}\sin\omega t)y° . \end{aligned}$$

Thus, $e^{tA}x° = Re\{e^{\lambda_o t} z°\}$ is, for all t, a linear combination of $x°$ and $y°$. Similarly,

$$e^{tA}y° = (e^{\mu t}\sin\omega t)x° + (e^{\mu t}\cos\omega t)y° .$$

Thus, for all $t \geqslant 0$, *a (zero-input) trajectory* which begins on the "plane" spanned by x° and y° must remain in that plane forever, spiraling about the point $0 \in \text{span}\{x^\circ, y^\circ\}$; the trajectory spirals outward when $\mu > 0$, and inward toward 0 when $\mu < 0$. The plane spanned by the real and imaginary parts of a complex eigenvector, along with the oscillatory "motion" of a zero-input trajectory which begins in that plane, constitute another *normal mode* or *natural mode* of the system. In Part IV, techniques which depend on the notion of natural mode will play a large part in helping us understand qualitative features of the behavior of linear systems, such as stability.

We return now to the systematic study of eigenvectors and eigenvalues for real $(n \times n)$ matrices A. We begin with an interesting technical lemma.

8.7 Lemma: *Let $T : V \to V$ be a linear transformation on an n-dimensional real or complex vector space V. Let* $\mathbf{v}$ *and* $\mathbf{w}$ *be two bases for V, and let $T_{\mathbf{v}}$ and $T_{\mathbf{w}}$ be the matrices of T with respect to* $\mathbf{v}$, $\mathbf{w}$. *Then*

$$\det(T_{\mathbf{v}}) = \det(T_{\mathbf{w}}) .$$

Proof: By the discussion preceding Definition 4.22, there exists an invertible $(n \times n)$ matrix P such that

$$T_{\mathbf{w}} = P^{-1} T_{\mathbf{v}} P$$

hence

$$\begin{aligned} \det(T_{\mathbf{w}}) &= \det(P^{-1} T_{\mathbf{w}} P) \\ &= \det(P^{-1}) \det(T_{\mathbf{v}}) \det(P) \\ &= \det(T_{\mathbf{v}}) , \end{aligned}$$

where the second equality follows from Fact 1.31, and the third from the fact that $\det(P^{-1})\det(P) = \det(P^{-1}P)$, $\det(I_n) = 1$. □

Observe that Lemma 8.7 says that the determinant of the matrix of a linear transformation with respect to a basis is *basis-independent*; thus, we might as well call that number the "determinant of the linear transformation." Many authors (for example, [Halmos]) define the determinant in such a "basis-invariant" fashion.

8.8 Theorem: *Let* **F** *be* **R** *or* **C**, *and let* $T : V \to V$ *be a linear transformation on an n-dimensional vector space* V *over* **F**. *Then* $\lambda_o \in \mathbf{F}$ *is an eigenvalue of* T *if and only if the determinant of the matrix*

$$T_{\mathbf{v}} - \lambda_o I_n$$

is zero, where $T_{\mathbf{v}}$ *is the matrix of* T *with respect to any basis* **v** *for* V.

Proof: Let $x \in V$ be an eigenvector of T corresponding to λ_o. Then $T(x) = \lambda_o x$; taking matrices and components with respect to **v** yields

$$T_{\mathbf{v}}[x]_{\mathbf{v}} = \lambda_o I_n [x_{\mathbf{v}}] ,$$

or

$$(T_{\mathbf{v}} - \lambda_o I_n)[x]_{\mathbf{v}} = 0 .$$

Since $x \neq 0$, $x_{\mathbf{v}} \neq 0$; Theorem 1.5, along with the discussion at the end of §1, show that $\det(T_{\mathbf{v}} - \lambda_o I_n) = 0$. Conversely, the same argument in reverse shows that if $\det(T_{\mathbf{v}} - \lambda_o I_n) = 0$, then there exists a nonzero $w \in \mathbf{F}^n$ such that $T_{\mathbf{v}} w = \lambda_o w$; find $y \in V$ such that $w = [y]_{\mathbf{v}}$, and obtain $T(y) = \lambda_o y$, implying that λ_o is an eigenvalue of T. □

In the setting of real $(n \times n)$ matrices, we deduce that λ_o is an eigenvalue of a real $(n \times n)$ matrix A (meaning, by Definition 8.3, λ_o is an eigenvalue of ${}_A\hat{T}$) if and only if

$$\det(A - \lambda_o I_n) = 0 . \qquad (*)$$

Equation (*) could have been derived directly using Definition 8.3 and the proof technique of Theorem 8.8, but Theorem 8.8 is an easily demonstrated and more general result, so we have included it.

Thus equation (*) must be satisfied by every eigenvalue λ_o of a real $(n \times n)$ matrix A. Observe from the definition of the determinant that the expression $\det(A - \lambda I_n)$, where λ is a "variable," is an n th degree polynomial in λ.

8.9 Definition: Let A be a real $(n \times n)$ matrix. The *characteristic polynomial* of A is the n th degree polynomial in λ given by $\det(A - \lambda I_n)$. The *characteristic equation* of A is the equation $\det(A - \lambda I_n) = 0$. □

In the terminology of Definition 8.9, λ_o is an eigenvalue of A if and only if λ_o is a *root* of the characteristic polynomial of A or, equivalently, is a *solution* to the characteristic equation of A.

It should be remarked that Observations 8.5 could have been deduced directly from Theorem 8.8 and the discussion surrounding Definition 8.9. Such a deduction would require some knowledge of elementary polynomial algebra, which the reader is assumed to have. Still, Observations 8.5, which were proved independently of polynomial algebra, may actually be used to *prove* that a real n th degree polynomial has at most n distinct roots and that complex roots come in conjugate pairs!

Observe that the characteristic polynomial of an $(n \times n)$ matrix A, like any n th degree polynomial, may be expressed as the product of first-degree terms as follows:

$$\det(A - \lambda I_n) = (-1)^n (\lambda_1 - \lambda)^{r_1} (\lambda_2 - \lambda)^{r_2} \ldots (\lambda_s - \lambda)^{r_s}$$

where λ_i, $1 \leq i \leq s$, are the s distinct eigenvalues of A and r_i, $1 \leq i \leq s$, are their respective multiplicities as roots of the characteristic polynomial. (Note that $r_1 + r_2 + \ldots + r_s = n$.)

8.10 Definition: Let λ_o be an eigenvalue of the real $(n \times n)$ matrix A. The *algebraic multiplicity* of λ_o is its multiplicity as a root of the characteristic polynomial of A. □

Definition 8.10 is important; it alludes to one type of multiplicity for eigenvalues. The other, *geometric multiplicity*, is closely related.

8.11 Definition: Let λ_o be an eigenvalue of a real $(n \times n)$ matrix A. The *eigenspace corresponding to* λ_o is the subspace of $\mathbf{C}^n$ spanned by all the eigenvectors for A corresponding to λ_o and is denoted by $E(\lambda_o)$. The *geometric multiplicity* of λ_o is the dimension of $E(\lambda_o)$ as a subspace of $\mathbf{C}^n$. □

8.12 Theorem: *Let A be a real $(n \times n)$ matrix and let λ_o be an eigenvalue of A. Then the geometric multiplicity of λ_o is less than or equal to the algebraic multiplicity of λ_o.*

Proof: Let r and m denote, respectively, the algebraic and geometric multiplicities of λ_o. Choose a basis $\mathbf{z}$ for $\mathbf{C}^n$ whose first m vectors form a basis for $E(\lambda_o)$. Let B be the matrix of ${}_A\hat{T}$ with respect to $\mathbf{z}$; then B takes the "block" form

$$B = \begin{bmatrix} \lambda_o I_m & B_2 \\ 0 & B_3 \end{bmatrix},$$

where 0 is $(n-m)\times m$, B_2 is $m\times(n-m)$, and B_3 is $(n-m)\times(n-m)$. By Lemma 8.7,

$$\det(B-\lambda I_n) = \det(A-\lambda I_n).$$

From Definition 1.28 of the determinant, or from Exercise 1.30(d), it is easily shown that

$$\begin{aligned}\det(B-\lambda I_n) &= \det[(\lambda_o-\lambda)I_m]\det(B_3-\lambda I_{n-m}) \\ &= (\lambda_o-\lambda)^m p(\lambda),\end{aligned}$$

where $p(\lambda)$ is a polynomial in λ of which λ_o may or may not be a root. Now,

$$\det(A-\lambda I_n) = (\lambda_o-\lambda)^r q(\lambda),$$

where $q(\lambda_o) \neq 0$. If $m > r$ were true, we could divide both of the last two expressions by $(\lambda_o-\lambda)^r$ to obtain

$$(\lambda_o-\lambda)^{m-r} p(\lambda) = q(\lambda),$$

with $(m-r) > 0$, which contradicts $q(\lambda_o) \neq 0$. Thus $m \leqslant r$ □

It is worthwhile at this point to step back and attempt to understand and interpret Theorem 8.12. To begin with, we have "justified" our search for eigenvectors of a real $(n\times n)$ matrix A by asserting that it is in some sense desirable to know which vectors in $\mathbf{C}^n$ are "operated upon" by ${}_A\hat{T}$ in a "simple" way. It would be convenient if *every* vector in $\mathbf{C}^n$ could be expressed as a linear combination of these eigenvectors; in that case, the "action" of ${}_A\hat{T}$ on any n-vector could be "decomposed" into a "sum" of scalings. Theorem 8.12 says that the dimension of each eigenspace for A is at most equal to the algebraic multiplicity of the corresponding eigenvalue. Since the algebraic multiplicities sum to n, we cannot hope to find an "eigenvector basis" for $\mathbf{C}^n$ unless each eigenvalue's geometric and algebraic multiplicities are equal. We give a somewhat more precise rendering of these rather vague statements in the definition and theorem which follow.

8.13 Definition: Let A be a real $(n\times n)$ matrix. A is said to be *semi-simple* if and only if there exists a basis for $\mathbf{C}^n$ consisting solely of eigenvectors for A. If this basis may be

chosen to be real, then A is said to be *diagonalizable.* □
[One cautionary note: some authors use the word "diagonalizable" to mean what we call "semi-simple."]

To understand the meaning of Definition 8.13, let A be a real $(n \times n)$ diagonalizable matrix; let $x^1, \ldots, x^n$ be n linearly independent (real) eigenvectors for A which correspond respectively to (necessarily real) eigenvalues $\lambda_1, \ldots, \lambda_n$. The λ_i, $1 \leqslant i \leqslant n$, are *not necessarily distinct* from each other. Now, $\mathbf{x} = \{x^1, \ldots, x^n\}$ is a basis for $\mathbf{R}^n$, as well as for $\mathbf{C}^n$; let us compute the matrix of the transformation ${}_AT : \mathbf{R}^n \to \mathbf{R}^n$ with respect to the basis $\mathbf{x}$. Since ${}_AT(x^i) = \lambda_i x^i$, $1 \leqslant i \leqslant n$, it is evident that

$$[({}_AT)_{\mathbf{x}}]ij = \lambda_i \delta_{ij} = \begin{cases} \lambda_i & i = j \\ 0 & i \neq j \end{cases}.$$

Thus, $({}_AT)_{\mathbf{x}}$ is a *diagonal* matrix; hence the name "diagonalizable" for such a real $(n \times n)$ matrix A. If A is semi-simple but not diagonalizable, we may take the matrix of ${}_A\hat{T}$ with respect to a basis $\mathbf{z} = \{z^1, \ldots, z^n\}$ consisting of eigenvectors corresponding to $\lambda_1, \ldots, \lambda_n$; this matrix will also be diagonal, although it will have complex numbers on its diagonal.

It might be said that the central result of this section is Theorem 8.15 below; in order to prove it, we need the following auxiliary proposition, which is really just a restatement Observation 8.5(b).

8.14 Lemma: *Let A be a real $(n \times n)$ matrix; let $\lambda_1, \ldots, \lambda_s$ be the distinct eigenvalues of A and $E(\lambda_i)$, $1 \leqslant i \leqslant s$, the corresponding eigenspaces. Then the $E(\lambda_i)$, $1 \leqslant i \leqslant s$, are mutually disjoint subspaces of $\mathbf{C}^n$.*

Proof: By Observation 4.28 (a), we merely need show that if $v^i \in E(\lambda_i)$, $1 \leqslant i \leqslant s$, and $v^1 + v^2 + \ldots + v^s = 0$, then $v^i = 0$, $1 \leqslant i \leqslant s$; this, however, is an immediate consequence of Observation 8.5(b). □

8.15 Theorem: *Let A be a real $(n \times n)$ matrix. A is semi-simple if and only if each of A's eigenvalues has geometric multiplicity equal to its algebraic multiplicity. Moreover, A is diagonalizable if and only if A is semi-simple and has only real eigenvalues.*

Proof: Suppose that A is semi-simple; then every complex n-vector may be expressed as a linear combination of eigenvectors of A. This means that the vector sum (Definition 4.26) of $E(\lambda_1), \ldots, E(\lambda_s)$ is $\mathbf{C}^n$. As usual, we denote by r_i and m_i, respectively, the algebraic multiplicity and geometric multiplicity of λ_i, $1 \leqslant i \leqslant s$. Note (by Observation 4.28(b)) that the vector sum of the $E(\lambda_i)$ has dimension at most $\sum_{i=1}^{s} m_i$; thus we need

$$\sum_{i=1}^{s} m_i = n ,$$

which, by Theorem 8.12, can occur only if $m_i = r_i$, $1 \leqslant i \leqslant s$.

Conversely, if for $1 \leqslant i \leqslant s$ we have $m_i = r_i$, then a basis for $\mathbf{C}^n$, by Observation 4.28(d), may be constructed by piecing together bases for all the eigenspaces. The set so constructed is indeed linearly independent, and must therefore be a basis for $\mathbf{C}^n$ (c.f. the discussion following Theorem 1.21).

Finally, A is diagonalizable precisely when a basis for $\mathbf{C}^n$ consisting solely of *real* eigenvectors for A may be found; this situation is clearly equivalent to semi-simplicity of A along with realness of each of A's eigenvalues. □

In §9, we consider in detail those real $(n \times n)$ matrices A which are *not* semi-simple; in the remainder of the present section, we concentrate on semi-simple matrices. We have already noted that if a real $(n \times n)$ matrix A is *diagonalizable*, then there exists a basis $\mathbf{x}$ for $\mathbf{R}^n$ with respect to which $({}_A T)_{\mathbf{x}}$ is a diagonal matrix. The converse of the last statement is also true; the general result is

8.16 Theorem: *Let A be a real $(n \times n)$ matrix. Then A is semi-simple if and only if there exists an invertible complex $(n \times n)$ matrix P such that $P^{-1}AP = \Lambda$ is a diagonal matrix. A is diagonalizable if and only if A is semi-simple and P may be chosen real; in either case, the diagonal elements of Λ are the eigenvalues of A, and each eigenvalue λ_i appears r_i times, where r_i is the algebraic (and geometric) multiplicity of λ_i.*

Proof: Suppose A is semi-simple; let $\mathbf{z} = \{z^1, \ldots, z^n\}$ be a basis for $\mathbf{C}^n$ consisting of eigenvectors of A. Write

$$z^j = \sum_{i=1}^{n} p_{ij} e^i , \qquad 1 \leqslant j \leqslant n ,$$

where e^i is the ith standard basis vector. Applying equation (**) from §4, we obtain

$$({}_A\hat{T})_{\mathbf{z}} = P^{-1}({}_A\hat{T})_{\mathbf{e}}P = P^{-1}AP ,$$

and $({}_A\hat{T})_{\mathbf{z}}$ is evidently (c.f. the discussion following Definition 8.13) a diagonal matrix Λ with each eigenvalue of A appearing on the diagonal the prescribed number of times.

Conversely, if such a P may be found, then we get

$$AP = P\Lambda .$$

Denote by z^j the jth column of P; the last equation says

$$Az^j = [\Lambda]_{jj} z^j ,$$

which implies that $z^1, \ldots, z^n$ are (linearly independent, by invertibility of P) eigenvectors for A corresponding respectively with eigenvalues $[\Lambda]_{jj}$. Since they form a basis for $\mathbf{C}^n$, A is semi-simple.

The statement about diagonalizability and realness of P is straightforward consequence of the definitions. □

Observe that the way in which P is constructed gives the formula

$$z^j = \begin{bmatrix} p_{1j} \\ p_{2j} \\ \cdot \\ \cdot \\ \cdot \\ p_{nj} \end{bmatrix} , \qquad 1 \leqslant j \leqslant n .$$

That is, the jth complex n-vector in an eigenvector basis for A is the jth column of P. In particular, if A is diagonalizable, we have a procedure for "diagonalizing" A:

8.17 Diagonalization Procedure: *Let A be a real $(n \times n)$ matrix which is diagonalizable.*

Step 1: Solve for the distinct eigenvalues $\lambda_1, \ldots, \lambda_s$ of A by solving (somehow) the characteristic equation

$$\det(A-\lambda I_n) = (-1)^n(\lambda_1-\lambda)^{r_1}\dots(\lambda_s-\lambda)^{r_s} = 0.$$

Step 2: Find a real basis $\{x^{i1}, \dots, x^{im_i}\}$ *for each eigenspace* $E(\lambda_i)$, $1 \leqslant i \leqslant s$, *by using the procedure in the proof of Theorem 1.25 to find a basis for the nullspace of* $(A-\lambda_i I_n)$, $1 \leqslant i \leqslant s$. *The diagonalizability assumption guarantees* $r_i = m_i$, $1 \leqslant i \leqslant s$.

Step 3: Let P *be the (real, invertible,* $(n\times n)$*) matrix whose first* r_1 *columns are* $x^{11}, x^{12}, \dots, x^{1r_1}$*; whose next* r_2 *columns are* $x^{21}, x^{22}, \dots, x^{2r_2}$*; etc.; and whose last* r_s *columns are* $x^{s1}, x^{s2}, \dots, x^{sr_s}$. *Then* $P^{-1}AP$ *is a diagonal matrix; the diagonal elements of* $P^{-1}AP$ *are given by*

$$[P^{-1}AP]_{ii} = \begin{cases} \lambda_1 & 1 \leqslant i \leqslant r_1 \\ \lambda_2 & r_1 < i \leqslant r_1 + r_2 \\ \cdot & \cdot \\ \cdot & \cdot \\ \lambda_s & r_1 + \dots + r_s - 1 < i \leqslant n. \end{cases}$$ □

If A is a real $(n\times n)$ matrix which is semi-simple but not diagonalizable, the same procedure may be used to find a *complex* invertible $(n\times n)$ matrix P such that $P^{-1}AP$ is diagonal; simply delete the word "real" from Steps 2 and 3. There is, however, another interesting manipulation which may be performed in the semi-simple, non-diagonalizable case.

We know from Observation 8.5(a) that the non-real eigenvalues and corresponding eigenvectors for A come in complex conjugate pairs. If λ, λ^* is such a pair of eigenvalues, with corresponding eigenvectors z, z^*, we saw in an earlier discussion that, if $x = \mathrm{Re}\{z\}$, $y = \mathrm{Im}\{z\}$, then $\{x, y\}$ is a pair of *real* n-vectors, and, moreover, that $\{x, y\}$ spans the same subspace of $\mathbb{C}^n$ as does $\{z, z^*\}$. Moreover, if $\lambda = \mu + i\omega$, then clearly

$$\begin{aligned} Ax &= \frac{1}{2}A(z+z^*) \\ &= \frac{1}{2}(\mu+i\omega)(x+iy) + \frac{1}{2}(\mu-i\omega)(x-iy) \qquad \text{(a)} \\ &= \mu x - \omega y, \end{aligned}$$

and also

$$Ay = \omega x + \mu y \ . \tag{b}$$

Now, let A be semi-simple; let $\lambda_1, \lambda_1^*, \ldots, \lambda_k, \lambda_k^*$, along with $\lambda_{2k+1}, \ldots, \lambda_s$, be the distinct eigenvalues of A, where λ_l is real for $l > 2k$. Choose a basis $\{z^{i1}, z^{i2}, \ldots, z^{ir_i}\}$ for $E(\lambda_i)$, $1 \leqslant i \leqslant k$; then $\{(z^{i1})^*, \ldots, (z^{ir_i})^*\}$ is a basis for $E(\lambda_i^*)$, $1 \leqslant i \leqslant k$. Construct a basis $\mathbf{x}$ for $\mathbf{C}^n$ as follows: the first $2r_1$ vectors are

$$x^{11}, y^{11}, x^{12}, y^{12}, \ldots, x^{1r_1}, y^{1r_1},$$

where $x^{1j} = \mathrm{Re}\{z^{1j}\}$, $y^{1j} = \mathrm{Im}\{z^{1j}\}$; the next r_2 vectors are $x^{21}, y^{21}, \ldots, x^{2r_2}, y^{2r_2}$; etc.; the last $(s-2k)$ vectors consist of bases, arranged consecutively as in Procedure 8.17, for the real eigenspaces $E(\lambda_i)$, $i > 2k$.

Observe that the basis $\mathbf{x}$ is a basis for $\mathbf{C}^n$ consisting of *real* n-vectors; hence $\mathbf{x}$ is also a basis for $\mathbf{R}^n$. If we construct the real $(n \times n)$ matrix P which has the basis vectors in $\mathbf{x}$ as its columns (arranged in order), then $P^{-1}AP$ will be the matrix of ${}_A\hat{T}$ (respectively of ${}_AT$) with respect to the basis $\mathbf{x}$ for $\mathbf{C}^n$ (respectively, for $\mathbf{R}^n$). Equations (a) and (b) above imply that $P^{-1}AP$ will have the following form: an array of $(r_1 + \ldots + r_k)$ blocks, each of size (2×2), will be strung out along the diagonal starting in the upper left; the last $r_{2k+1} + r_{2k+2} + \ldots + r_s$ diagonal elements of $P^{-1}AP$ will be the real eigenvalues, arrayed as in Procedure 8.17.

The (2×2) blocks will be as follows: the first r_1 of them will all be

$$\begin{bmatrix} \mu_1 & \omega_1 \\ -\omega_1 & \mu_1 \end{bmatrix},$$

where $\lambda_1 = \mu_1 + i\omega_1$; the next r_2 will be

$$\begin{bmatrix} \mu_2 & \omega_2 \\ -\omega_2 & \mu_2 \end{bmatrix},$$

where $\lambda_2 = \mu_2 + i\omega_2$; etc. For example, if A is (6×6) and has eigenvalues 4, 5, and $1 \pm i$ (each of multiplicity two), then $P^{-1}AP$ might be

$$\begin{bmatrix} 1 & 1 & 0 & 0 & 0 & 0 \\ -1 & 1 & 0 & 0 & 0 & 0 \\ 0 & 0 & 1 & -1 & 0 & 0 \\ 0 & 0 & 1 & 1 & 0 & 0 \\ 0 & 0 & 0 & 0 & 5 & 0 \\ 0 & 0 & 0 & 0 & 0 & 4 \end{bmatrix}.$$

8.18 Exercise: Let A be a real $(n \times n)$ matrix; suppose that each eigenvalue of A has algebraic multiplicity 1, i.e., that there exist n different eigenvalues for A.

(a) Prove that A is semi-simple.

(b) Prove that A is diagonalizable if all the eigenvalues are real. □

Exercise 8.18 shows that the family of real $(n \times n)$ semi-simple matrices includes those matrices which have n different eigenvalues. In §10, it is shown that every real symmetric matrix is also semi-simple (in fact, diagonalizable!). Although the *whole* class of semi-simple matrices is difficult to characterize, semi-simplicity is seen to be a *generic* property of $(n \times n)$ matrices; if A is $(n \times n)$ and has repeated eigenvalues, then an arbitrarily small perturbation of A will yield a matrix which has n different eigenvalues. See [Hirsch and Smale] for a more detailed discussion.

We return now to the study of Equations (I) and (II) at the start of the section; techniques involving eigenvalues and eigenvectors will be the key to computing $e^{(t-s)A}$ and $A^{(k-l)}$, which are the transition matrices for (I) and (II), respectively. The key observation is that if A is $(n \times n)$, P is $(n \times n)$ invertible, and $\Lambda = P^{-1}AP$, then

$$A^k = P\Lambda^k P^{-1}$$

for every $k > 0$. If Λ is diagonal, then Λ^k is quite easy to compute. Indeed,

$$[\Lambda^k]_{ij} = \begin{cases} ([\Lambda]_{ii})^k & i = j \\ 0 & i \neq j \end{cases}$$

Taking the argument one step further, observe that in the same context

$$
\begin{aligned}
e^{(t-s)A} &= \sum_{k=1}^{\infty} \frac{(t-s)^k (P\Lambda P^{-1})^k}{k!} \\
&= P[\sum_{k=0}^{\infty} \frac{(t-s)^k}{k!} \Lambda^k]P^{-1} \\
&= P\, e^{(t-s)\Lambda} P^{-1} ;
\end{aligned}
$$

once again, if Λ is diagonal, $e^{(t-s)\Lambda}$ is easy to compute:

$$
[\exp(t-s)\Lambda]_{ij} = \begin{cases} e^{(t-s)[\Lambda]_{ii}} & i = j \\ 0 & i \neq j \end{cases}
$$

Hence we have a technique for finding $e^{(t-s)A}$ and A^{k-l}, $k \geqslant l$, given that A is semi-simple: use Procedure 8.17 or its complex variant to find an invertible $(n \times n)$ matrix P such that $P^{-1}AP = \Lambda$ is diagonal; then use the above formulas to find $e^{(t-s)A}$ and $A^{(k-l)}$ in terms of P and Λ.

It should be remarked that this procedure, while theoretically pleasing, is not computationally efficient; see, for example, [Golub and van Loan]. Step 1 of Procedure 8.17 which entails finding the eigenvalues of A, is itself a nontrivial task. In §11, the results of the present section and those of §§9-10 will be viewed in a somewhat different light. For now, it suffices to regard them as a means by which, *in principle*, one may compute $e^{(t-s)A}$ and $A^{(k-l)}$ $k > l$, and thereby "solve" the differential equation in (I) and the difference equation in (II).

9. *The M + N Decomposition for Matrices Which Are Not Semi-Simple*

In this section, we continue the eigenvector/eigenvalue analysis of real $(n \times n)$ matrices A. We saw in §8 how such an analysis leads not only to insights about the linear mappings $_A T : \mathbf{R}^n \to \mathbf{R}^n$ and $_A \hat{T} : \mathbf{C}^n \to \mathbf{C}^n$, but also to procedures which may, at least in principle, be used in computing $e^{(t-s)A}$ and $A^{(k-l)}$. In the pages which follow, we focus attention on matrices A which are *not* semi-simple (hence, in particular, not diagonalizable); the techniques we shall be discussing should be viewed as *generalizations* of the techniques we employed in analyzing semi-simple A's.

The essential idea is the following: if a real $(n \times n)$ matrix A is semi-simple (or diagonalizable), then a basis for $\mathbf{C}^n$ (or $\mathbf{R}^n$)

may be found which consists solely of eigenvectors for A. If A is *not* semi-simple, then there are in some sense "not enough" eigenvectors for A; that is, there exist vectors in $\mathbf{C}^n$ which cannot be expressed as linear combinations of A's eigenvectors. We are prompted to make the following definition.

9.1 Definition: Let A be a real $(n \times n)$ matrix; let λ_o be an eigenvalue of A. A *generalized eigenvector* of A *corresponding to eigenvalue* λ_o is a nonzero vector $z \in \mathbf{C}^n$ such that, for some $k > 0$, $(A - \lambda_o I_n)^k z = 0$. Furthermore, the *generalized eigenspace corresponding to* λ_o (notation: $F(\lambda_o)$) is the subspace of $\mathbf{C}^n$ spanned by all the generalized eigenvectors of A corresponding to the eigenvalue λ_o. □

Clearly, every eigenvector of A corresponding to eigenvalue λ_o is also a generalized eigenvector of A corresponding to λ_o; the converse, however, is not in general true. The central result of this section is that, given a real $(n \times n)$ matrix A, *any vector in* $\mathbf{C}^n$ *may be expressed as a linear combination of generalized eigenvectors of* A. We shall see that A is semi-simple precisely when every *generalized* eigenvector of A is also an eigenvector of A. The argument leading to these results is somewhat intricate; we have broken it down into a series of propositions along the lines of the development in §8.

9.2 Proposition: *Let A be a real $(n \times n)$ matrix; let z° be a generalized eigenvector of A corresponding to eigenvalue λ_o, i.e., $z^\circ \neq 0$ and $z^\circ \in F(\lambda_o)$. Then $A^l z^\circ \in F(\lambda_o)$, as well, for every $l > 0$.*

Proof: Suppose $(A - \lambda_o I_n)^k z^\circ = 0$. Then, since A^l commutes with $(A - \lambda_o I_n)$, we have

$$(A - \lambda_o I_n)^k A^l z^\circ = A^l (A - \lambda_o I_n)^k z^\circ = 0 .$$

Hence $A^l z^\circ \in F(\lambda_o)$ by definition. □

A direct consequence of Proposition 9.2 is that if z° is in the generalized eigenspace of A corresponding to eigenvalue λ_o, then so is $(A - \alpha I_n)^k z^\circ$ for any complex number α. To see this, write $(A - \alpha I_n)^k$ as a polynomial in A; each term in the resulting expansion of $(A - \alpha I_n)^k z^\circ$ is of the form $cA^l z^\circ$, where c is a scalar, and is therefore in $F(\lambda_o)$ by Proposition 9.2.

9.3 Proposition: *Let A be a real $(n \times n)$ matrix; let $\lambda_1, \ldots, \lambda_s$ be the distinct eigenvalues of A with corresponding generalized eigenspaces $F(\lambda_1), \ldots, F(\lambda_s)$. Then, if $i \neq j$;*

$$F(\lambda_i) \cap F(\lambda_j) = \{0\}.$$

Proof: Suppose $z \neq 0$ and $z \in F(\lambda_i) \cap F(\lambda_j)$; let k and l be the smallest positive integers satisfying

$$(A - \lambda_i I_n)^k z = (A - \lambda_j I_n)^l z = 0.$$

Then $(A - \lambda_j I_n)^{l-1} z$ is an eigenvector of A corresponding to eigenvalue λ_j. Multiply the first equation by $(A - \lambda_j I_n)^{l-1}$ to obtain

$$\begin{aligned} 0 &= (A - \lambda_i I_n)^k [(A - \lambda_j I_n)^{l-1} z] \\ &= (\lambda_j - \lambda_i)^k (A - \lambda_j I_n)^{l-1} z = 0 . \end{aligned}$$

Since $\lambda_i \neq \lambda_j$, we get $(A - \lambda_j I_n)^{l-1} z = 0$, contradicting choice of l. Thus $z = 0$. □

We shall require the following stronger version of Proposition 9.3, which generalizes Lemma 8.14.

9.4 Proposition: *Let A be a real $(n \times n)$ matrix; let $\lambda_1, \ldots, \lambda_s$ be the distinct eigenvalues of A with corresponding generalized eigenspaces $F(\lambda_i)$, $1 \leqslant i \leqslant s$. Then $F(\lambda_i)$, $1 \leqslant i \leqslant s$, are mutually disjoint subspaces of $\mathbb{C}^n$.*

Proof: By Observation 4.28(a), we merely need to show that if $z^i \in F(\lambda_i)$, $1 \leqslant i \leqslant s$, and $\sum_{i=1}^{s} z^i = 0$, then $z^i = 0$, $1 \leqslant i \leqslant s$. We use an induction argument: we show that, for each m, if $z^i \in F(\lambda_i)$, $1 \leqslant i \leqslant m$, and

$$z^1 + \ldots + z^m = 0 ,$$

then $z^i = 0$, $1 \leqslant i \leqslant m$. This is clearly true for $m = 1$; suppose now that it is true for $m = l-1$, and suppose that $z^i \in F(\lambda_i)$, $1 \leqslant i \leqslant l$, are such that

$$\sum_{i=1}^{l} z^i = 0 .$$

Since $z^l \in F(\lambda_l)$, there exists a $k > 0$ such that $(A - \lambda_l I_n)^k z^l = 0$. Multiplying both sides of the second-to-last equation by $(A - \lambda_l I_n)^k$, we obtain

$$\sum_{i=1}^{l-1} (A-\lambda_l I_n)^k z^i = 0 .$$

By Proposition 9.2 and the discussion which follows it, $(A-\lambda_l I_n)^k z^i \in F(\lambda_i)$, $1 \leqslant i \leqslant l-1$; by induction hypothesis, $(A-\lambda_l I_n)^k z^i = 0$, $1 \leqslant i \leqslant l-1$. Thus, for $1 \leqslant i \leqslant l$, $z^i \in F(\lambda_l)$, which implies (since $z^i \in F(\lambda_i)$) that $z^i = 0$, $1 \leqslant i \leqslant l-1$, by Proposition 9.3; since, therefore, $z^l = 0$ as well, the proof is complete. □

9.5 Proposition: *Let A be a real $(n \times n)$ matrix; let $\lambda_1, \ldots, \lambda_s$ be the distinct eigenvalues of A, with respective algebraic multiplicities r_i, $1 \leqslant i \leqslant s$, and generalized eigenspaces $F(\lambda_i)$, $1 \leqslant i \leqslant s$. Then the dimension of $F(\lambda_i)$ is r_i, $1 \leqslant i \leqslant s$.*

Proof: Choose i, and let k be the dimension of $F(\lambda_i)$; we'll show that $k = r_i$. Choose a basis $\mathbf{z} = \{z^1, \ldots, z^n\}$ for $\mathbf{C}^n$ such that $\{z^1, \ldots, z^k\}$ is a basis for $F(\lambda_i)$. Let B be the $(n \times n)$ matrix of ${}_A\hat{T} : \mathbf{C}^n \to \mathbf{C}^n$ with respect to the basis $\mathbf{z}$; then B takes the form

$$B = \begin{bmatrix} A_1 & A_2 \\ 0 & A_3 \end{bmatrix},$$

where A_1, A_2, A_3, and 0 are of respective sizes $(k \times k)$, $(k \times (n-k))$, $((n-k) \times (n-k))$, and $((n-k) \times k)$. This special form for B follows from Proposition 9.2, since each ${}_A\hat{T}(z^i)$, $1 \leqslant i \leqslant k$, is a linear combination of $z^1, \ldots, z^k$.

Consider now the eigenvalues of A_1 and A_3. First, we claim that λ_i is the only eigenvalue of A_1. If, for some $j \neq i$, λ_j were also an eigenvalue of A_1, then there would exist a nonzero $y \in \mathbf{C}^k$ such that $A_1 y = \lambda_j y$; then $y_1 z^1 + \ldots + y_k z^k$ would be an eigenvector of A corresponding to eigenvalue λ_j as well as a generalized eigenvector corresponding to λ_i, which is impossible by Proposition 9.3. Moreover, λ_i *cannot* be an eigenvalue of A_3; if λ_i were an eigenvalue of A_3, then there would exist a nonzero $y \in \mathbf{C}^{n-k}$ such that $(A_3 - \lambda_i I_{n-k})y = 0$. Then

$$z = \sum_{i=k+1}^{n} y_i z^i$$

would be a generalized eigenvector of A corresponding to eigenvalue λ_i, since by taking coordinates with respect to $\mathbf{z}$ we obtain

$$[(A-\lambda_i I_n)z]_{\mathbf{z}} = (B-\lambda_i I_n)\begin{bmatrix}0\\ y\end{bmatrix}$$
$$= \begin{bmatrix}A_2 y\\ 0\end{bmatrix}$$
$$= [\sum_{i=1}^{k} [A_2 y]_i z^i]_{\mathbf{z}},$$

so that $(A-\lambda_i I_n)z \in F(\lambda_i)$, implying that $(A-\lambda_i I_n)^k z = 0$ for some $k > 0$; therefore, $x \in F(\lambda_i)$. Since z is by construction a sum of vectors linearly independent of every vector in $F(\lambda_i)$, this situation cannot arise.

We see, then, that λ_i is the only eigenvalue of A_1 and is not an eigenvalue of A_3. By Lemma 8.7, we have

$$\det(A-\lambda I_n) = \det(B-\lambda I_n)$$
$$= \det(A_1-\lambda I_k)\det(A_3-\lambda I_{n-k})$$
$$= (\lambda_i-\lambda)^k q(\lambda),$$

where $q(\lambda)$ does not have λ_i as a root. Since λ_i has multiplicity r_i as a root of $\det(A-\lambda_i I_n)$, we see that $k = r_i$. □

Propositions 9.4 and 9.5, taken together with Observations 4.28, yield a quick proof of the following fundamental result.

9.6 Theorem: *Let A be a real $(n \times n)$ matrix; let $\lambda_1, \ldots, \lambda_s$ be the distinct eigenvalues of A with respective algebraic multiplicities $r_1, \ldots, r_s$ and generalized eigenspaces $F(\lambda_1), \ldots, F(\lambda_s)$. Then*

(a) There exists a basis for $\mathbf{C}^n$ consisting solely of generalized eigenvectors of A; and

(b) If all eigenvalues of A are real, there exists a basis for $\mathbf{R}^n$ consisting solely of (real) generalized eigenvectors of A.

Proof: By Proposition 9.4 and Observation 4.28(c), we know that the vector sum of the $F(\lambda_i)$ — that is, the subspace of $\mathbf{C}^n$ consisting of all sums $z^1 + \ldots + z^s$ with $z^i \in F(\lambda_i)$ — has dimension equal to the sum of the dimensions of $F(\lambda_i)$. By Proposition 9.5, this sum is $r_1 + \ldots + r_s = n$. Since the only n-dimensional subspace of $\mathbf{C}^n$ is $\mathbf{C}^n$ itself, this shows that

$F(\lambda_1) + \ldots + F(\lambda_s) = \mathbf{C}^n$. By stringing together bases for the $F(\lambda_i)$ (as in Observation 4.28 (d)), we obtain a basis for $\mathbf{C}^n$ consisting solely of generalized eigenvectors of A; this proves (a).

Assume now that A has only real eigenvalues. By replacing $\mathbf{C}^n$ with $\mathbf{R}^n$ in Definition 9.1 and in the statements and proofs of Propositions 9.4 and 9.5, while leaving Propositions 9.2 and 9.3 unaltered, we obtain a chain of results which, by precisely the same argument used in proving (a), leads to a proof of (b). □

The relationship between Theorem 9.6 and the discussion of semi-simple matrices in §8 should now be apparent. In the notation §8 and Theorem 9.6, we know that for each i, $1 \leqslant i \leqslant s$,

$$E(\lambda_i) \subset F(\lambda_i).$$

That is, the eigenspace corresponding to each eigenvalue of an arbitrary real $(n \times n)$ matrix is a subspace of the *generalized* eigenspace corresponding to the same eigenvalue. By Proposition 9.5, the dimension of $F(\lambda_i)$ is r_i; hence $E(\lambda_i) = F(\lambda_i)$ for every i precisely when the dimension of each $E(\lambda_i)$ is r_i. In that case, the geometric multiplicity of each eigenvalue λ_i is the same as its algebraic multiplicity r_i; this last statement is equivalent to A's being semi-simple. As a consequence, another way of saying that A is semi-simple is by saying that *every generalized eigenvector of A is also an eigenvector of A*. If A is not semi-simple, then there "aren't enough" eigenvectors of A to span $\mathbf{C}^n$; by Theorem 9.6, however, the *generalized* eigenvectors will always span $\mathbf{C}^n$.

We turn now to the problem of applying Theorem 9.6 to the computation of e^{tA} and A^k for matrices A which are not semi-simple. We shall have occasion to choose bases for the generalized eigenspaces of A; choosing arbitrary such bases will give us a method for computing e^{tA} and A^k, while certain special choices of basis will lead in §11 to the Jordan canonical form for A. To begin with, we require an auxiliary result which is of central importance in what follows.

9.7 Lemma: *Let λ_0 be an eigenvalue of a real $(n \times n)$ matrix A; let r_0 be the algebraic multiplicity of λ_0 and $F(\lambda_0)$ the corresponding generalized eigenspace. Then*

$$(A-\lambda_o I_n)^{r_o} z = 0$$

for every $z \in F(\lambda_o)$. *Hence* $F(\lambda_o)$ *is precisely the nullspace of* $(A-\lambda_o I_n)^{r_o}$.

Proof: Pick a nonzero $z \in F(\lambda_o)$, and let k be the smallest positive integer such that $(A-\lambda_o I_n)^k z = 0$; we'll show that $k \leqslant r_o$. Denote by N the matrix $(A-\lambda_o I_n)$; we claim that $\{z, Nz, N^2 z, \ldots, N^{k-1} z\}$ is a linearly independent set of vectors contained in $F(\lambda_o)$. These vectors are indeed in $F(\lambda_o)$, by Proposition 9.2; to see that they are linearly independent, multiply a relation of the form

$$c_1 z + c_1 Nz + \ldots + c_{k-1} N^{k-1} z = 0$$

by N^{k-1} to obtain

$$c_o N^{k-1} z = 0,$$

implying that $c_o = 0$ since $N^{k-1} z \neq 0$. We are left with a new relation having one term fewer:

$$c_1 Nz + \ldots + c_{k-1} N^{k-1} z = 0.$$

Multiplying by N^{k-2}, we find $c_1 = 0$; repeating this procedure, we may deduce that $c_i = 0, 0 \leqslant i \leqslant k-1$.

Since, by Proposition 9.5, $F(\lambda_o)$ has dimension r_o, it is seen that $k \leqslant r_o$ is required, since $k > r_o$ would imply the existence of too large a linearly independent set of vectors in $F(\lambda_o)$. Since z was arbitrary, $(A-\lambda_o I_n)^{r_o} z = 0$ must hold for every $z \in F(\lambda_o)$. The last statement in the theorem is then an immediate consequence. □

A corollary of Theorem 9.6 and Lemma 9.7 is a famous result which is known as the *Cayley-Hamilton Theorem.*

9.8 Theorem (Cayley and Hamilton): *Let A be a real* $(n \times n)$ *matrix; let the characteristic equation of A be given by*

$$\begin{aligned} 0 &= p(\lambda) = \det(A-\lambda I_n) \\ &= (-1)^n \lambda^n + a_1 \lambda^{n-1} + \ldots + a_{n-1}\lambda + a_n . \end{aligned}$$

Then A "satisfies its own characteristic equation;" i.e.:

$$p(A) = (-1)^n A^n + a_1 A^{n-1} + \ldots + a_{n-1} A + a_n I_n = 0.$$

Proof: Let λ_i, $1 \leqslant i \leqslant s$, be the distinct eigenvalues of A with respective algebraic multiplicities r_i, $1 \leqslant i \leqslant s$. Observe then that since

$$p(\lambda) = (\lambda_1 - \lambda)^{r_1} \ldots (\lambda_s - \lambda)^{r_s} ,$$

we have

$$p(A) = (\lambda_1 I_n - A)^{r_1} \ldots (\lambda_s I_n - A)^{r_s} .$$

If z is a generalized eigenvector of A corresponding to eigenvalue λ_i, then $p(A)z = 0$, since the factors in $p(A)$ commute; indeed by letting $N_j = (\lambda_j I_n - A)$, we get

$$p(A)z = N_1^{r_1} \ldots N_{i-1}^{r_{i-1}} N_{i+1}^{r_{i+1}} \ldots N_s^{r_s} (N_i^{r_i} z) = 0 ,$$

where the last equality follows from Lemma 9.7. Since by Theorem 9.6 every $z \in \mathbf{C}^n$ may be expressed as a sum of generalized eigenvectors for A, we obtain $p(A)z = 0$ for every $z \in \mathbf{C}^n$, which implies that $p(A) = 0$. □

The Cayley-Hamilton Theorem may be viewed as providing a means by which high powers of A may be expressed in terms of lower powers of A. With notation as in the statement of Theorem 9.8, observe that

$$A^n = (-1)^{n+1}[a_1 A^{n-1} + a_2 A^{n-2} + \ldots + a_n I_n] .$$

Thus A^n may be expressed as a linear combination of I_n, $A, \ldots, A^{n-1}$. In fact, we may express any power A^k of A as a linear combination of A's first $(n-1)$ powers; to see this inductively, first multiply the above equation by A^{k-n}, and obtain a formula for A^k in terms of $A^{k-1}, \ldots, A^{k-n}$; then use the (already obtained) formulas for $A^{k-1}, \ldots, A^{k-n}$ in terms of $I_n, A, \ldots, A^{n-1}$ to obtain A^k as a linear combination of the first $(n-1)$ powers of A. In principle, the Cayley-Hamilton Theorem could also be used to get a formula for e^{tA} in terms of the first $(n-1)$ powers of A; the formula would be

$$e^{tA} = f_0(t) I_n + f_1(t) A + \ldots + f_{n-1}(t) A^{n-1} ;$$

the $t \to f_i(t)$ would be scalar functions of t obtained by substituting linear combinations of $I_n, A, \ldots, A^{n-1}$ for $A^n, A^{n+1}, \ldots$ in the series for e^{tA} and then factoring out the $A^k, 0 \leqslant k \leqslant n-1$.

The method for computing e^{tA} which we shall discuss presently is somewhat different, but still relies heavily on

Lemma 9.7. Our aim will be to express any real $(n \times n)$ matrix A as the sum of two $(n \times n)$ (possibly complex) matrices M and N such that:

(*i*) M and N commute; i.e., $MN = NM$; and

(*ii*) M is semi-simple, and N is *nilpotent*, i.e., $N^r = 0$ for some $r \leqslant n$.

To understand why the $A = M + N$ decomposition is significant, consider the following expression for A^k derived using the *binomial expansion*:

$$A^k = (M+N)^k = \sum_{l=0}^{k} \begin{pmatrix} k \\ l \end{pmatrix} M^{k-l} N^l ; \qquad (*)$$

here, we are using the standard notation

$$\begin{pmatrix} k \\ l \end{pmatrix} = \frac{k!}{l!(k-l)!} .$$

The last equality will not hold in general unless M and N commute; indeed, $(M+N)^2 = M^2 + MN + NM + N^2$ for arbitrary (possibly non-commuting) M and N. Observe that for large values of k, all but r terms in the expansion (*) will vanish due to nilpotence of N, and hence only a few terms will have to be computed in order to find A^k. Observe in addition that, since M is semi-simple, M^k may be computed using the methods of §8.

An even more convincing reason for attempting to write $A = M + N$ for two commuting matrices M and N is foreshadowed by the following important technical result.

9.9 Lemma: *Let* M *and* N *be* $(n \times n)$ *matrices. Then* $MN = NM$ *if and only if for every* $t \in \mathbf{R}$,

$$e^{t(M+N)} = e^{tM} e^{tN} .$$

Proof: Let $L(t)$ and $R(t)$ denote, respectively, the left- and right-hand sides of the above identity. Suppose now that $MN = NM$. Then clearly $e^{tM} N = N e^{tM}$ for all t, since each term in the series expansion for e^{tM} is a power of M. Consequently, upon taking derivatives, we obtain

$$\dot{L}(t) = (M+N)L(t), \qquad L(0) = I_n;$$

along with

$$\dot{R}(t) = Me^{tM}e^{tN} + e^{tM}Ne^{tN}$$
$$= (M+N)R(t), \qquad R(0) = I_n.$$

Hence $t \to L(t)$ and $t \to R(t)$ satisfy the same linear matrix differential equation and initial condition, hence must be equal for all $t \in \mathbf{R}$.

Conversely, assume $L(t) = R(t)$ for all t. Taking *two* derivatives with respect to t gives

$$\frac{d^2L}{dt^2} = (M+N)^2L(t)$$

$$\frac{d^2R}{dt^2} = M^2e^{tM}e^{tN} + 2Me^{tM}Ne^{tN} + e^{tM}N^2e^{tN}.$$

Now, these two expressions must be equal for all $t \in \mathbf{R}$; evaluating at $t = 0$ yields

$$(M+N)^2 = M^2 + 2MN + N^2,$$

or

$$M^2 + MN + NM + N^2 = M^2 + 2MN + N^2,$$

implying that $MN = NM$. □

Suppose, then, that $A = M+N$, where M is semi-simple, $N^r = 0$ for some $r > 0$, and $MN = NM$. By Lemma 9.9, it is seen that

$$e^{tA} = e^{t(M+N)} = e^{tM}e^{tN}.$$

Since $N^r = 0$, $N^k = 0$ for all $k \geqslant r$; thus the series for e^{tN} contains only finitely many nonzero terms. Moreover, e^{tM} may be computed using the methods of §8, since M is semi-simple; by expressing A as $A = M+N$, we have reduced the problem of finding e^{tA} to a series of computations which we know how to perform.

Let us derive the "$(M+N)$ decomposition" for an arbitrary real $(n \times n)$ matrix A. Suppose A has s distinct eigenvalues $\lambda_1, \ldots, \lambda_s$ with corresponding algebraic multiplicities and generalized eigenspaces r_i and $F(\lambda_i)$, $1 \leqslant i \leqslant s$. We know from Proposition 9.5 and Lemma 9.7 that the dimension of $F(\lambda_i)$ is r_i and that $F(\lambda_i)$ is precisely the nullspace of $(A-\lambda_i I_n)^{r_i}$. Let z_i be a basis for $F(\lambda_i)$, $1 \leqslant i \leqslant s$; note that z_i contains r_i vectors, and that the ordered set $z = z_1 \cup z_2 \cup \ldots \cup z_s$ is a basis for $\mathbf{C}^n$.

Denote by B the $(n \times n)$ matrix of the linear transformation ${}_A\hat{T} : \mathbf{C}^n \to \mathbf{C}^n$ with respect to the basis z; by Proposition 9.2, since ${}_A\hat{T}$ operates on any vector in $F(\lambda_i)$ to give another vector in $F(\lambda_i)$, B takes the special form

$$B = \begin{bmatrix} A_1 & 0 & 0 & 0 & 0 & 0 \\ 0 & A_2 & 0 & . & . & 0 \\ . & 0 & . & . & . & . \\ . & . & . & . & . & . \\ . & . & . & 0 & A_{s-1} & 0 \\ 0 & 0 & 0 & 0 & 0 & A_s \end{bmatrix},$$

where A_i, $1 \leqslant i \leqslant s$, is an $(r_i \times r_i)$ complex matrix. Observe (see equation (**) in §4 and the discussion which precedes 8.17) that $B = P^{-1}AP$, where P is the matrix whose jth column is the jth vector in the basis z.

Define the $(n \times n)$ diagonal matrix Λ by

$$\Lambda = \begin{bmatrix} \lambda_1 I_{r_1} & 0 & . & . & . \\ 0 & \lambda_2 I_{r_2} & . & . & . \\ . & . & . & . & . \\ . & . & . & . & . \\ 0 & 0 & . & . & \lambda_s I_{r_s} \end{bmatrix}.$$

Thus Λ is a matrix in which the eigenvalues of A appear in order along the diagonal; each appears as many times as its algebraic multiplicity. It is clear that $B\Lambda = \Lambda B$; that is, B *commutes* with Λ. Moreover, $B - \Lambda$ is *nilpotent*.

9.10 Observation: *In the above notation,* $(A_i - \lambda_i I_{r_i})^{r_i} = 0$. *Hence* $(B - \Lambda)^r = 0$, *where r is the largest of the* r_i.

Proof: Choose i, $1 \leqslant i \leqslant s$, and y be an arbitrary complex r_i-vector. Define $z \in \mathbf{C}^n$ by

$$z = \sum_{j=1}^{r_i} y_j z^{ij},$$

where z^{ij} is the jth vector in the basis z_i for $F(\lambda_i)$. This means by definition of P that $z = P\hat{y}$, where $\hat{y}$ is the complex n-vector with $r_1 + \ldots + r_{i-1}$ zeros at the top, y in the next r_i positions, and $r_{i+1} + \ldots + r_s$ zeros at the bottom. By Lemma 9.7, since $z \in F(\lambda_i)$, we know that $(A - \lambda_i I_n)^{r_i} z = 0$. Hence

$$\begin{aligned} 0 &= P^{-1}(A - \lambda_i I_n)^{r_i} z \\ &= (P^{-1}AP - \lambda_i I_n)^{r_i} P^{-1} z \\ &= (B - \lambda_i I_n)^{r_i} \hat{y} , \end{aligned}$$

where the second equality follows from the discussion at the end of §8 and the last equality from the definitions of B, z, and $\hat{y}$.

Note that $(B - \lambda_i I_n)^{r_i} \hat{y}$ is the n-vector which has $r_1 + \ldots + r_{i-1}$ zeros at the top, the vector $(A_i - \lambda_i I_{r_i})^{r_i} y$ in the next r_i positions, and $r_{i+1} + \ldots + r_s$ zeros at the bottom. We conclude that $(A_i - \lambda_i I_n)^{r_i} y = 0$; since $y \in \mathbf{C}^{r_i}$ was arbitrary, $(A_i - \lambda_i I_n)^{r_i} = 0$. Since the last identity is true for every i, $1 \leqslant i \leqslant s$, $(B - \Lambda)^r = 0$, where r is the maximum of the r_i, $1 \leqslant i \leqslant s$. □

The $(M + N)$ decomposition for A is now at hand. Define P, B, and Λ as above; set

$$M = P\Lambda P^{-1} ; \qquad N = P(B - \Lambda)P^{-1} = A - P\Lambda P^{-1} .$$

Then $M + N = PBP^{-1} = A$; moreover, M is semi-simple by Theorem 8.16. Finally $N^r = P(B - \Lambda)^r P^{-1} = 0$, where $r = \max\{r_i : 1 \leqslant i \leqslant s\}$. We summarize the foregoing discussion as follows:

9.11 Decomposition Procedure: *Let A be a real $(n \times n)$ matrix.*

Step 1: Compute the distinct eigenvalues λ_i, $1 \leqslant i \leqslant s$, and their corresponding algebraic multiplicities r_i, $1 \leqslant i \leqslant s$, by solving $\det(A - \lambda I_n) = 0$.

Step 2: For each i, find a basis $\mathbf{z}_i$ for $F(\lambda_i)$, where $F(\lambda_i)$ is the generalized eigenspace corresponding to λ_i. To find these bases, use the fact that $F(\lambda_i)$ is the nullspace of $(A - \lambda_i I_n)^{r_i}$ along with the procedure in the proof of Theorem 1.25.

Step 3: Form the matrices P and Λ as above; i.e., the jth column of P is the jth vector in the basis for $\mathbf{C}^n$ obtained by stringing together $\mathbf{z}_1, \ldots, \mathbf{z}_s$ in order; and Λ is the diagonal matrix with diagonal elements given by

$$[\Lambda]_{jj} = \lambda_i \, , \quad r_1 + \ldots + r_{i-1} < j \leqslant r_1 + \ldots + r_i \, .$$

Step 4: Set $M = P\Lambda P^{-1}$ and $N = A - M$. Then $A = M + N$, $MN = NM$, M is semi-simple, and N is nilpotent; in fact, $N^r = 0$, where $r = \max\{r_i : 1 \leqslant i \leqslant s\}$. □

After Procedure 9.11 has been accomplished, we may conclude from equation (*) and Lemma 9.9 that

$$A^k = \sum_{l=0}^{r-1} \binom{k}{l} M^{k-l} N^l \, , \qquad k > 0;$$

and

$$e^{tA} = e^{tM} e^{tN} = e^{tM} \Big(\sum_{k=0}^{r-1} \frac{t^k N^k}{k!} \Big) \, , \qquad t \in \mathbf{R} \, .$$

Here, M^{k-l} and e^{tM} may be computed using the methods of §8, since M is semi-simple. It is easy to see that another formula for e^{tA} is

$$e^{tA} = P e^{t\Lambda} e^{t\tilde{N}} P^{-1} \, ,$$

and that another formula for A^k is

$$A^k = P\Big[\sum_{l=0}^{r-1} \binom{k}{l} \Lambda^{k-l} \tilde{N}^l \Big] P^{-1} \, ;$$

here, $\tilde{N} = P^{-1}NP = B - \Lambda$. Thus, once P, B, and Λ have been computed, e^{tA} and A^k may be found immediately.

The reader who wishes to apply Procedure 9.11 to a given $(n \times n)$ A should be careful to check that A is not semi-simple before figuring out $(A - \lambda_i I_n)^{r_i}$ in Step 2; in fact, if $E(\lambda_i) = F(\lambda_i)$ for any particular i, then every nonzero vector in $F(\lambda_i)$ is an eigenvector of A, and hence $F(\lambda_i)$ is equal to the nullspace of $(A - \lambda_i I_n)$. Moreover, if r_i is large, one could shorten Step 2 by successively computing the nullspaces of $(A - \lambda_i I_n)$, $(A - \lambda_i I_n)^2$, etc., until a k is reached for which the nullspace of $(A - \lambda_i I_n)^k$ has dimension r_i.

10. Complex Matrices and the Unitary Diagonalizability of Hermitian Matrices

In this section, we seek to tie up some loose ends. First, although we have stated all the definitions and proved all the results in §§8-9 as if they applied only to *real* $(n \times n)$ matrices A, the truth is that just about everything said in §§8-9 about real matrices has its analogue in the theory of *complex* matrices. Let us go through §§8-9 point by point and isolate the (few) instances where realness of A makes a difference.

Definition 8.1 and Examples 8.2 remain unaltered for complex matrices; in the discussion following Examples 8.2, note that a complex A does *not* necessarily define a linear transformation ${}_AT : \mathbf{R}^n \to \mathbf{R}^n$, but that ${}_AT : \mathbf{C}^n \to \mathbf{C}^n$ arises from a complex A in exactly the same fashion as for real A. Definition 8.3 stands verbatim for complex A if the word "real" is omitted; similarly, Example 8.4.

Observation 8.5(a) and Corollary 8.6(a) are the first results which hinge on the realness of A; Observation 8.5(b) and Corollary 8.6(b), however, hold equally for complex A, as do Lemma 8.7 and Theorem 8.8. Definitions 8.9, 8.10, and 8.11 are fine for complex A, as well; Theorem 8.12 holds for complex A, too.

The description of semi-simplicity in Definition 8.13 is also good for complex A; the word "diagonalizable," however, is synonymous with "semi-simple" when A is complex. Lemma 8.14 and Theorem 8.15 (minus the last sentence) are both equally true when a complex A is involved. Theorem 8.16 applies also to complex A, except for the statement about diagonalizability and realness of P. Procedure 8.17, modified according the comment which immediately follows it, is an equally good procedure for diagonalizing a semi-simple *complex* matrix A. The statement in Exercise 8.18(a) may also be shown to hold for complex A, although the discussion preceding it applies only to real A. The methods for computing A^k and e^{tA} which arise from Theorems 8.15 and 8.16 along with Procedure 8.17 work equally well for semi-simple complex matrices A. Finally, the treatment in §9 of matrices A which are not semi-simple stands unchanged if the word "real" is deleted from the description of A in every definition and theorem statement.

We turn now to the discussion of an extremely important result, which is often called the *spectral theorem* for $(n \times n)$ Hermitian matrices. Recall from 1.1(f) that a *Hermitian matrix* A is a complex matrix satisfying

$$A^{\dagger} = A \ ,$$

where the dagger superscript represents the operation of "complex conjugate transposition," i.e.

$$A^{\dagger} = (A^*)^T \ ;$$

here, a star superscript denotes (element by element) complex conjugate. A *real symmetric matrix* A is a real matrix which is equal to its own transpose; every real symmetric matrix A is therefore also Hermitian, since $A^* = A$. The following result is central in the theory of Hermitian matrices.

10.1 Lemma: *If A be a Hermitian $(n \times n)$ matrix, then the eigenvalues of A are real.*

Proof: Let $\lambda \in \mathbf{C}$ be an eigenvalue of A; let $z \in \mathbf{C}^n$ be a corresponding eigenvector. Then

$$Az = \lambda z \ ,$$

so that

$$z^{\dagger} A z = \lambda \, \| z \|^2 \, .$$

Take conjugate transpose of both sides to obtain

$$z^{\dagger} A^{\dagger} z = \lambda^* \, \| z \|^2 \, .$$

Since $A^{\dagger} = A$, we may subtract the third equation from the second, and obtain

$$0 = (\lambda - \lambda^*) \| z \|^2 \ ,$$

from which it follows that $\lambda = \lambda^*$ — i.e., $\lambda \in \mathbf{R}$ — since by definition $z \neq 0$, and $\| z \|^2 = |z_1|^2 + \ldots + |z_n|^2$. □

The next result says that every Hermitian matrix A is semi-simple; if A is real and symmetric, this implies by Theorem 8.16 and Lemma 10.1 that A is also diagonalizable.

10.2 Theorem: *If A be an $(n \times n)$ Hermitian matrix, then A is semi-simple.*

Proof: By Theorem 8.15, it suffices to show that the algebraic multiplicity of each eigenvalue of A is equal to its geometric multiplicity. Equivalently, in view of the discussion in §9, we need to show only that every generalized eigenvector z of A is also an eigenvector of A.

Accordingly, let the generalized eigenvector z correspond to the (necessarily real) eigenvalue λ. Thus for some $k > 0$ we have

$$(A-\lambda I_n)^k z = 0 .$$

Suppose $k = 2$; then multiply on the left by $z^\dagger$, use realness of λ, and use Hermitian-ness of A to conclude

$$\begin{aligned} 0 &= z^\dagger(A-\lambda I_n)(A-\lambda I_n)z \\ &= z^\dagger(A-\lambda I_n)^\dagger(A-\lambda I_n)z \\ &= \|(A-\lambda I_n)z\|^2 . \end{aligned}$$

Hence $(A-\lambda I_n)z = 0$, and z is an eigenvector of A corresponding to eigenvalue λ.

Suppose, inductively, that we have shown that every generalized eigenvector z satisfying $(A-\lambda I_n)^l z = 0$ for $l < k$ is an eigenvector of A corresponding to λ. If k is even, proceed as follows:

$$\begin{aligned} (A-\lambda I_n)^k z &= 0 ; \\ z^\dagger(A-\lambda I_n)^{k/2}(A-\lambda I_n)^{k/2}z &= 0 ; \\ \|(A-\lambda I_n)^{k/2}z\|^2 &= 0 ; \end{aligned}$$

from which, by induction, it follows that z is an eigenvector. If k is odd, simply multiply the first equation by $(A-\lambda I_n)$, and perform the same maneuver; conclude by induction that z is an eigenvector of A corresponding to eigenvalue λ. □

We turn now to our main result; infinite-dimensional analogues of this very fundamental theorem form the fulcrum for the spectral theory of linear operators on Hilbert Space, which in turn lies at the heart of modern functional analysis.

10.3 Theorem: *Let A be an $(n \times n)$ Hermitian matrix. Then there exists an* **orthonormal** *basis* $\mathbf{w}$ *for* $\mathbf{C}^n$ *consisting solely of eigenvectors for A. Moreover, if A is real and Hermitian (hence symmetric), then this basis may be chosen to be* *real.*

Proof: Let $\lambda_1, \ldots, \lambda_s$ be the distinct eigenvalues of A; the λ_i are all real by Lemma 10.1. Let $E(\lambda_i)$, $1 \leqslant i \leqslant s$, be the corresponding eigenspaces; by Theorem 10.2 along with Lemma 1.18 and Observation 4.28(c), the vector sum $E(\lambda_1) + \ldots + E(\lambda_s)$ of the (disjoint) subspaces is $\mathbf{C}^n$. Choosing a basis $\mathbf{z}_i = \{z^{i1}, \ldots, z^{ir_i}\}$ for each $E(\lambda_i)$, we may obtain as in Theorem 8.15 a basis for $\mathbf{C}^n$ by piecing together the $\mathbf{z}_i$:

$$\mathbf{z} = \mathbf{z}_1 \cup \mathbf{z}_2 \cup \ldots \cup \mathbf{z}_s .$$

This basis, however, need not be an orthonormal basis for $\mathbf{C}^n$.

We construct a new basis $\mathbf{w}_i$ for each $E(\lambda_i)$ by applying the Gram-Schmidt Procedure 5.18 to the basis $\mathbf{z}_i$ for $E(\lambda_i)$. The bases $\mathbf{w}_i = \{w^{i1}, \ldots, w^{ir_i}\}$ for each $E(\lambda_i)$ will then be orthonormal; that is, for every i, $1 \leqslant i \leqslant s$:

$$(w^{ij})^\dagger(w^{ik}) = \delta_{jk} = \begin{cases} 1 & j = k \\ 0 & j \neq k . \end{cases}$$

We piece together the $\mathbf{w}_i$ to make $\mathbf{w}$, which is again a basis for $\mathbf{C}^n$; the claim is that $\mathbf{w}$ is an *orthonormal* basis for $\mathbf{C}^n$. To prove this, it remains to be shown that for every j, $1 \leqslant j \leqslant r_i$ and k, $1 \leqslant k \leqslant r_l$:

$$(w^{ij})^\dagger w^{lk} = 0 \quad \text{if} \quad i \neq l .$$

In other words, every eigenvector in $E(\lambda_i)$ is orthogonal to every eigenvector in $E(\lambda_l)$ when $i \neq l$.

To prove the last equation, proceed as follows. First, observe that, since $w^{lk} \in E(\lambda_l)$ and $w^{ij} \in E(\lambda_i)$,

$$Aw^{lk} = \lambda_l w^{lk}$$

and

$$Aw^{ij} = \lambda_i w^{ij} .$$

By Hermitian-ness of A and consequent realness of λ_i and λ_l, we have:

$$(w^{ij})^\dagger A = \lambda_i (w^{ij})^\dagger .$$

Hence

$$\begin{aligned} (w^{ij})^\dagger A w^{lk} &= \lambda_l (w^{ij})^\dagger w^{lk} \\ &= \lambda_i (w^{ij})^\dagger w^{lk} . \end{aligned}$$

Subtracting, we obtain

$$(\lambda_l - \lambda_i)(\mathbf{w}^{ij})^\dagger \mathbf{w}^{lk} = 0 ,$$

from which the result follows since $\lambda_l \neq \lambda_i$ by construction. The statement about realness of $\mathbf{w}$ for real A is obvious. □

There exists a "matrix alternative" which re-states Theorem 10.3 in the same way that Theorem 8.16 re-states Theorem 8.15. First, we must introduce some terminology.

10.4 Definition: Let U be an $(n \times n)$ complex matrix. U is said to be *unitary* if and only if it is invertible and has its conjugate transpose as its inverse; i.e. $U^{-1} = U^\dagger$, or

$$U^\dagger U = UU^\dagger = I_n .$$

If U is unitary and real, U is said to be an *orthogonal* matrix. □

10.5 Exercise: Let U be an $(n \times n)$ complex matrix.

(a) Show that U is unitary if and only if its columns form an orthonormal set of complex n-vectors.

(b) Show that U is unitary if and only if the transposes of its *rows* form an orthonormal set of complex n-vectors. [*Suggestions:* Let $\mathbf{w}^j$ be the jth column of U. Then $[U^\dagger U]_{ij} = (\mathbf{w}^i)^\dagger \mathbf{w}^j$, $1 \leqslant i, j \leqslant n$.] □

10.6 Theorem: *Let A be an $(n \times n)$ Hermitian matrix. Then there exists an $(n \times n)$ unitary matrix U such that $U^\dagger A U = \Lambda$ is a diagonal matrix whose diagonal elements are the eigenvalues of A, each appearing as many times as its algebraic multiplicity. If, in addition, A is real, then U may be chosen orthogonal.*

Proof: Suppose A is Hermitian; let $\mathbf{w}$ be the orthonormal eigenvector basis for $\mathbf{C}^n$ guaranteed by Theorem 10.3. Let U be the $(n \times n)$ matrix whose jth column is the jth vector in the basis $\mathbf{w}$; then, exactly as in the proof of Theorem 8.16, $U^{-1} A\, U = \Lambda$, where Λ is the diagonal "eigenvalue matrix." Since $\mathbf{w}$ is an orthonormal set, from Exercise 10.5 we conclude that U is unitary; hence $U^{-1} = U^\dagger$ and $U^\dagger A U = \Lambda$. The statement about realness of A and the ability to choose U orthogonal is an immediate consequence of the last statement in Theorem 10.3. □

10.7 Exercise: Show that if A is an $(n \times n)$ Hermitian matrix, and if $Az = 0$ for some nonzero $z \in \mathbf{C}^n$, then the range of A is the orthogonal complement (see Definition 5.13) of the nullspace of A. [*Suggestion:* Use Theorem 10.3, recalling that the nullspace of A is merely the eigenspace of A corresponding to the eigenvalue zero.] □

11. Nilpotent Matrices and the Jordan Canonical Form

In this section, we derive the Jordan canonical form for an arbitrary $(n \times n)$ real or complex matrix A. The Jordan canonical form of A is simply the matrix of the linear transformation ${}_A\hat{T} : \mathbf{C}^n \to \mathbf{C}^n$ with respect to a special basis for $\mathbf{C}^n$. We saw in §9 that if the distinct eigenvalues of A are $\lambda_1, \ldots, \lambda_s$, with respective generalized eigenspaces $F(\lambda_1), \ldots, F(\lambda_s)$, and if $\mathbf{z}_i$ is a basis for $F(\lambda_i)$, $1 \leqslant i \leqslant s$, then the matrix of A with respect to the ordered basis $\mathbf{z}_1 \cup \ldots \cup \mathbf{z}_s$ for $\mathbf{C}^n$ takes the special form

$$B = \begin{bmatrix} A_1 & 0 & \cdot & \cdot & \cdot \\ 0 & A_2 & \cdot & \cdot & \cdot \\ \cdot & 0 & \cdot & \cdot & \cdot \\ \cdot & \cdot & \cdot & A_{s-1} & \cdot \\ \cdot & \cdot & \cdot & 0 & A_s \end{bmatrix}, \qquad (*)$$

where each A_i is an $(r_i \times r_i)$ matrix (r_i is the algebraic multiplicity of the eigenvalue λ_i) which satisfies

$$(A_i - \lambda_i I_{r_i})^{r_i} = 0 .$$

For each i, then, $(A_i - \lambda_i I_{r_i})$ is a nilpotent matrix; observe that if the eigenspace $E(\lambda_i)$ *coincides* with $F(\lambda_i)$, then $A_i - \lambda_i I_{r_i} = 0$, since ${}_A\hat{T}(z) = \lambda_i z$ for every $z \in F(\lambda_i)$. This situation occurs for all i, $1 \leqslant i \leqslant s$, precisely when A is semi-simple. In what follows, after a preliminary discussion of nilpotent matrices, we shall see how to choose a basis for $\mathbf{C}^n$ with respect to which the matrix of ${}_A\hat{T}$ has the form

$$J = \Lambda + \hat{N} ,$$

where Λ is as in Observation 9.10 and $\hat{N}$ is an $(n \times n)$ matrix whose every element is zero except (possibly) for some 1's which appear in some of the $(i+1, i)$-positions just below the main diagonal. J in the last equation is the *Jordan canonical form* for A. We begin with a few definitions.

11.1 Definition: Let N be an $(r \times r)$ complex matrix; let $k \geqslant 1$ be an integer. N is said to be *nilpotent of order* k if and only if $N^{k-1} \neq 0$ and $N^k = 0$. □

Observe that the $(r \times r)$ zero matrix is the only $(r \times r)$ matrix which is nilpotent of order 1. Note also that, for each A_i in equation (*) above, $(A_i - \lambda_i I_{r_i})$ is nilpotent of order *at most* r_i.

11.2 Observation: *Let N be an $(r \times r)$ nilpotent matrix. Then zero is N's only eigenvalue, and (therefore) has algebraic multiplicity equal to r.*

Proof: Clearly, zero is an eigenvalue of N; its algebraic multiplicity is r (from which we conclude that zero is N's *only* eigenvalue) since, by definition of nilpotence, every nonzero vector $w \in \mathbf{C}^r$ is a generalized eigenvector of N corresponding to eigenvalue zero; now, use Proposition 9.5. □

11.3 Definition: Let $w \neq 0$, $w \in \mathbf{C}^r$, and let N be an $(r \times r)$ matrix nilpotent of order k. The *cyclic subspace* $Z(w; N)$ *generated by* w *under* N is the subspace of $\mathbf{C}^r$ given by

$$Z(w; N) = \operatorname{span}\{w, Nw, \ldots, N^{k-1}w\}. \qquad \square$$

The following technical result is central to the ensuing development.

11.4 Lemma: *Let N be an $(r \times r)$ complex matrix which is nilpotent of order k. If that $w \in \mathbf{C}^r$ and l is the smallest positive integer which satisfies $N^l w = 0$, then*

$$\{w, Nw, \ldots, N^{l-1}w\}$$

is a linearly independent set of vectors, and therefore forms a basis for the cyclic subspace $Z(w; N)$.

Proof: Assume a relation of the form

$$c_0 w + c_1 N w + \ldots + c_{l-1} N^{l-1} w = 0 ;$$

multiply the relation by N^{l-1} to get $c_0 = 0$. Now take the resulting new relation between the vectors $Nw, \ldots, N^{l-1}w$ and proceed similarly to show $c_1 = 0$. Continue in this fashion until all the c_j's have been shown to be zero, and deduce that the indicated set of vectors is linearly independent.

It is evident that the set forms a spanning set for $Z(w; N)$, since $N^l w = 0$; since it is also a linearly independent set, it forms a *basis* for $Z(w; N)$, as well. □

In what follows, we demonstrate that given a nilpotent $(r \times r)$ matrix, a family $w^1, w^2, \ldots, w^m$ of r-vectors may be chosen so that the cyclic subspaces $Z(w^i; N)$, $1 \leqslant i \leqslant m$, are *mutually disjoint* (c.f. Definition 4.28) and satisfy

$$Z(w^1; N) + \ldots + Z(w^m; N) = \mathbf{C}^r ,$$

where "+" in the last equation denotes "vector sum" (see Definition 4.26). Once having chosen $w^1, w^2, \ldots, w^m$, we shall use Observation 4.28(d) to construct a basis for $\mathbf{C}^r$ by piecing together bases of the form $\{w^i, Nw^i, \ldots, N^{l_i - 1} w^i\}$ for each $Z(w^i; N)$.

The result we are about to describe is very difficult to prove rigorously in a way which is "algorithmically satisfying" in the sense that it indicates how the special vectors w^i may be found; many rigorous proofs proceed by induction on the overall vector space dimension r. The interested reader should consult [Hirsch and Smale, pp. 334-336] for such a proof. We attempt to describe the procedure in a way which indicates (at least to an extent) how each step follows from the one before it.

Accordingly, let N be an $(r \times r)$ complex matrix which is nilpotent of order $k > 0$. For each $j \geqslant 0$, define R_j to be the range of N^j, where N^0 is understood to stand for the $(r \times r)$ identity matrix I_r. Then

$$\mathbf{C}^r = R_0 \supset R_1 \supset \ldots \supset R_{k-1} \supset R_k = \{0\} ,$$

where the inclusion relations follow because everything in the range of N^{j+1} is clearly in the range of N^j, $j \geqslant 0$. Observe, for each j, that

$$\dim(R_j) \geqslant \dim(R_{j+1}) .$$

We now detail the procedure for choosing the "special" basis for C^r to which we have alluded. Begin by selecting $w^{11}, w^{12}, \ldots, w^{1d_1}$ in C^r so that $\{N^{k-1}w^{11}, \ldots, N^{k-1}w^{1d_1}\}$ is a basis for R_{k-1}. Note that each w^{1i}, $1 \leqslant i \leqslant d_1$, generates a k-dimensional cyclic subspace $Z(w^{1i}; N)$ of C^r given by

$$Z(w^{1i}) = \text{span}\{w^{1i}, Nw^{1i}, \ldots, N^{k-1}w^{1i}\} .$$

These subspaces are *mutually disjoint;* to see this, assume a relation of the form

$$\sum_{j=1}^{k} \sum_{i=1}^{d_1} c_j N^{k-j} w^{1i} = 0 ;$$

multiply on the left by N^{k-1}, and use $N^k = 0$ to conclude that

$$\sum_{i=1}^{d_1} c_{i1} N^{k-1} w^{1i} = 0 ,$$

which implies that $c_{i1} = 0$, $1 \leqslant i \leqslant d_1$, by linear independence of $\{w^{1i} : 1 \leqslant i \leqslant d_1\}$. Continue as in the proof of Lemma 11.4, and the linear independence of $\{N^{k-j}w^{1i}\}$, $1 \leqslant j \leqslant k$, $1 \leqslant i \leqslant d_1$ follows. The mutual disjointness of the cyclic subspaces $Z(w^{1i})$, $1 \leqslant i \leqslant d_1$, is then an easy consequence of Observations 4.28.

If $k = 1$, stop; observe then that $w^{11}, \ldots, w^{1d_1}$ is a basis for $C^r = R_0$. (In this case, of course, the result is trivial, since $N = 0$!)

If $k > 1$, look at R_{k-2}; note that

$$\{N^{k-1}w^{11}, N^{k-1}w^{12}, \ldots, N^{k-1}w^{1d_1}, N^{k-2}w^{11}, \ldots, N^{k-2}w^{1d_1}\} \quad (**)$$

is a linearly independent subset of R_{k-2}. We would now like to *extend* the linearly independent set (**) to a basis for R_{k-2} in a particular way; the key to the construction is that we may choose the additional basis vectors so that they all lie in the *nullspace* of N.

First, pick $\hat{w}^{21}, \ldots, \hat{w}^{2d_2}$ in C^r so that

$$\{N^{k-1}w^{11}, \ldots, N^{k-1}w^{1d_1}, N^{k-2}w^{11}, \ldots,$$

$$N^{k-2}w^{1d_1}, N^{k-2}\hat{w}^{21}, \ldots, N^{k-2}\hat{w}^{2d_2}\}$$

is a basis for R_{k-2}. Implicit in the last choice is the fact that the dimension of R_{k-2} is $2d_1+d_2$. Multiplying any of the new basis vectors with N yields a vector in R_{k-1}; thus, for each j, $1 \leqslant j \leqslant d_2$, we may find constants c_{ij} such that

$$N^{k-1}\hat{w}^{2j} = \sum_{i=1}^{d_1} c_{ij} N^{k-1} w^{1i} .$$

Set

$$w^{2j} = \hat{w}^{2j} - \sum_{i=1}^{d_1} c_{ij} w^{1i} , \qquad 1 \leqslant j \leqslant d_2 ;$$

we claim that

$$\{N^{k-1}w^{11}, \ldots, N^{k-1}w^{1d_1}, N^{k-2}w^{11}, \ldots,$$

$$N^{k-2}w^{1d_1}, N^{k-2}w^{21}, \ldots, N^{k-2}w^{2d_2}\}$$

is a basis for R_{k-2}, and that $N^{k-1}w^{2j} = 0$ for each j, $1 \leqslant j \leqslant d_2$.

11.5 Exercise: (a) Show that the last set of vectors is linearly independent by using an argument such as that in the proof of Lemma 11.4.

(b) Prove that the d_1+d_2 cyclic subspaces $Z(w^{1i}; N)$, $1 \leqslant i \leqslant d_1$, and $Z(w^{2j}; N)$, $1 \leqslant j \leqslant d_2$, are mutually disjoint; use an argument similar to the one in (a). □

A bit of thought should convince the reader that, in the notation and terminology of Definition 11.3, we have simply chosen $w^{11}, \ldots, w^{1d_1}, w^{21}, \ldots, w^{2d_2}$ so that

$$R_{k-2} = Z(N^{k-2}w^{11}; N) + \ldots + Z(N^{k-2}w^{1d_1}; N)$$

$$+ Z(N^{k-2}w^{21}; N) + \ldots + Z(N^{k-2}w^{2d_2}; N);$$

that is, *we have expressed R_{k-2} as the vector sum of mutually disjoint cyclic subspaces*. The subspaces are disjoint by Exercise 11.5(b), since each is a subspace of $Z(w^{1i}; N)$ or $Z(w^{2j}; N)$ for some i or j, $1 \leqslant i \leqslant d_1$, $1 \leqslant j \leqslant d_2$. Observe that $d_2 = 0$ is, indeed, a possibility; the idea is that the $2d_1$ vectors $\{N^{k-1}w^{1i}, N^{k-2}w^{1i} : 1 \leqslant i \leqslant d_1\}$ might already constitute a basis for R_{k-2}.

The next step in the procedure (provided, of course, that $k > 2$) is to construct a basis for R_{k-3} by extending the following linearly independent subset of R_{k-3}:

$$\{N^{k-1}w^{1i}, N^{k-2}w^{1i}, N^{k-3}w^{1i}, N^{k-2}w^{2j},$$

$$N^{k-3}w^{2j} : 1 \leqslant i \leqslant d_1, 1 \leqslant j \leqslant d_2\}.$$

As in the construction of the basis for R_{k-2}, we extend the last set to a basis for R_{k-3} by adding d_3 new vectors $\hat{w}^{31}, \ldots, \hat{w}^{3d_3}$, and then *modify* the new vectors to obtain $w^{31}, \ldots, w^{3d_3}$ which lie in the nullspace of N^{k-2}. Accordingly, R_{k-2} has dimension $3d_1 + 2d_2 + d_3$, and R_{k-3}, like R_{k-2}, may be exhibited as the following vector sum of disjoint cyclic subspaces:

$$\underset{1 \leqslant i \leqslant d_1}{+} Z(N^{k-3}w^{1i}; N) \underset{1 \leqslant j \leqslant d_2}{+} Z(N^{k-3}w^{2j}; N) \underset{1 \leqslant q \leqslant d_3}{+} Z(N^{k-3}w^{3q}; N),$$

where the big "plus" signs signify vector sums of the indicated subspaces.

Continue in this fashion; that is, for each l, $1 \leqslant l \leqslant k$, choose d_l new vectors $w^{l1}, w^{l2}, \ldots, w^{ld_l}$ so that R_{k-l} may be expressed as the vector sum of disjoint cyclic subspaces:

$$\underset{1 \leqslant j \leqslant l}{+} \; \underset{1 \leqslant i \leqslant d_j}{+} Z(N^{k-l}w^{ji}; N).$$

By the time the procedure has been completed for $l = k$, we have expressed $R_o = \mathbf{C}^r$ as the vector sum of disjoint cyclic subspaces

$$\underset{1 \leqslant l \leqslant k}{+} \; \underset{1 \leqslant i \leqslant d_l}{+} Z(w^{li}; N).$$

Observe that the dimension of each of these subspaces is given by

$$dim(Z(w^{li}; N)) = k - l + 1, \qquad 1 \leqslant i \leqslant d_l,$$

since by construction we have $N^{k-l+1}w^{li} = 0$ for each l, $1 \leqslant l \leqslant k$, and each i, $1 \leqslant i \leqslant d_l$. Observe also that some terms in the sum may be absent; i.e., if some $d_l = 0$, then there are no vectors w^{li} selected along the way.

The procedure just described might seem somewhat baffling to the reader who has never before seen at least some version of it. As is often the case, a picture proves to be illuminating. Suppose that N is (31×31) and is nilpotent of order 7; that is, $N^7 = 0$ and $N^6 \neq 0$. Then we have the sequence of subspaces $\mathbf{C}^{31} = R_o \supset R_1 \supset \ldots \supset R_6 \supset R_7 = \{0\}$, where R_j is range of N^j for each j, $0 \leqslant j \leqslant 7$. Suppose in the selection procedure outlined above that $d_1 = 2$, $d_2 = d_3 = 0$, $d_4 = 1$, $d_5 = 2$, $d_6 = 2$, and $d_7 = 3$. Consider the following table:

0	1	2	3	4	5	6
w^{11}	Nw^{11}	N^2w^{11}	N^3w^{11}	N^4w^{11}	N^5w^{11}	N^6w^{11}
w^{12}	Nw^{12}	N^2w^{12}	N^3w^{12}	N^4w^{12}	N^5w^{12}	N^6w^{12}
w^{41}	Nw^{41}	N^2w^{41}	N^3w^{41}			
w^{51}	Nw^{51}	N^2w^{51}				
w^{52}	Nw^{52}	N^2w^{52}				
w^{61}	Nw^{61}					
w^{62}	Nw^{62}					
w^{71}						
w^{72}						
w^{73}						

The entries in the table are the vectors in the basis for $\mathbf{C}^{31}$ chosen according to the inductive procedure we have described. The vectors in each *row* of the table starting with w^{li} form a basis for the cyclic subspace $Z(w^{li}; N)$. Meanwhile, the vectors to the right of the jth column of the table, including those in the jth column, together form a basis for R_j. (N.b.: the columns are numbered 0 through 6 rather than 1 through 7.) In this example, $d_2 = d_3 = 0$ means that no new vectors were added during the procedure when bases were constructed for R_5 and R_4.

Note that each row terminates (on the right) with a vector in the nullspace of N. Indeed, the right-hand end vectors in the $(k-l)$th column, $1 \leqslant l \leqslant k$, are the d_l new vectors selected (during the lth step of the procedure) in the course of finding a basis for R_{k-l}; each of these vectors, by construction, lies in the nullspace of N. In fact, the vectors at the right-hand ends of the rows taken together form a *basis* for the nullspace of N. These vectors are linearly independent by construction; to see that they span the nullspace of N, let w be any vector in the nullspace of N. Express w as a linear combination of the table entries:

$$w = \sum_{l=1}^{7} \sum_{j=1}^{k-l} \sum_{i=1}^{d_l} c_{ijl} \, N^{j-1} w^{li} .$$

Now multiply through by N, and use the facts that $Nw = 0$ and that N annihilates the far right-hand table entries to conclude that

$$\sum_{l=1}^{7} \sum_{j=1}^{k-l-1} \sum_{i=1}^{d_l} c_{ijl}\, N^j w^{li} = 0.$$

This means that $c_{ijl} = 0$, $1 \leqslant l \leqslant 7$, $1 \leqslant i \leqslant d_l$, and $1 \leqslant j \leqslant k-l-1$, by linear independence of the table entries. Hence, w may be expressed as a linear combination of the vectors which lie at the right-hand ends of the rows in the table, which correspond in the above sums to $j = k-l$. which is an expansion of w as a linear combination of vectors at the right-hand ends of the rows of the table. The above calculation is easily generalized to arbitrary $(r \times r)$ complex matrices N which are nilpotent of order k. The general result is summarized in the following theorem.

11.6 Theorem: *Let N be an $(r \times r)$ complex matrix which is nilpotent of order k; i.e., $N^k = 0$ and $N^{k-1} \neq 0$. Then for each l, $1 \leqslant l \leqslant k$, there exist nonnegative integers d_l and (when $d_l > 0$) r-vectors w^{li}, $1 \leqslant i \leqslant d_l$ such that:*

(a) The cyclic subspaces $Z(w^{li}; N)$, $1 \leqslant l \leqslant k$, $1 \leqslant i \leqslant d_l$ are $(k-l+1)$-dimensional, disjoint, and span $\mathbf{C}^r$ in the sense that $\mathbf{C}^r$ is their vector sum;

(b) The sum $m = d_1 + d_2 + \ldots + d_l$, which is the total number of cyclic subspaces so defined, is equal to the dimension of the nullspace of N; in fact, a basis for the nullspace of N is given by

$$\{N^{k-l} w^{li} : 1 \leqslant l \leqslant k,\ 1 \leqslant i \leqslant d_l\};$$

(c) Accordingly, there exists an $(r \times r)$ invertible complex matrix P such that $\hat{N} = P^{-1}NP$ has the following form: $[\hat{N}]_{ij} = 0$ for all i and j except possibly when $i = j+1$, that is, at positions just below the diagonal. In those positions, the following sequence of zeros and ones appears, counting down and to the right, beginning at the $(2, 1)$ position: d_1 sequences of $(k-1)$ ones, with a single zero following each sequence; d_2 sequences of $(k-2)$ ones, with a zero following each sequence; etc.; d_{k-1} ones, with a zero following each; and $d_k - 1$ zeroes. (If $d_k = 0$, there is no zero following the last sequence of ones.) □

During the course of the foregoing discussion, we proved essentially all of Theorem 11.6 except for part (c). Part (c) follows if we choose P as the $(r \times r)$ matrix whose columns are the basis vectors chosen during the above sequence of operations, ordered as follows: each w^{li} is followed by Nw^{li}, then by $N^2 w^{li}$, and so on up through $N^{k-l} w^{li}$; these "i-chains," each containing $k-l+1$ vectors, are in turn ordered according to i, $1 \leqslant i \leqslant d_l$; these compound strings are then ordered according to l, $1 \leqslant l \leqslant k$. The "abstract" interpretation of Theorem 11.6(c) is that there exists a basis $\mathbf{w}$ for $\mathbf{C}^r$ with respect to which the matrix of the linear transformation ${}_N T : \mathbf{C}^r \to \mathbf{C}^r$ has the indicated form. An important observation is that the *lengths* of the sequences of ones and zeroes are *uniquely determined by* N. To see this, recall that d_1 is merely the dimension of the range of N^{k-1}; d_2 is given by

$$d_2 = \dim(\mathrm{R}(N^{k-2})) - 2d_1;$$

and so on, where R, as in §1, denotes range.

In the (31x31) example which we considered previously, the matrix P would be constructed by taking the vectors from the first row of the table in left-to-right order, followed by those from the second row in order, etc., followed by those in the last three rows. In any case, the form of $\hat{N}$ given in Theorem 11.6(c) is not hard to deduce; N times the jth column of P is either the $(j+1)$th column of P or zero, according to whether the jth column is a vector which lies "inside" or "at the end" of one of the rows in the table. Thus, in the (31x31) example, $NP = P\hat{N}$, where $\hat{N}$ is as described in Theorem 11.6(c), so that $P^{-1}NP = \hat{N}$.

With Theorem 11.6 in hand, we are ready to describe the Jordan canonical form for an $(n \times n)$ matrix A. Given such an A, let $\lambda_1, \ldots, \lambda_s$ be the distinct eigenvalues of A; let r_i, $1 \leqslant i \leqslant s$, and m_i, $1 \leqslant i \leqslant s$ be, respectively, their algebraic and geometric multiplicities; and let $F(\lambda_i)$, $1 \leqslant i \leqslant s$, be the corresponding generalized eigenspaces. We showed in §9 how to construct an $(n \times n)$ invertible matrix P such that $B = P^{-1}AP$ took the form

$$B = \begin{bmatrix} A_1 & 0 & \cdot & \cdot & \cdot \\ 0 & A_2 & \cdot & \cdot & \cdot \\ \cdot & 0 & \cdot & \cdot & \cdot \\ \cdot & \cdot & \cdot & A_{s-1} & \cdot \\ \cdot & \cdot & \cdot & 0 & A_s \end{bmatrix},$$

where A_i is an $(r_i \times r_i)$ matrix satisfying $(A_i - \lambda_i I_{r_i})^{r_i} = 0$.

Define

$$N_i = A_i - \lambda_i I_{r_i}, \qquad 1 \leqslant i \leqslant s;$$

then N_i is an $(r_i \times r_i)$ matrix which is nilpotent. By Theorem 11.6, we may find P_i, an $(r_i \times r_i)$ invertible matrix, such that $\hat{N}_i = P_i^{-1} N_i P_i$ has the standard form described in Theorem 11.6(c). Define the $(n \times n)$ matrices $\tilde{P}$, $\hat{N}$, and Λ by

$$\tilde{P} = \begin{bmatrix} P_1 & 0 & . & . & . \\ 0 & P_2 & . & . & . \\ . & 0 & . & . & . \\ . & . & . & P_{s-1} & . \\ . & . & . & 0 & P_s \end{bmatrix};$$

$$\Lambda = \begin{bmatrix} \lambda_1 I_{r_1} & 0 & . & . & . \\ 0 & \lambda_2 I_{r_2} & . & . & . \\ . & 0 & . & . & . \\ . & . & . & \lambda_{s-1} I_{r_{s-1}} & . \\ . & . & . & 0 & \lambda_s I_{r_s} \end{bmatrix};$$

and

$$\hat{N} = \begin{bmatrix} \hat{N}_1 & 0 & . & . & . \\ 0 & \hat{N}_2 & . & . & . \\ . & 0 & . & . & . \\ . & . & . & \hat{N}_{s-1} & . \\ . & . & . & 0 & \hat{N}_s \end{bmatrix}.$$

Then

$$\begin{aligned} \tilde{P}^{-1} B \tilde{P} &= \tilde{P}^{-1}(B - \Lambda)\tilde{P} + \tilde{P}^{-1}\Lambda\tilde{P} \\ &= \hat{N} + \Lambda, \end{aligned}$$

where the last equality follows by construction of $\tilde{P}$ and the fact that $\tilde{P}\Lambda = \Lambda\tilde{P}$.

Finally, observe that if we set $Q = P\tilde{P}$, then

$$\begin{aligned} Q^{-1}AQ &= \tilde{P}^{-1}P^{-1}AP\tilde{P} \\ &= \tilde{P}^{-1}B\tilde{P} \\ &= \hat{N} + \Lambda = J. \end{aligned}$$

The matrix J is defined as the *Jordan canonical form* for A. Its diagonal elements $[J]_{qq}$ are a replica of Λ's; that is,

$[J]_{qq} = \lambda_i$ whenever q is such that

$$r_1 + \ldots + r_{i-1} < q \leqslant r_1 + \ldots + r_i \ .$$

The only other (possibly) nonzero elements in J are 1's (from $\hat{N}$) lying just below the main diagonal. The matrix J is simply the matrix of the linear transformation ${}_AT : \mathbf{C}^n \to \mathbf{C}^n$ with respect to the basis for $\mathbf{C}^n$ given by the columns of the $(n \times n)$ matrix Q.

We have spoken of "the" Jordan canonical form for A even though there is nonuniqueness built into our definition of it. It is worthwhile at this point to isolate those features of the Jordan canonical form for A which *are* uniquely specified by A. Certainly, the distinct eigenvalues $\lambda_1, \ldots, \lambda_s$, along with their algebraic and geometric multiplicities, are determined uniquely by A. If λ_o is one of A's eigenvalues, then, the *total length* of the "λ_o block" in J is uniquely determined, since this number is the algebraic multiplicity of λ_o. Evidently, the *ordering* of these blocks along the diagonal of J is *not* uniquely specified.

The lengths of the various sequences of ones and zeroes appearing below the main diagonal in J are determined by the $\hat{N}_i$, $1 \leqslant i \leqslant s$, through Theorem 11.6. Furthermore, $\hat{N}_i$, $1 \leqslant i \leqslant s$, is the matrix (with respect to a particular basis) of the linear transformation $(A - \lambda_i I_n)$ *restricted* to the generalized eigenspace $F(\lambda_i)$; since the lengths of the zero- and one-chains depend only on *basis-invariant properties* (namely, dimensions of ranges) of this restriction, they are uniquely determined by the restriction, and hence by A.

It should be remarked that the matrix N which appears in the $M+N$ decomposition (c.f. §9) for A is amenable to the the same analysis as that which led to Theorem 11.6. The matrix of A with respect to the special basis for $\mathbf{C}^n$ which puts N in the form of $\hat{N}$ in Theorem 11.6 is another "Jordan-like" canonical form for A; in this new canonical form, however, an eigenvalue λ_i of A does not, in general, have associated with it an $(r_i \times r_i)$ block; instead, the "mini-blocks" arising from the structure of $\hat{N}$ are arrayed along the diagonal in decreasing order of size.

The reader might (justifiably) wonder about the usefulness of the Jordan canonical form; the computation involved in obtaining the Jordan canonical form for an arbitrary matrix is arduous and is difficult to explain. Under many circumstances, one is not primarily interested in actually *finding* the Jordan form for A; rather one's purposes are served merely because one *knows that the Jordan form for A exists*. In other words, one may often draw important conclusions about A simply by letting Q be an (unspecified) matrix such that $Q^{-1}AQ = J$ is the Jordan form for A; then properties of A may be deduced from results about J, which has a very elementary structure.

11.7 Exercise: Let A be a (3×3) real matrix whose eigenvalues all have magnitudes less than one. Use the *existence* of the Jordan canonical form to show that

(a) $A^k \to 0$ as $k \to \infty$; and

(b) $(I_3 - A)^{-1}$ is given by the *convergent* series

$$(I_3 - A)^{-1} = \sum_{k=0}^{\infty} A^k . \qquad \square$$

11.8 Exercise: Find the Jordan canonical form of the following (5×5) matrix A.

$$A = \begin{bmatrix} 1 & 1 & 0 & 0 & 0 \\ 1 & 1 & 0 & 0 & 0 \\ 0 & 0 & 0 & 1 & 2 \\ 0 & 0 & 0 & 0 & 0 \\ 0 & 0 & 0 & 0 & 0 \end{bmatrix} \qquad \square$$

We note in closing this section that if A is semi-simple, then the Jordan canonical form for A is simply the diagonal matrix Λ. As we said at the end of §9, for a semi-simple matrix A the matrices A_i in $B = P^{-1}AP$ all satisfy

$$(A_i - \lambda_i I_{r_i}) = N_i = 0;$$

thus $\hat{N}_i = 0$, $1 \leqslant i \leqslant s$, and consequently $\hat{N} = 0$, meaning that $J = \Lambda$.

12. Positive Definiteness, Matrix Factorization, and an Imperfect Analogy

The results of this section are purely algebraic, but they will prove to be of great significance in §§13-15 when we discuss linear matrix equations, reachability, and observability, and again in Part III when realization theory for time-invariant input-output systems is treated. The "imperfect analogy" to which the section title alludes is a rough parallel which we draw between the family of real $(n \times n)$ matrices and the family of complex numbers; the reader may judge independently whether this correspondence is a useful or interesting one.

Recall (Lemma 10.1) that every eigenvalue of an $(n \times n)$ Hermitian matrix A is real; in particular, *every real symmetric* $(n \times n)$ *matrix* A *possesses real eigenvalues only*. It may be shown, by an argument exactly parallel to the argument used in the proof of Lemma 10.8, that all the eigenvalues of any *skew*-Hermitian matrix (see Example 1.1(f)) are purely imaginary; in particular, this holds for every real skew-symmetric matrix. We are therefore tempted to draw the first correspondences in our imperfect analogy: *real symmetric* $(n \times n)$ matrices (respectively, *real skew-symmetric* $(n \times n)$ matrices) play a similar role in the family of *all* real $(n \times n)$ matrices to the role which real numbers (respectively, imaginary numbers) play in the family of all complex numbers.

To carry the analogy one step further, note that a real number is a complex number which is equal to its own conjugate, whereas an imaginary number is one which is equal to the negative of its own conjugate. Meanwhile, a symmetric $(n \times n)$ matrix A is equal to its own transpose, while a skew symmetric A satisfies $A = -A^T$. We are tempted to draw a parallel between conjugation (of complex numbers) and transposition (of real $(n \times n)$ matrices).

The temptation is reinforced by the following facts: first, every complex number z may be decomposed in a unique way as the sum of a real number $\mathrm{Re}\{z\}$ and an imaginary number $i\,\mathrm{Im}\{z\}$; moreover,

$$\begin{aligned} \mathrm{Re}\{z\} &= \tfrac{1}{2}(z + z^*) \\ i\,\mathrm{Im}\{z\} &= \tfrac{1}{2}(z - z^*) \\ z &= \mathrm{Re}\{z\} + i\,\mathrm{Im}\{z\} \,. \end{aligned}$$

Furthermore, every real $(n \times n)$ matrix A may be decomposed *uniquely* as the sum of a real symmetric $(n \times n)$ matrix $\mathrm{Sym}(A)$

and a real skew-symmetric $(n \times n)$ matrix Skew(A); in fact

$$\begin{aligned} \mathrm{Sym}(A) &= \frac{1}{2}(A + A^T) \\ \mathrm{Skew}(A) &= \frac{1}{2}(A - A^T) \\ A &= \mathrm{Sym}(A) + \mathrm{Skew}(A)\,. \end{aligned}$$

12.1 Exercise: Show that if $A = M + N$, with M symmetric and N skew-symmetric, then $M = \mathrm{Sym}(A)$ and $N = \mathrm{Skew}(A)$. Conclude that the above decomposition is unique. □

We turn now to the study of real positive definite $(n \times n)$ matrices. In the context of our proposed analogy between real $(n \times n)$ matrices and complex numbers, real positive definite $(n \times n)$ matrices will play the role of positive real numbers.

12.2 Definition: Let P be a real $(n \times n)$ matrix. Then P is said to be *positive definite* if and only if

(a) P is symmetric, and

(b) $x^T Px \geqslant 0$ for every $x \in \mathrm{R}^n$, and $x^T Px > 0$ except when $x = 0$.

A matrix satisfying (a) and (b) except for the last clause of (b) is said to be *nonnegative definite*, or *positive semidefinite*. □

To see how Definition 12.2 fits in with the foregoing discussion, view 12.2(a) as saying that P corresponds with a real number, and 12.2(b) as saying that, in addition, P corresponds with a *positive* real number. Many authors define positive definite matrices as those satisfying Definition 12.2(b) without the symmetry assumption 12.2(a); such a convention would correspond precisely with a definition of "positive complex number" as "a complex number with positive real part." To see this, observe that if A is any $(n \times n)$ matrix, then

$$\begin{aligned} x^T Ax &= x^T[\mathrm{Sym}(A) + \mathrm{Skew}(A)]x \\ &= x^T \mathrm{Sym}(A)x\ , \end{aligned}$$

where the last equality follows from the fact that $x^T \mathrm{Skew}(A)x$ is a scalar, hence equal to its own transpose, but is also equal to minus its own transpose by skew-symmetry of

Skew(A). Thus $x^T Ax \geqslant 0$ for all $x \in \mathbf{R}^n$ precisely when $x^T \text{Sym}(A)x \geqslant 0$ for all $x \in \mathbf{R}^n$. The "positivity" property 12.2(b) of $x^T Ax$, then, depends only on the corresponding property for Sym(A).

12.3 Examples: (a) The $(n \times n)$ identity matrix I_n is positive definite for any n, since it is symmetric and for any non-zero $x \in \mathbf{R}^n$, we have $x^T I_n x = \|x\|^2 > 0$. (Here, $\| \ \|$ denotes the Euclidean norm.) Observe also that the $(n \times n)$ zero matrix is nonnegative definite for any n.

(b) Any real diagonal $(n \times n)$ matrix Λ with positive (nonnegative) diagonal elements is positive (nonnegative) definite, since it is symmetric and

$$x^T \Lambda x = \sum_{i=1}^{n} [\Lambda]_{ii} x_i^2 \geqslant 0 .$$

(c) Let B be any real $(m \times n)$ matrix; then the $(m \times m)$ matrix BB^T and the $(n \times n)$ matrix $B^T B$ are both nonnegative definite. To see this, observe that if $x \in \mathbf{R}^n$, then

$$x^T B^T Bx = (Bx)^T Bx = \|Bx\|^2 \geqslant 0 ,$$

while if $y \in \mathbf{R}^m$ then

$$y^T BB^T y = (B^T y)^T (B^T y) = \|B^T y\|^2 \geqslant 0 . \qquad \square$$

12.4 Exercise: Show that if B is $(m \times n)$, with $m > n$, then BB^T is never positive definite, even though it is always nonnegative definite. [*Suggestion*: Use Lemma 1.12.] □

Soon, we shall find that Example 12.3(c) is, in some sense, *universal* with respect to nonnegative definite matrices. As might be expected, there are many properties of an $(n \times n)$ matrix P which are equivalent to positive definiteness of P. An important criterion for positive definiteness is given in terms of *eigenvalues* by the following result.

12.5 Theorem: *Let P be a real symmetric $(n \times n)$ matrix. Then P is nonnegative definite if and only if every eigenvalue of P is nonnegative. P is positive definite if and only if every eigenvalue of P is strictly positive.*

Proof: By Theorem 10.6, we may write

$$P = U \Lambda U^T ,$$

where U is an orthogonal matrix ($UU^T = I_n$) and Λ is a diagonal matrix whose diagonal elements are the eigenvalues of P. Now, $x^T Px \geqslant 0$ for every $x \in \mathbf{R}^n$ precisely when $x^T U \Lambda U^T x \geqslant 0$ for every $x \in \mathbf{R}^n$; thus P is nonnegative definite if and only if $y^T \Lambda y \geqslant 0$ for every $y \in \mathbf{R}^n$ which is of the form $U^T x$ for some $x \in \mathbf{R}^n$. Every y, however, may be written that way since U is invertible. Moreover,

$$y^T \Lambda y = \sum_{i=1}^{n} [\Lambda]_{ii} y_i^2 ,$$

which is nonnegative for *every* y if and only if all the $[\Lambda]_{ii}$, which are the eigenvalues of P, are nonnegative. Furthermore, $y^T \Lambda y$ is strictly positive for all nonzero y precisely when all the eigenvalues are strictly positive, and the proof is complete. □

One conclusion which may be drawn from Theorem 12.5 is that an $(n \times n)$ nonnegative definite matrix P is positive definite if and only if zero is not an eigenvalue of P; this situation, however, is equivalent to P's being invertible, since zero fails to be an eigenvalue of P precisely when $\det(P) \neq 0$. Equivalently, if P had zero as an eigenvalue, P would have an eigenvector corresponding to eigenvalue zero; such a vector would be a nonzero vector in the nullspace of P, which would preclude invertibility of P by Theorem 1.5. Consequently, the positive definite $(n \times n)$ matrices are simply those nonnegative definite $(n \times n)$ matrices which are also invertible.

To continue the development, we need an important technical lemma.

12.6 Lemma: *If A is any real symmetric $(n \times n)$ matrix which has zero as an eigenvalue, then the dimension of the nullspace of A is the algebraic multiplicity of zero as an eigenvalue of A.*

Proof: By Theorem 10.2, A is semi-simple; thus, by Theorem 8.15, the eigenspace corresponding to eigenvalue zero has dimension equal to the algebraic multiplicity of zero as an eigenvalue; but the eigenspace corresponding to eigenvalue zero is exactly the nullspace of A. □

By Theorem 1.25, the rank of *any* real $(n \times n)$ matrix A is equal to n minus the *geometric* multiplicity of zero as an eigenvalue of A; Lemma 12.6 shows that if A is *symmetric*, then the rank of A is n minus the *algebraic* multiplicity of zero as an eigenvalue of A. We are now ready to demonstrate the "universality" of Example 12.3(c).

12.7 Theorem: *Let P be a real $(n \times n)$ nonnegative definite matrix which has rank $r \leqslant n$. Then there exists an $(r \times n)$ matrix B such that*

$$P = B^T B$$

Moreover, any other $(r \times n)$ matrix C for which $P = C^T C$ is related to B as follows: $C = \hat{U} B$, where $\hat{U}$ is an $(r \times r)$ orthogonal matrix.

Proof: By Theorem 10.6, we may write

$$P = U \Lambda U^T ,$$

where U is $(n \times n)$ orthogonal, and Λ has the eigenvalues of P on its diagonal, each appearing as many times as its algebraic multiplicity. By Lemma 12.6, zero appears $(n-r)$ times on the diagonal of Λ; assume without loss of generality that the last $(n-r)$ diagonal entries of Λ are zero. Suppose, then, that

$$[\Lambda]_{ii} = \begin{cases} \lambda_i & 1 \leqslant i \leqslant r \\ 0 & r < i \leqslant n , \end{cases}$$

where the λ_i are positive and are not necessarily distinct. Define the $(r \times n)$ matrix Γ by

$$\Gamma = \begin{bmatrix} \sqrt{\lambda_1} & 0 & . & . & . & 0 & . & 0 \\ 0 & \sqrt{\lambda_2} & . & . & . & 0 & . & . \\ . & 0 & . & 0 & . & . & . & . \\ . & . & . & \sqrt{\lambda_{r-1}} & 0 & . & . & . \\ 0 & . & . & 0 & \sqrt{\lambda_r} & 0 & . & 0 \end{bmatrix},$$

where the zero block on the right is of size $(r \times (n-r))$. Then $\Gamma^T \Gamma = \Lambda$; if we set $B = \Gamma U^T$, then $B^T B = U \Lambda U^T = P$. Moreover, B has rank r; it has rank at most r, since it has only r rows, and has rank equal to r since $B^T B = P$ and P has rank r (c.f. Theorem 1.27).

Now let C be any other $(r \times n)$ matrix satisfying $C^T C = P$. By Theorem 1.28, the matrix B has a right inverse

G; i.e., there exists an $(n \times r)$ matrix G such that $BG = I_r$. Set

$$\hat{U} = CG \ .$$

Then $\hat{U}$ is $(r \times r)$; moreover,

$$\hat{U}^T \hat{U} = G^T C^T CG = G^T B^T BG = I_r \ ,$$

so that $\hat{U}$ is an orthogonal matrix. We claim that $C = \hat{U}B$; to see this, note that

$$\hat{U}^T C = G^T C^T C = G^T B^T B = I_r B \ ,$$

so that, upon multiplying by $\hat{U}$, we obtain

$$C = \hat{U}B \ . \qquad \square$$

12.8 Exercise: Show that the rank of B equals the rank of BB^T equals the rank of $B^T B$ when B is an *arbitrary* $(m \times n)$ matrix. [This exercise is not as easy as it may look.] $\square$

In applications, it is often very important to obtain a "full rank" factorization $P = B^T B$ (whose existence is guaranteed by Theorem 12.7) for a given $(n \times n)$ nonnegative definite matrix P. The factorization technique given in the proof of Theorem 12.7 is not an efficient one; for one thing, it requires a computation of the eigenvalues and eigenvectors of P. A more computationally economical technique, known as *Cholesky factorization*, may be used when P is positive definite.

12.9 Cholesky Factorization Procedure: *Let P be an $(n \times n)$ positive definite matrix. Then an $(n \times n)$ matrix B satisfying $B^T B = P$ may be constructed as follows:*

Step 1: Find C such that $CP = M$ is the echelon form of P, by Gauss elimination (c.f. 1.8).

Step 2: Define B as $\Gamma^{-1}M$, where Γ is the diagonal matrix whose diagonal elements are the square roots of the corresponding (necessarily positive) diagonal elements of M. It will be found that $C^{-1}\Gamma = B^T$; thus $B^T B = P$. Moreover, B will be upper triangular, i.e. $[B]_{ij} = 0, i > j$. $\square$

We present Procedure 12.9 without proving the assertions which are implicit in its description; for example, the diagonal elements of M will be positive, $C^{-1}\Gamma = B$, and so on. A good reference is [Strang]. An additional result, which concerns yet another criterion for positive definiteness, is also proven in [Strang].

12.10 Criterion for Positive Definiteness: *If P is an $(n \times n)$ symmetric matrix, then P is positive definite (respectively, nonnegative definite) if and only if all the principal minor determinants of P are positive (respectively, nonnegative).* □

Recall that a *principal minor determinant of P* is the determinant of one of the $(k \times k)$ matrices obtained from P by deleting all entries except those which lie in both the first k rows and the first k columns of P.

We return now to our proposed analogy between real $(n \times n)$ matrices and complex numbers. Recall that we have pointed out (and justified, to an extent) a correspondence which relates positive definite $(n \times n)$ matrices with positive real numbers. Two operations which may be performed on positive real numbers are the taking of square roots and of natural logarithms; that is, if p is a positive real number, there exists a unique *positive real number* q such that $q^2 = p$ and a unique *real number* x such that $e^x = p$. If Λ is an $(n \times n)$ diagonal matrix which is positive definite (i.e., has positive diagonal elements), then by defining *diagonal* matrices $\Lambda^{1/2}$ and $\log(\Lambda)$ by

$$[\Lambda^{1/2}]_{ii} = ([\Lambda]_{ii})^{1/2}, \qquad 1 \leqslant i \leqslant n ,$$

and

$$[\log(\Lambda)]_{ii} = \ln([\Lambda]_{ii}), \qquad 1 \leqslant i \leqslant n ,$$

we see that there exist matrices $\Lambda^{1/2}$ and $\log(\Lambda)$ satisfying respectively $(\Lambda^{1/2})^2 = \Lambda$ and $\exp[\log(\Lambda)] = \Lambda$. The extension to general $(n \times n)$ positive definite matrices is straightforward.

12.11 Theorem: *Let P be an $(n \times n)$ positive definite matrix. Then*

(a) There exists a unique $(n \times n)$ positive definite matrix Q, called the square root of P, such that $Q^2 = P$. [Notation: $Q = P^{1/2}$.]

(b) There exists a unique $(n \times n)$ real symmetric matrix X, called the natural log of P, such that $e^X = P$. [Notation: $X = \log(P)$.]

Proof: The hard part is uniqueness, so we'll treat that first. Let P have (necessarily positive) distinct eigenvalues $\lambda_1, \ldots, \lambda_s$ with algebraic multiplicities $r_1, \ldots, r_s$; suppose also that Q is positive definite, X is symmetric, and we have $Q^2 = e^X = P$. Observe that if $z \in \mathbf{C}^n$ is an eigenvector of Q corresponding to eigenvalue σ, then z is also an eigenvector of P, since $Q^2 z$ is just $\sigma^2 z$. Thus every eigenvector of Q is an eigenvector of P; furthermore, every eigenvalue σ of Q must be positive by positive definiteness of Q along with Theorem 12.5, and must satisfy $\sigma^2 = \lambda_i$ for some i, $1 \leqslant i \leqslant s$.

By Theorem 10.6, we may find an $(n \times n)$ orthogonal matrix U such that UPU^T (respectively, UQU^T) is a diagonal matrix with eigenvalues of P (respectively, of Q) on the diagonal; since

$$(UQU^T)^2 = UPU^T ,$$

we see that each $\sqrt{\lambda_i}$, $1 \leqslant i \leqslant s$, is an eigenvalue of Q of algebraic multiplicity r_i, $1 \leqslant i \leqslant s$. Similarly, by showing that if $z \in \mathbf{C}^n$ is an eigenvector of X corresponding to eigenvalue of σ then z is an eigenvector of P corresponding with e^σ, we can prove using symmetry of X and Theorem 10.6 that the natural logarithm of each λ_i, $1 \leqslant i \leqslant s$, is an eigenvalue of X with algebraic multiplicity r_i.

From this we may conclude that if $z \in \mathbf{C}^n$ is an eigenvector of P corresponding to λ_i, then z is an eigenvector of Q corresponding to $\sqrt{\lambda_i}$ and is also an eigenvector of X corresponding to $\ln(\lambda_i)$. The idea is that we may, by Theorem 10.3, expand z as a linear combination of linearly independent eigenvectors of Q (or of X) corresponding to the various $\sqrt{\lambda_j}$ (or $\ln(\lambda_j)$), multiply by Q (or X), subtract equations, and conclude that the terms in the expansion corresponding to $\sqrt{\lambda_j}$ (or $\ln(\lambda_j)$) when $j \neq i$ are zero.

Choose, therefore, an orthogonal matrix U such that U's columns form an orthonormal basis for $\mathbf{C}^n$ consisting of eigenvectors for P. Then $U^T Q U$, $U^T PU$, and $U^T XU$ will all be diagonal; moreover, $U^T XU$ and $U^T Q U$ are uniquely determined by the conditions

$$[U^T Q U]_{ii}^2 = [U^T PU]_{ii} , \qquad 1 \leqslant i \leqslant n ,$$

and

$$\exp([U^T XU]_{ii}) = [U^T PU]_{ii} , \qquad 1 \leqslant i \leqslant n .$$

If R (respectively, Y) is any other positive definite (respectively, symmetric) matrix such that $R^2 = P$ (respectively, $e^Y = P$), then the same U determines $U^T RU$ and $U^T YU$ as being equal to $U^T QU$ and $U^T XU$; thus $R = Q$ and $Y = X$.

The proof of existence for A and X is buried in the above uniqueness argument; given P, choose U so that $U^T PU = \Lambda$ is diagonal. Then if

$$\Lambda = \begin{bmatrix} \lambda_1 I_{r_1} & 0 & \cdot & \cdot & \cdot \\ 0 & \lambda_2 I_{r_2} & \cdot & \cdot & \cdot \\ \cdot & 0 & \cdot & \cdot & \cdot \\ \cdot & \cdot & \cdot & \lambda_{s-1} I_{r_{s-1}} & \cdot \\ \cdot & \cdot & \cdot & 0 & \lambda_s I_{r_s} \end{bmatrix},$$

then Q and X are given by

$$Q = U \begin{bmatrix} \sqrt{\lambda_1} I_{r_1} & 0 & \cdot & \cdot & \cdot \\ 0 & \sqrt{\lambda_2} I_{r_2} & \cdot & \cdot & \cdot \\ \cdot & \cdot & \cdot & \cdot & \cdot \\ \cdot & \cdot & \cdot & \sqrt{\lambda_{s-1}} I_{r_{s-1}} & \cdot \\ 0 & \cdot & \cdot & 0 & \sqrt{\lambda_s} I_{r_s} \end{bmatrix} U^T$$

and

$$X = U \begin{bmatrix} \ln\lambda_1 I_{r_1} & 0 & \cdot & \cdot & \cdot \\ 0 & \ln\lambda_2 I_{r_2} & \cdot & \cdot & \cdot \\ \cdot & \cdot & \cdot & \cdot & \cdot \\ \cdot & \cdot & \cdot & \ln\lambda_{s-1} I_{r_{s-1}} & \cdot \\ 0 & \cdot & \cdot & 0 & \ln\lambda_s I_{r_s} \end{bmatrix} U^T . \qquad \square$$

One final operation which may be performed on complex numbers and has a matricial analogy is the *polar decomposition*. If z is a nonzero complex number, we may write z uniquely as

$$z = rw ,$$

where r is positive and $|w|^2 = ww^* = 1$. The formulas giving r and w are

$$r = (zz^*)^{1/2}$$

and

$$w = \frac{z}{r} .$$

The result for real $(n \times n)$ matrices bears a striking resemblance.

12.12 Theorem: *Let A be an invertible real $(n \times n)$ matrix. Then there exist unique $(n \times n)$ matrices R and W such that $A = RW$, R is positive definite, and W is orthogonal; that is, $WW^T = W^T W = I_n$.*

Proof: By Example 12.3(c) AA^T is nonnegative definite; by invertibility of A and the discussion following Theorem 12.5, AA^T is positive definite. Let R be the unique positive definite square root of AA^T given by Theorem 12.11; let $W = R^{-1}A$. Then $A = RW$; moreover

$$\begin{aligned} WW^T &= R^{-1}AA^T R^{-1} \\ &= R^{-1}RRR^{-1} = I_n . \end{aligned}$$

Hence $W^T W = I_n$, as well; moreover, R and W are unique, since if $\hat{R}$ is positive definite, $\hat{W}$ is orthogonal, and $\hat{R}\hat{W} = A$, then

$$AA^T = \hat{R}\hat{W}\hat{W}^T \hat{R} = (\hat{R})^2$$

and $\hat{R}$ must be equal to R by the uniqueness part of Theorem 12.11(a). Conclude that $\hat{W} = W$ since $\hat{W} = R^{-1}A$. □

Observe that Theorem 12.12 points up yet another correspondence: $(n \times n)$ orthogonal matrices U behave like complex numbers w of magnitude 1. Indeed, keeping in mind the analogy between transposition and conjugation, we see that $ww^* = 1$ just as $UU^T = I_n$.

While Theorems 12.11 and 12.12 reinforce the imperfect analogy between real $(n \times n)$ matrices and complex numbers, the analogy has some serious deficiencies. Perhaps the most obvious flaw is that complex numbers commute under multiplication, whereas real $(n \times n)$ matrices do not. More interesting is the fact that it is impossible for a (real) complex number to be nonnegative and not positive without being zero, whereas a real (symmetric) $(n \times n)$ matrix may be nonnegative definite, nonzero, and still not be positive definite. The rank of a real $(n \times n)$ matrix A, then, is in some sense a measure of the "zeroness" of A. If A is invertible, then A is "nonzero;" the more A's rank is deficient, the more "zero-like" A is.

12.13 Exercise: (a) Let A be a real $(n \times n)$ matrix; consider the linear transformation ${}_AT : \mathbf{R}^n \to \mathbf{R}^n$ defined by A as in §4. Recall (c.f. Definition 5.12, Exercise 5.13, and the discussion which follows) that the *norm* $\| {}_AT \|_2$ of the transformation ${}_AT$ with respect to the Euclidean norm on $\mathbf{R}^n$ is given by

$$\| {}_AT \|_2 = \sup\{ \| Ax \|_2 : \| x \|_2 = 1\} .$$

Show that $\| {}_AT \|_2$ is equal to the square root of the maximum eigenvalue of the (nonnegative definite) $(n \times n)$ matrix $A^T A$. ($\| {}_AT \|_2$ is also known as the *spectral radius* of A.) [*Suggestion:* Given $x \in \mathbf{R}^n$ with $\| x \|_2 = 1$, expand x as a linear combination of orthonormal eigenvectors for $A^T A$.]

(b) Show that if $\| \ \|_a$ is *any* norm on $\mathbf{R}^n$, then the induced norm $\| {}_AT \|_a$ of the transformation ${}_AT$ is bounded below by the spectral radius of A. □

Recall from §5 that the norm of a linear transformation $T : \mathbf{R}^n \to \mathbf{R}^n$ induced by a norm on $\mathbf{R}^n$ may be interpreted as describing how much the transformation can *stretch* vectors in $\mathbf{R}^n$ with respect to the given norm on $\mathbf{R}^n$. Thus, if A is a real $(n \times n)$ matrix, the induced norm $\| {}_AT \|_2$ may be thought of as indicating how *expansive* the matrix A can be when A is viewed as defining a transformation $x \to Ax$ on $\mathbf{R}^n$, which is endowed with its standard Euclidean notion of distance. It follows that a "reasonable" measure of the *bigness* of the matrix A might be the square root of the largest eigenvalue of $A^T A$.

In the discussion preceding Exercise 12.13, we alluded to the fact that the *rank* of a real $(n \times n)$ matrix A may be viewed as measuring the "nonzeroness" of A; as it happens, the ideas which underlie the calculation in Exercise 12.13 enable us to measure A's "distance from zero" in a more *quantitative* fashion. We begin with a very general definition which encompasses even non-square matrices.

12.14 Definition: Let A be a real $(m \times n)$ matrix. The *singular values* of A are the (positive) square roots of the nonzero eigenvalues of $A^T A$. □

In terms of Definition 12.14, the largest singular value of a real $(n \times n)$ matrix A is equal to the induced Euclidean norm of the transformation ${}_AT : \mathbf{R}^n \to \mathbf{R}^n$. It will be seen in what follows that the *smallest* singular value of a real *full-rank* $(m \times n)$ matrix A is a reasonable measure of "how close A is to zero." We require the following technical lemma.

12.15 Lemma: *If A be a real $(m \times n)$ matrix which has rank r, then there exist an $(m \times m)$ orthogonal matrix U and an $(n \times n)$ orthogonal matrix V such that*

$$U^T AV = \begin{bmatrix} \Sigma & 0_{r \times (n-r)} \\ 0_{(m-r) \times r} & 0_{(m-r) \times (n-r)} \end{bmatrix},$$

where Σ is the $(r \times r)$ diagonal matrix with positive diagonal entries $\sigma_1, \sigma_2, \ldots, \sigma_r$ arranged in decreasing order on the diagonal.

Proof: By Exercise 12.8, the $(n \times n)$ matrix $A^T A$ has rank r; by Example 12.3(c), $A^T A$ is nonnegative definite. By Lemma 12.6, $A^T A$ has r positive eigenvalues $\sigma_1^2 \geqslant \ldots \geqslant \sigma_r^2$; the rest of the eigenvalues of $A^T A$ are zero. By the results of §10, we may find an $(n \times n)$ orthogonal matrix V such that $V^T A^T AV$ is a diagonal matrix whose (i, i) element is σ_i^2 for $1 \leqslant i \leqslant r$ and zero for $i > r$. Let v^j denote the jth column of V; observe that v^j, $1 \leqslant j \leqslant r$, is an eigenvector of $A^T A$ corresponding to eigenvalue σ_i^2. It is *also* true that

$$\{Av^1, \ldots, Av^r\}$$

is an *orthogonal* set of real m-vectors; to see this, note that

$$(Av^i)^T Av^j = (v^i)^T \sigma_j^2 v^j = \sigma_j^2 \delta_{ij}, \quad 1 \leqslant i, j \leqslant r,$$

because the v^k are orthonormal eigenvectors for $A^T A$.

It follows that

$$\{\sigma_1^{-1} Av^1, \ldots, \sigma_r^{-1} Av^r\}$$

is an orthonormal set of real m-vectors. Extend this last set to an orthonormal basis for $\mathbf{R}^m$ (perhaps by first extending the set to an arbitrary basis, and then applying the Gram-Schmidt procedure 5.18); let the $(m \times m)$ (orthogonal) matrix U have as its jth column u^j, the jth basis vector so constructed, for $1 \leqslant j \leqslant m$. It is left to the reader to complete the proof by showing that

$$(u^i)^T Av^j = \begin{cases} \sigma_j \delta_{ij} & 1 \leqslant i, j \leqslant r \\ 0 & \text{otherwise}; \end{cases}$$

from these formulas, the conclusion of the theorem follows

immediately. □

Observe that the positive numbers σ_i, $1 \leqslant i \leqslant r$, are precisely the *singular values of* A. We may "rotate" the conclusion of Lemma 12.15 so as to obtain the following result.

12.16 Corollary: *Let A be a real $(m \times n)$ matrix having rank r; let $\sigma_1 \geqslant \sigma_2 \geqslant \ldots \geqslant \sigma_r$ be the singular values of A. Then there exists an orthonormal set of m-vectors $\{u^1, \ldots, u^r\}$ and an orthonormal set of n-vectors $\{v^1, \ldots, v^r\}$ such that A may be decomposed as follows:*

$$A = \sigma_1 u^1 (v^1)^T + \sigma_2 u^2 (v^2)^T + \ldots + \sigma_r u^r (v^r)^T .$$

Proof: Let the u^i and the v^j be as in the proof of Lemma 12.15; the corollary follows if we multiply the equation in the statement of the lemma on the left by U and on the right by V^T, and remember that they are orthogonal matrices. □

The expansion of A given in the statement of Corollary 12.16 is called a *singular value decomposition* for A; the decomposition has associated with it a certain amount of uniqueness, but we shall not dwell on that issue. The singular value decomposition is particularly important in the situation where the $(m \times n)$ matrix A has *full* rank $r = \min(m, n)$. In that case, the *smallest* singular value σ_r turns out to provide a reasonable measure of *how close A is to being rank-deficient*. In other words, one may say that the *distance from zero* of a full rank $(m \times n)$ matrix A is given by A's *smallest singular value*.

If A is a real $(n \times n)$ (i.e., *square*) matrix, then A has full rank if and only if A is invertible; in this case, A has n positive singular values $\sigma_1 \geqslant \ldots \geqslant \sigma_n$. It turns out that the smallest singular value σ_n has a great deal of significance with regard to the stability of *numerical computations* involving the matrix A. To understand this last comment, consider the simple scalar equation $ax = b$, where $a \neq 0$ and b are given real numbers and x is to be solved for. If b is changed by a small amount, say ϵ, then the solution to the equation changes by $a^{-1}\epsilon$. Hence, if a is *close to zero*, a very small change in the *parameters* of the computational problem of determining x from $ax = b$ might translate into a very large change in the solution.

Similarly, if A is an invertible real $(n \times n)$ matrix, then we may always solve the equation $Ax = b$ for a unique real n-vector x, no matter what $b \in \mathbf{R}^n$ is. Nonetheless, the *computational* problem of determining x given b is quite sensitive to variations in b if the smallest singular value σ_n of A is very small. To see this, write A as $A = U \Sigma V^T$, as in the conclusion of Lemma 12.15; here, Σ is the $(n \times n)$ diagonal matrix of singular values of A. The solution to $Ax = b$ is given by $x = A^{-1}b$; meanwhile, A^{-1} is given by $V \Sigma^{-1} U^T$. Suppose that b is perturbed slightly in the "direction" of u^n, the last column of U; more specifically, set $\tilde{b} = b + \epsilon u^n$. The solution $\tilde{x}$ to the equation $A\tilde{x} = \tilde{b}$ is $\tilde{x} = x + \sigma_n^{-1}\epsilon$, which might be quite far away from x if σ_n is much smaller than ϵ. In other words, errors in the parameters of the problem may be *amplified* by as great an amount as the inverse of the smallest singular value of A.

12.17 Exercise: (a) Let A be an invertible real $(n \times n)$ matrix; show that the largest singular value of A may be approximated asymptotically as follows:

$$\sigma_1 = \lim_{k \to \infty} \frac{\| A^{k+1} x \|_2}{\| A^k x \|_2}$$

whenever $x \in \mathbf{R}^n$ does *not* lie in the orthogonal complement of the eigenspace of $A^T A$ corresponding to σ_1.

(b) In the same context, show that so long as $x \in \mathbf{R}^n$ is not in the orthogonal complement of the eigenspace of $A^T A$ corresponding to A's smallest singular value σ_n, it is true that

$$\sigma_n^{-1} = \lim_{k \to \infty} \frac{\| (A^{-1})^{k+1} x \|_2}{\| (A^{-1})^k x \|_2} . \qquad \square$$

12.18 Exercise: Use Lemma 12.15 to show that if A is a real $(m \times n)$ matrix and B is a real $(n \times m)$ matrix, then the $(m \times m)$ matrix AB and the $(n \times n)$ matrix BA have the same nonzero eigenvalues; in fact,

$$\det(\lambda I_m - AB) = \begin{cases} \lambda^{m-n} \det(\lambda I_n - BA) & m \geqslant n \\ \lambda^{n-m} \det(\lambda I_n - BA) & m \leqslant n . \end{cases}$$

[*Suggestion:* Assume $m \geqslant n$, and apply Lemma 12.15 to A;

multiply the matrix $(\lambda I_m - AB)$ on the left by U^T and on the right by U, preserving the determinant, and proceed; recall that $AB = AVV^T B$.] □

Observe that Exercise 12.18 enables us, with impunity, to "flip" Definition 12.14 of singular values; that is, the singular values of A are *also* the square roots of the positive eigenvalues of AA^T, and hence are the same as the singular values of A^T.

13. *Reachability and Controllability for Time-Invariant Continuous-Time Systems*

It is worthwhile at this point to look back at the six previous sections and determine the roles they play in our study of state space linear systems. In §6, we presented a formal definition of *state space linear system;* such a system is a composite mathematical object with various components: the space U of m-vector input functions, the spaces $\mathbf{Y}_{t_o}$ of p-vector output functions, the (finite-dimensional) state space X, the state transition mapping ϕ, and the readout mapping ρ. Fundamental examples of linear systems were given in Examples 6.5 and 6.6; in §7, it was shown that an understanding of these examples leads in an essential way to a deep understanding of "almost all" finite-dimensional state space linear systems. The principle underlying the development in §7 is that *realizations* may be constructed for "most" continuous-time systems and all discrete-time systems; the crucial step in the realization process is a *choice of basis* for the state space X.

The results of §7 motivated the study of time-invariant realizations which was undertaken in §§8-11. Specifically, we used the realization process to boil down the study of arbitrary real, *time-invariant* state space linear systems to the study of the time-invariant systems with state space $\mathbf{R}^n$ defined as in Examples 6.5 and 6.6 by the equations

$$\dot{x}(t) = Ax(t) + Bu(t), \quad t \geqslant t_o$$

$$x(t_o) = x^\circ \qquad \text{(I)}$$

$$y(t) = Cx(t) + Du(t)$$

in continuous time and

$$x(k+1) = Ax(k) + Bu(k), \qquad k \geqslant k_o$$
$$x(k_o) = x^\circ \tag{II}$$
$$y(k) = Cx(k) + Du(k)$$

in discrete time. The matrices A, B, C, and D are real, constant, and have respective sizes $(n \times n)$, $(n \times m)$, $(p \times n)$, and $(p \times m)$. The development in §§8-11 constitutes only the *first step* in analyzing (I) and (II) above. In §§8-11, methods are described by which, in principle, e^{tA} and A^k, $k > 0$, may be computed and/or better understood; knowing all about e^{tA} and A^k is essentially equivalent to knowing all about the *state-transition mappings* for the systems described by (I) and (II), since these mappings are given respectively by

$$\phi(t, t_o, x^\circ, u) = e^{(t-t_o)A}x^\circ + \int_{t_o}^{t} e^{(t-\tau)A}Bu(\tau)d\tau, \quad t \geqslant t_o,$$

and

$$\phi(k, k_o, x^\circ, u) = A^{(k-k_o)}x^\circ + \sum_{l=k_o}^{k-1} A^{(k-l-1)}Bu(l), \quad k \geqslant k_o.$$

The discussion at the beginning of §6 should alert the reader to the fact that the state transition behavior of a linear state space system is only *one of several system features* which are of interest to us. The other two aspects of system behavior which merit further investigation are the way in which *input functions* determine (and may be used to influence) the system's *state*, and the way in which the *state* in turn determines the system's *output*. It is our goal in the present section and in the three which follow to investigate these two "connections:" input to state, and state to output.

The study of reachability and controllability which we are about to undertake may be regarded as marking the point at which "modern" linear system theory becomes something more than "classical linear dynamics with inputs." In the classical dynamics of systems of particles, one is concerned with a *descriptive* analysis of how a given physical system behaves when a given forcing function is applied to the system with a set of initial conditions (position and velocity for each particle, say) having been given. The system is described by a (possibly quite complicated) set of differential equations; if the dynamics are linear, these equations will resemble the linear differential equations of §2 and (I) above.

In "modern" linear system theory, on the other hand, one is often principally interested in a *prescriptive* analysis of systems such as those determined by (I) and (II). The question which one asks is not so much, "How does the system behave given the input?" but rather, "How may the input be used to *influence* the behavior of the system?" Consider the following simple example.

13.1 Example: Controllable instability versus uncontrollable instability.

Consider the equations

$$\dot{x}(t) = Ax(t) + Bu(t)\,; \qquad x(0) = \begin{bmatrix} 1 \\ 0 \end{bmatrix}.$$

The differential equation might be that arising in a realization of a state space linear system with 2-dimensional state space. Suppose first that

$$A = \begin{bmatrix} 1 & 0 \\ 1 & 0 \end{bmatrix}; \quad B = \begin{bmatrix} 0 \\ 1 \end{bmatrix}.$$

It is easy to check that, since $A^k = A$ for every $k > 0$,

$$\begin{aligned} e^{tA} &= I_2 + A\Big(\sum_{k=1}^{\infty} \frac{t^k}{k!}\Big) \\ &= \begin{bmatrix} e^t & 0 \\ e^t - 1 & 1 \end{bmatrix}. \end{aligned}$$

Thus, for $t > 0$, $x(t)$ is given by

$$\begin{aligned} x(t) &= e^{tA}x(0) + \int_0^t e^{(t-\tau)A}Bu(\tau)d\tau \\ &= \begin{bmatrix} e^t \\ e^t - 1 + \int_0^t u(\tau)d\tau \end{bmatrix}. \end{aligned}$$

Observe that *no matter how* $u(t)$, $t \geqslant 0$, *is chosen*, $x_1(t) \to \infty$ as $t \to \infty$.

Suppose now that A is as above but that B is given by

$$B = \begin{bmatrix} 1 \\ 0 \end{bmatrix}.$$

Then $x(t)$, $t > 0$, is given by

$$x(t) = \begin{bmatrix} e^t + \int_0^t e^{(t-\tau)} u(\tau) d\tau \\ e^t - 1 + \int_0^t (e^{(t-\tau)} - 1) u(\tau) d\tau \end{bmatrix}.$$

If we choose, say,

$$u(t) = -2e^{-t}, \qquad t \geqslant 0,$$

then

$$x(t) = \begin{bmatrix} e^{-t} \\ 1-e^{-t} \end{bmatrix}, \qquad t \geqslant 0,$$

so that $x(t)$ remains bounded as $t \to \infty$. Thus a seemingly minor change in B results in a rather large increase in our ability to "control" the state of the linear system whose time evolution is described by the above differential equation. □

Example 13.1, for one thing, shows that the "B-matrix," in addition to the "A-matrix," has much to do with how the state of a linear system realized by (A, B, C, D) in (I) may be influenced by inputs to the system.

The notions of *reachability* and *controllability*, which we shall define presently, turn out to provide a useful approach to formalizing the foregoing intuitive discussion. Following along the lines of §§6-7, we define these notions both for "abstract" linear systems and in the context of realizations. Our definitions apply only to time-invariant systems; there are, however, analogous definitions in the setting of time-varying systems. The interested reader should consult [Desoer], [Brockett] [Kailath], or [Chen] for a treatment of time-varying reachability and controllability.

13.2 Definition: Let there be given, as in Definition 6.1, a time-invariant real, m-input, p-output continuous-time state space linear system with input function space $\mathbf{U}$, state space X, and state transition mapping ϕ. A state $x^1 \in X$ is said to be a *reachable state* if and only if there exist a real $T > 0$ and a $u \in \mathbf{U}$ such that

$$\phi(T, 0, 0, u) = x^1.$$

Moreover, the system is said to be *reachable*, or to have a *reachable state space*, if and only if every $x^1 \in X$ is a reachable state. □

13.3 Definition: Let A and B be real matrices of respective sizes $(n \times n)$ and $(n \times m)$. A vector $x^1 \in \mathbf{R}^n$ is said to be *reachable with respect to (the pair)* (A, B) if and only if there exist a real $T > 0$ and a continuous function $u : [0, T] \to \mathbf{R}^m$ such that

$$x^1 = \int_0^T e^{(T-\tau)A} Bu(\tau) d\tau \ .$$

Moreover, (A, B) is said to be a *reachable pair* if and only if every $x^1 \in \mathbf{R}^n$ is reachable with respect to (A, B). □

In other words, a state $x^1 \in X$ of a time-invariant continuous-time linear system is reachable precisely when the initial state $x^\circ = 0$ at initial time $t_o = 0$ may be "steered" to x^1 in finite time if a suitable input function is applied. Observe also that $x^1 \in \mathbf{R}^n$ is reachable with respect to a pair (A, B) if and only if x^1 is, in the sense of Definition 13.2, a reachable state of the continuous-time linear system with state space $\mathbf{R}^n$ defined as in Example 6.5 by equations (I) above. A solid connection between Definitions 13.2 and 13.3 is given by the following result.

13.4 Theorem: *Let there be given a real, time-invariant, m-input, p-output continuous-time state space linear system having n-dimensional state space X and satisfying the "realizability" Assumptions 7.3. A state $x^1 \in X$ is reachable if and only if, for every basis* $\mathbf{x}$ *for X, $[x^1]_{\mathbf{x}}$ is reachable with respect to (A, B), where A and B arise as in Theorem 7.8 in the realization of the system provided by* $\mathbf{x}$. *Consequently, such a system is reachable if and only if the matrices (A, B) arising in any realization for the system constitute a reachable pair.*

Proof: Suppose $x^1 \in X$; let $\mathbf{x}$ be any basis for X, and let the matrix pair (A, B) arise as in Theorem 7.8. Now, x^1 is reachable if and only if there exist a $T > 0$ and a $u \in \mathbf{U}$ such that

$$x^1 = \phi(T, 0, 0, u) \ .$$

The function u must be continuous, by Assumption 7.3(b); moreover, taking coordinates of the last equation with respect to $\mathbf{x}$ allows one to conclude that x^1 is reachable if and only if there exist T and u such that

$$[x^1]_{\mathbf{x}} = [\phi(T, 0, 0, u)]_{\mathbf{x}}$$
$$= \int_0^T e^{(T-\tau)A} Bu(\tau)d\tau ,$$

where the last equality follows from equations in the proof of Theorem 7.8 since $\Phi(T, \tau) = e^{(T-\tau)A}$. That is, x^1 is reachable if and only if $[x^1]_{\mathbf{x}}$ is reachable with respect to (A, B). Since $\mathbf{x}$ was an arbitrary basis, all but the last sentence of the theorem statement follows. This last sentence, however, is an obvious consequence of the definitions. □

Theorem 13.4 enables us to focus our discussion on the "concrete" manifestation of reachability, namely, reachability of real n-vectors with respect to pairs (A, B). Our ultimate objective is to find simple criteria, given in terms of A and B, for the reachability of n-vectors with respect to (A, B). To this end, we define the *reachability Gramian* and *reachability matrix* of a pair (A, B) as follows.

13.5 Definitions: Let A and B be real matrices having respective sizes $(n \times n)$ and $(n \times m)$.

(a) Given $T > 0$, the *T-reachability Gramian* $M_r(T; A, B)$ of (A, B) is the $(n \times n)$ matrix given by

$$M_r(T; A, B) = \int_0^T e^{(T-\tau)A} BB^T e^{(T-\tau)A^T} d\tau .$$

(b) The *reachability matrix* $Q_r(A, B)$ is the $(n \times nm)$ matrix given by

$$Q_r(A, B) = [B \;\; AB \;\; A^2B \; \dots \; A^{n-1}B] .$$ □

It will happen that $x^1 \in \mathbf{R}^n$ is reachable with respect to (A, B) if and only if x^1 lies in the range of the matrix $Q_r(A, B)$; it will turn out that this range coincides with the range of $M_r(T; A, B)$ for *every* $T > 0$. We begin with some crucial observations.

13.6 Observations: *(a)* $M_r(T; A, B)$ *is nonnegative definite for every* $T > 0$.

(b) The nullspace of $M_r(T\,;A\,,B)$ is the same for every $T>0$, and equals the set of all $w \in \mathbf{R}^n$ such that

$$w^T Q_r(A\,,B) = 0\,.$$

(c) The range of $M_r(T\,;A\,,B)$ is the same for every $T>0$, and equals the range of $Q_r(A\,,B)$.

Proof: (a) $M_r(T\,;A\,,B)$ is clearly symmetric; moreover, given $x \in \mathbf{R}^n$, observe that

$$x^T M_r(T\,;A\,,B)x = \int_0^T x^T e^{(t-\tau)A} BB^T e^{(T-\tau)A^T} x d\tau$$

$$= \int_0^T \| B^T e^{(T-\tau)A^T} x \|^2 d\tau \geqslant 0\,.$$

(b) Let $T>0$ be given. Suppose that $M_r(T\,;A\,,B)w = 0$; then $w^T M_r(T\,;A\,,B)w = 0$, as well. As in the proof of (a), we have

$$\int_0^T \| B^T e^{(T-\tau)A^T} w \|^2 d\tau = 0\,.$$

As the integrand is nonnegative, continuous, and integrates to zero, it must be identically zero on the interval $[0,T]$. Hence

$$B^T e^{(T-\tau)A^T} w = 0\,, \qquad 0 \leqslant \tau \leqslant T\,.$$

Taking derivatives and evaluating at $\tau = T$, we get

$$\begin{aligned} B^T w = -B^T A^T w &= B^T (A^T)^2 w \\ &= \ldots. \\ &= (-1)^k B^T (A^T)^k w = 0 \end{aligned}$$

for every $k>0$. Hence $w^T A^k B = 0$ for every $k \geqslant 0$; clearly, then, $w^T Q_r(A\,,B) = 0$.

Conversely, if $w^T Q_r(A\,,B) = 0$, then $w^T A^k B = 0$ for *every* $k>0$ since by the Cayley-Hamilton Theorem (our Theorem 9.8) $w^T A^k B$ may be expressed as a linear combination of $\{w^T A^l B\}$ with $0 \leqslant l \leqslant n-1$. We conclude that, for every $\tau \in [0,T]$,

$$B^T e^{(T-\tau)A^T} w = \sum_{k=0}^{\infty} \frac{(T-\tau)^k}{k!} B^T (A^T)^k w = 0\,.$$

Hence

$$M_r(T\ ;A\,,B)w = \int_0^T e^{(T-\tau)A}BB^T e^{(T-\tau)A^T} w\, d\tau = 0\ ,$$

and w is in the nullspace of $M_r(T\ ;A\,,B)$. Since T was arbitrary, (b) is proven.

(c) Let $T>0$ be given; since $M_r(T\ ;A\,,B)$ is symmetric, its range is the orthogonal complement of its nullspace by Exercise 10.7. Now, w is in the nullspace of $M_r(T\ ;A\,,B)$ if and only if (by (b)) w is orthogonal to every column of $Q_r(A\,,B)$; or, equivalently, if and only if w is in the orthogonal complement of the subspace of $\mathbf{R}^n$ spanned by the columns of $Q_r(A\,,B)$. This last subspace is the range of $Q_r(A\,,B)$; hence the range of $Q_r(A\,,B)$ is the orthogonal complement of the nullspace of $M_r(T\ ;A\,,B)$ and therefore equals the range of $M_r(T\ ;A\,,B)$. Since T was arbitrary, the result follows. □

The promised test for reachability is now at hand.

13.7 Theorem: *Let A and B be real matrices of respective sizes $(n\times n)$ and $(n\times m)$. Then $x^1\in\mathbf{R}^n$ is reachable with respect to $(A\,,B)$ if and only if x^1 is in the range of $Q_r(A\,,B)$, which is the same for every $T>0$ as the range of $M_r(T\ ;A\,,B)$. Moreover, $(A\,,B)$ is a reachable pair if and only if $Q_r(A\,,B)$ has rank n; equivalently, $(A\,,B)$ is a reachable pair if and only if $M_r(T\ ;A\,,B)$ is invertible (hence positive definite) for some (and consequently all) $T>0$.*

Proof: Suppose x^1 is reachable with respect to $(A\,,B)$; we may then find a $T>0$ and a continuous $u:[0,T]\to\mathbf{R}^m$ such that

$$x^1 = \int_0^T e^{(T-\tau)A}Bu(\tau)d\tau\ .$$

Hence

$$x^1 = \sum_{k=0}^{\infty} A^k B \int_0^T \frac{(T-\tau)^k}{k!}u(\tau)d\tau\ .$$

The right-hand side is a linear combination of the columns of $B\,,AB\,,A^2B\,,\ldots,A^kB\,,\ldots$; by the Cayley-Hamilton Theorem 9.8, it may be written as a linear combination of the columns of $B\,,AB\,,\ldots,A^{n-1}B$, i.e., the columns of $Q_r(A\,,B)$. Thus x^1 is in the range of $Q_r(A\,,B)$ (hence also in the range of $M_r(T\ ;A\,,B)$ for *every* $T>0$) if x^1 is reachable with respect to (A, B).

Conversely, if x^1 is in the range of $Q_r(A, B)$, it is in the range of $M_r(T; A, B)$ for every $T > 0$; to see this, pick a T and find a $z \in \mathbf{R}^n$ so that

$$x^1 = M_r(T; A, B)z ;$$

then

$$x^1 = \int_0^T e^{(T-\tau)A} Bu(\tau)d\tau ,$$

where

$$u(\tau) = B^T e^{(T-\tau)A^T} z , \qquad 0 \leqslant \tau \leqslant T ,$$

and x^1 is therefore reachable with respect to (A, B).

Now, (A, B) is a reachable pair if and only if every $x^1 \in \mathbf{R}^n$ is reachable with respect to (A, B); thus (A, B) is a reachable pair if and only if the range of $Q_r(A, B)$ is $\mathbf{R}^n$, meaning that the rank of $Q_r(A, B)$ is n. By Observation 13.6(c), then, (A, B) is a reachable pair if and only if, for every $T > 0$, $M_r(T; A, B)$ has range equal to $\mathbf{R}^n$, which occurs precisely when $M_r(T; A, B)$ is invertible. Since $M_r(T; A, B)$ is nonnegative definite, we conclude that (A, B) is a reachable pair if and only if $M_r(T; A, B)$ is positive definite for some (hence all) $T > 0$ (c.f. the Theorem 12.5 and what follows). □

One rather startling conclusion which may be drawn from Observations 13.6 and Theorem 13.7 is that *an* $x^1 \in \mathbf{R}^n$ *which is reachable with respect to* (A, B) *may be "reached" in an arbitrarily small time by suitable choice of input*. This assertion has practical implications with regard to control systems. Suppose, for instance, that we wish to "steer" the trajectory of the differential equation

$$\dot{x}(t) = Ax(t) + Bu(t)$$

from $x(0) = 0$ to some $x^1 \in \mathbf{R}^n$ which is reachable with respect to (A, B). By Theorems 13.6 and 13.7, such a transfer may be accomplished in an arbitrarily small amount of time; a *very rapid transfer*, however, might require a *very large input*. If A and B are positive scalars, say, and we choose u so that

$$x^1 = \int_0^\epsilon e^{(\epsilon-\tau)A} Bu(\tau)d\tau ,$$

then clearly the smaller ϵ is, the larger $|u|$ must be on $[0, \epsilon]$.

The reader might wonder how reachability is related to the situation encountered in Example 13.1. There, the problem was not one of choosing an input to *reach* a desired state starting from zero, but rather to choose an input so as to *prevent* the trajectory starting at a given initial condition from "blowing up." The concept of *controllability*, which is equivalent to reachability for *continuous-time* systems, turns out to be the key.

13.8 Definition: Let there be given a real, time-invariant, m-input, p-output continuous-time state space linear system as in Definition 6.1. The system is said to be *controllable* if and only if for every initial state $x^\circ \in X$ there exists a time $T > 0$ and an input function $u \in \mathbf{U}$ such that

$$\phi(T, 0, x^\circ, u) = 0 . \qquad \square$$

13.9 Definition: Let A and B be real matrices of respective sizes $(n \times n)$ and $(n \times m)$. (A, B) is said to be a *controllable pair* if and only if for every $x^\circ \in \mathbf{R}^n$ there exist a $T > 0$ and a continuous function $u : [0, T] \rightarrow \mathbf{R}^m$ such that

$$0 = e^{tA} x^\circ + \int_0^T e^{(T-\tau)A} Bu(\tau) d\tau . \qquad \square$$

A system is controllable, then, if an arbitrary initial condition may be "zeroed out" in finite time by suitable choice of input function. Controllability would certainly appear to be a desirable property for a system to have. Observe also that (A, B) is a controllable pair precisely when the system with state space $\mathbf{R}^n$ constructed from equations (I) as in Example 6.5 is a controllable system in the sense of Definition 13.8. We leave it to the reader to prove the following analogue to Theorem 13.4.

13.10 Theorem: *Let there be given a real, time-invariant, m-input, p-output continuous-time state space linear system having n-dimensional state space X and satisfying the "realizabilty" Assumptions 7.3. The system is controllable if and only if for every choice of basis* $\mathbf{x}$ *for X*, (A, B) *is a controllable pair, where A and B arise as in Theorem 7.8 in the realization of the system provided by* $\mathbf{x}$. $\square$

Our final result, which is *not* true for discrete-time systems, asserts the equivalence for continuous-time systems of reachability and controllability. We state the result first in terms of pairs (A, B).

13.11 Theorem: *Let A and B be real matrices of respective sizes $(n \times n)$ and $(n \times m)$. Then (A, B) is a reachable pair if and only if (A, B) is a controllable pair.*

Proof: Assume that (A, B) is a reachable pair. Given $x^\circ \in \mathbf{R}^n$, pick any $T > 0$. Let $x^1 = -e^{TA}x^\circ$; choose a continuous $u : [0, T] \to \mathbf{R}^n$ so that

$$x^1 = \int_0^T e^{(T-\tau)A} Bu(\tau) d\tau .$$

Then

$$0 = e^{TA} x^\circ + \int_0^T e^{(T-\tau)A} Bu(\tau) d\tau ,$$

and (A, B) is therefore a controllable pair since x° was arbitrary.

Conversely, assume that (A, B) is a controllable pair. If (A, B) were not reachable, there would exist by Theorem 13.7 a nonzero vector $w \in \mathbf{R}^n$ satisfying $w^T Q_r(A, B) = 0$. Thus $w^T A^k B = 0$ for every $k \geqslant 0$ by the Cayley-Hamilton Theorem 9.8. Suppose we could find $T > 0$ and a continuous $u : [0, T] \to \mathbf{R}^m$ such that

$$0 = e^{TA} w + \int_0^T e^{(T-\tau)A} Bu(\tau) d\tau .$$

Multiply both sides on the left by $w^T e^{-TA}$, and conclude that

$$0 = w^T w + w^T \left(\int_0^T e^{-\tau A} Bu(\tau) d\tau\right) .$$

Now, w^T times the integral is zero since the integral is a linear combination of columns of $B, AB, \ldots, A^k B, \ldots$ (c.f. the proof of Theorem 13.7); consequently, we get $w^T w = 0$, which is a contradiction. □

It turns out that reachability *implies* controllability for discrete time systems, but that controllability does not imply reachability. The crucial step in the proof of Theorem 13.11 which has no discrete-time analogue is the *inversion of the state transition matrix* e^{TA} to give e^{-TA}. Once again, we have found a manifestation of the "temporal asymmetry" in discrete-time systems which was alluded to in §3.

For completeness, we conclude the section with the statement of an "abstract counterpart" for Theorem 13.11; the proof is a straightforward application of Theorems 13.4, 13.10, and 13.13.

13.12 Theorem: *A real, time-invariant, m-input, p-output continuous-time state space linear system satisfying the "realizability" Assumptions 7.3 has a reachable state space if and only if it is controllable.* □

14. Reachability and Controllability for Time-Invariant Discrete-Time Systems

In this section, we extend the development in §13 to cover time-invariant discrete-time state space systems. The proofs of some of the results are a bit easier than the proofs of their continuous-time analogues; much of the technical machinery which we shall require is already at hand. As in §13, various definitions and results are stated and proved both in the context of "abstract" discrete-time linear systems, as per Definition 6.2, and in the setting of realizations.

14.1 Definition: Let there be given, as in Definition 6.2, a real, time-invariant, m-input, p-output discrete-time state space linear system with input function space $\mathbf{U}$, state space X, and state transition mapping ϕ. A state $x^1 \in X$ is said to be a *reachable state* if and only if there exists an integer $K > 0$ and a $u \in \mathbf{U}$ such that

$$\phi(K, 0, 0, u) = x^1 .$$

Moreover, the system is said to be *reachable*, or to have a *reachable state space*, if and only if every $x^1 \in X$ is a reachable state. □

14.2 Definition: Let A and B be real matrices of respective sizes $(n \times n)$ and $(n \times m)$. A vector $x^1 \in \mathbf{R}^n$ is said to be *discrete-time reachable with respect to (the pair)* (A, B) if and only if there exists an integer $K > 0$ and a sequence

$u(0), u(1), \ldots, u(K-1)$ of real m-vectors such that

$$x^1 = \sum_{l=0}^{K-1} A^{K-l-1} u(l) .$$

Moreover, (A, B) is said to be a *discrete-time reachable pair* if and only if every $x^1 \in \mathbf{R}^n$ is discrete-time reachable with respect to (A, B). □

Definitions 14.1 and 14.2 have the same intuitive meaning with regard to discrete-time systems as Definitions 13.2 and 13.3 have with regard to continuous-time systems. Specifically, a state $x^1 \in X$ of a time-invariant discrete-time linear system is reachable precisely when the initial state $x^\circ = 0$ at initial time $k_o = 0$ may be "steered" to x^1 in finite time if a suitable input sequence is applied. Note also that $x^1 \in \mathbf{R}^n$ is reachable with respect to (A, B) if and only if x^1 is, in the sense of Definition 14.1, a reachable state of the discrete-time linear system with state space $\mathbf{R}^n$ defined as in Example 6.6 by the equations

$$x(k+1) = Ax(k) + Bu(k), \quad k \geqslant k_o$$

$$x(k_o) = x^\circ \tag{II}$$

$$y(k) = Cx(k) + Du(k) .$$

The discrete-time analogue of Theorem 13.4 is

14.3 Theorem: *Let there be given a real, time-invariant, m-input, p-output discrete-time state space linear system having n-dimensional state space X. A state $x^1 \in X$ is reachable if and only if, for every basis $\mathbf{x}$ for X, $[x^1]_{\mathbf{x}}$ is discrete-time reachable with respect to (A, B), where A and B arise as in Theorem 7.1 in the realization of the system provided by $\mathbf{x}$. Consequently, such a system is reachable if and only if the matrices (A, B) arising in any realization for the system constitute a discrete-time reachable pair.*

Proof: The proof is exactly the same as that of Theorem 13.4, except that we are working in discrete time. We need the fact that, for time-invariant discrete-time systems (c.f. the proof of Theorem 7.1), we have

$$[\phi(K, 0, 0, u)]_{\mathbf{x}} = \sum_{l=0}^{K-1} A^{(K-l-1)} Bu(l)$$

for every $K > 0$ and $u \in \mathbf{U}$. Thus $x^1 \in X$ is reachable if and only if there exists a $u \in \mathbf{U}$ such that

$$[x^1]_{\mathbf{x}} = [\phi(K, 0, 0, u)]_{\mathbf{x}} = \sum_{l=0}^{K-1} A^{(K-l-1)} Bu(l)$$

where $\mathbf{x}$ is an arbitrary basis for X and A and B arise from the realization for the system given by $\mathbf{x}$. That is, x^1 is reachable if and only if $[x^1]_{\mathbf{x}}$ is reachable with respect to (A, B). The last sentence of the theorem is (as in continuous time) an immediate consequence of the definitions. □

We now introduce the discrete-time analogues of the special matrices which appeared in Definitions 13.5.

14.4 Definitions: Let A and B be real matrices having respective sizes $(n \times n)$ and $(n \times m)$.

(a) Given an integer $K > 0$, define the *K-reachability Gramian* $M_r(K ; A, B)$ of (A, B) as the $(n \times n)$ matrix given by

$$M_r(K ; A, B) = \sum_{l=0}^{K-1} A^{(K-l-1)} BB^T (A^T)^{(K-l-1)} .$$

(b) The *reachability matrix* $Q_r(A, B)$ is (exactly as in Definition 13.5) defined to be the $(n \times nm)$ matrix given by

$$Q_r(A, B) = [B \;\; AB \;\; A^2B \;\; \ldots \;\; A^{n-1}B] . \qquad \square$$

It will happen, as in the continuous-time situation, that $x^1 \in \mathbf{R}^n$ is reachable with respect to (A, B) if and only if x^1 lies in the range of $Q_r(A, B)$. It will also turn out that this range coincides with the range of $M_r(K ; A, B)$ for every $K \geqslant n$. We prove these results in a slightly different order than their continuous-time counterparts in §13.

14.5 Theorem: *Let A and B be real matrices having respective sizes $(n \times n)$ and $(n \times m)$. A vector $x^1 \in \mathbf{R}^n$ is discrete-time reachable with respect to (A, B) if and only if x^1 lies in the range of $Q_r(A, B)$. Consequently, (A, B) is a reachable pair if and only if $Q_r(A, B)$ has (full) rank n.*

Proof: If x^1 is in the range of $Q_r(A, B)$, we may find a real nm-vector z such that

$$x^1 = Q_r(A, B)z \ .$$

For $0 \leqslant i \leqslant n-1$, define the m-vector $u(n-i-1)$ by

$$u(n-i-1) = \begin{bmatrix} z_{mi+1} \\ z_{mi+2} \\ \cdot \\ \cdot \\ \cdot \\ z_{m(i+1)} \end{bmatrix} .$$

That is,

$$z^T = [u(0)^T \ u(1)^T \ \ldots \ u(n-1)^T] .$$

Then

$$x^1 = \sum_{l=0}^{n-1} A^{(n-l-1)} Bu(l) ,$$

and we conclude that x^1 is reachable with respect to (A, B).

Conversely, if we may find $K > 0$ and an m-vector sequence $u(0), u(1), \ldots, u(K-1)$ satisfying

$$x^1 = \sum_{l=0}^{K-1} A^{(K-l-1)} Bu(l) ,$$

then x^1 is in the range of $Q_r(A, B)$ by the Cayley-Hamilton Theorem 9.8. To see this, write each A^k in the above expansion as a linear combination of A^l for $0 \leqslant l \leqslant n-1$; then

$$x^1 = \sum_{l=0}^{n-1} A^{(n-l-1)} B\hat{u}(l)$$

for another sequence of m-vectors $\hat{u}(0), \ldots, \hat{u}(n-1)$. Clearly, then, x^1 is in the range of $Q_r(A, B)$; precisely, $x^1 = Q_r(A, B)z$, where z is the nm-vector defined by

$$z^T = [\hat{u}(n-1)^T \ \hat{u}(n-2)^T \ \ldots \ \hat{u}(0)^T] .$$

The last sentence in the theorem statement is an obvious consequence of the definition of rank. □

We now state and prove the discrete-time versions of Observations 13.6.

14.6 Observations: *(a) $M_r(K\,;A\,,B)$ is nonnegative definite for every $K > 0$.*

(b) The nullspace of $M_r(K\,;A\,,B)$ is the same for every $K \geqslant n$, and equals the set of all $w \in \mathbf{R}^n$ which satisfy

$$w^T Q_r(A\,,B) = 0\,.$$

(c) The range of $M_r(K\,;A\,,B)$ is the same for every $K \geqslant n$, and equals the range of $Q_r(A\,,B)$.

Proof: (a) $M_r(K\,;A\,,B)$ is obviously symmetric; moreover, given $x \in \mathbf{R}^n$ we have

$$\begin{aligned} x^T M_r(K\,;A\,,B)x &= \sum_{l=0}^{K-1} x^T A^{(K-l-1)} B B^T (A^T)^{(K-l-1)} x \\ &= \sum_{l=0}^{K-1} \| B^T (A^T)^{(K-l-1)} x \|^2 \geqslant 0\,. \end{aligned}$$

(b) Let $K \geqslant n$ be given; then $M_r(K\,;A\,,B)w = 0$ implies that

$$\begin{aligned} 0 &= w^T M_r(K\,;A\,,B)w \\ &= \sum_{l=0}^{K-1} w^T A^{(K-l-1)} B B^T (A^T)^{(K-l-1)} w \\ &= \sum_{l=0}^{K-1} \| B^T (A^T)^{(K-l-1)} w \|^2\,. \end{aligned}$$

Hence if $M_r(K\,;A\,,B)w = 0$, then $w^T A^k B = 0$ at least for every k, $0 \leqslant k \leqslant n-1$; we conclude that $w^T Q_r(A\,,B) = 0$. Conversely, if $w^T Q_r(A\,,B) = 0$, then (once again by the Cayley-Hamilton Theorem 9.8) $w^T A^k B = 0$ for *every* $k \geqslant 0$; thus for every $K \geqslant n$ (in fact, for every $K \geqslant 0$):

$$M_r(K\,;A\,,B)w = \sum_{l=0}^{K-1} A^{(K-l-1)} B B^T (A^T)^{(K-l-1)} w = 0\,.$$

We conclude that the nullspace of $M_r(K\,;A\,,B)$ for every $K \geqslant n$ coincides with the set of $w \in \mathbf{R}^n$ such that $w^T Q_r(A\,,B) = 0$.

(c) Let $K \geqslant n$ be given; since $M_r(K\,;A\,,B)$ is symmetric, its range is the orthogonal complement of its nullspace by Exercise 10.7. A vector w is in the nullspace of $M_r(K\,;A\,,B)$ if and only if (by (b)) w is orthogonal to every column of $Q_r(A\,,B)$; or, equivalently, if and only if w is in the orthogonal complement of the subspace of $\mathbf{R}^n$ spanned by the columns of $Q_r(A, B)$. This very last subspace is the range of $Q_r(A, B)$; thus the range of $Q_r(A, B)$ is the orthogonal complement of the nullspace of $M_r(K; A, B)$, and therefore equals the range of $M_r(K; A, B)$. Since $K \geqslant n$ was arbitrary, the result follows. □

The test for discrete-time reachability which corresponds with Theorem 13.7 is

14.7 Theorem: *Let A and B be real matrices having respective sizes $(n \times n)$ and $(n \times m)$. Then (A, B) is a discrete-time reachable pair if and only if $Q_r(A, B)$ has rank n, or, equivalently, $M_r(K; A, B)$ is invertible (hence positive definite) for some (and consequently all) $K \geqslant n$.*

Proof: We have already proven the first part of the theorem statement (c.f. Theorem 14.5); by Observation 14.6(c), $Q_r(A, B)$ has rank n precisely when $M_r(K; A, B)$ has rank n; since the matrix $M_r(K; A, B)$ is $(n \times n)$, it has rank n if and only if it is invertible. By Observation 14.6(a) (along with the discussion following Theorem 12.5), we conclude that (A, B) is a reachable pair if and only if $M_r(K; A, B)$ is positive definite for some (and consequently all, by Observation 14.6(c)) $K \geqslant n$. □

The reader may have noticed a slight difference between the foregoing results and those of §13. The continuous-time reachability Gramian $M_r(T; A, B)$ has the same rank for every $T > 0$, by Observation 13.6(c); as a consequence, a vector $x^1 \in \mathbf{R}^n$ reachable with respect to (A, B) may be "reached" in an arbitrarily small time $T > 0$ by appropriate choice of $u: [0, T] \rightarrow \mathbf{R}^m$. In the discrete time situation, however, the rank $M_r(K; A, B)$ does not necessarily "stabilize" at its full value until $K \geqslant n$. We may conclude that a vector $x^1 \in \mathbf{R}^n$ which is *discrete-time* reachable with respect to (A, B) may be reached in time $K = n$ or less, but perhaps not earlier than $K = n$.

We turn now to the question of discrete-time controllability. Following Definitions 13.8 and 13.9, we have

14.8 Definition: Let there be given a real, time-invariant, m-input, p-output discrete-time state space linear system as in Definition 6.2. The system is said to be *discrete-time controllable* if and only if for every initial state $x^\circ \in X$ there exist an integer time $K > 0$ and an input function $u \in \mathrm{U}$ such that

$$\phi(K, 0, x^\circ, u) = 0 . \qquad \square$$

14.9 Definition: Let A and B be real matrices having respective sizes $(n \times n)$ and $(n \times m)$. (A, B) is said to be a *discrete-time controllable pair* if and only if for every $x^\circ \in \mathbf{R}^n$ there exist an integer $K > 0$ and a sequence of m-vectors $u(0), u(1), \ldots, u(K-1)$ such that

$$0 = A^K x^\circ + \sum_{l=0}^{K-1} A^{(K-l-1)} B u(l) . \qquad \square$$

As in continuous time, a discrete-time system is controllable if and only if an arbitrary initial condition may be "zeroed out" in finite time by suitable choice of (discrete-time) input function. Observe that (A, B) is a discrete-time controllable pair precisely when the system with state space $\mathbf{R}^n$ constructed from equations (II) in §13 as in Example 6.6 is a controllable system in the sense of Definition 14.8. The reader may easily verify the following analogue of Theorem 13.10.

14.10 Theorem: *Let there be given a real, time-invariant, m-input, p-output discrete-time state space linear system with n-dimensional state space X. The system is controllable if and only if, for every choice of basis* $\mathbf{x}$ *for X, (A, B) is a controllable pair, where A and B arise as in Theorem 7.1 in the realization of the system provided by* $\mathbf{x}$. $\square$

Our final major result marks a point at which the theories of continuous- and discrete-time systems differ. As we remarked in §13, reachability of a pair (A, B) implies controllability of (A, B), both in continuous and discrete time. The converse, however, is not in general true for discrete-time systems. Consider the pair

$$A = \begin{bmatrix} 0 & 1 \\ 0 & 0 \end{bmatrix}; \quad B = \begin{bmatrix} 1 \\ 0 \end{bmatrix}.$$

Note that

$$Q_r(A, B) = \begin{bmatrix} 1 & 0 \\ 0 & 0 \end{bmatrix};$$

thus, by Theorem 14.5, (A, B) is *not* a discrete-time reachable pair because $Q_r(A, B)$ fails to have rank 2.

On the other hand, (A, B) *is* a controllable pair; to see this, first observe that $A^k = 0$ for every $k \geqslant 2$. Thus *any* initial condition $x(0) = x^\circ$ for the difference equation in (II) from §13 "dies out" in time $K = 2$ if the inputs $u(0) = u(1) = 0$ are applied. We remark once again that the key difference between time-invariant continuous- and discrete-time systems is that *the continuous-time state-transition matrix $\Phi(t, \tau) = e^{(t-\tau)A}$ is always invertible, while the discrete-time state-transition matrix $\Phi(k, l) = A^{(k-l)}$ may not be invertible.* This distinction causes the failure, in discrete time, of the statement "controllability implies reachability."

14.11 Theorem: *Let A and B be real matrices having respective sizes $(n \times n)$ and $(n \times m)$. If (A, B) is a discrete-time reachable pair, then (A, B) is a discrete-time controllable pair.*

Proof: Assume that (A, B) is a discrete-time reachable pair. Given $x^\circ \in \mathbf{R}^n$, pick any $K \geqslant n$. Let $x^1 = -A^K x^\circ$; choose a sequence of m-vectors $u(0), u(1), \ldots, u(K-1)$ so that

$$x^1 = \sum_{l=0}^{K-1} A^{(K-l-1)} B u(l).$$

Then

$$0 = A^K x^\circ + \sum_{l=0}^{K-1} A^{(K-l-1)} B u(l),$$

and (A, B) is therefore a discrete-time controllable pair since x° was arbitrary. □

The following "abstract analogue" of Theorem 14.11 is as easily proved as Theorem 13.12.

14.12 Theorem: *A real, time-invariant, m-input, p-output discrete-time* state space system is controllable if it has a reachable state space. □

15. Observability for Time-Invariant Continuous-Time Systems

We remarked in §13 that the notions of reachability and controllability serve to enhance our understanding of how the inputs to a state space linear system may be used to influence the state of the system. A particular system state is reachable if it is, in a sense, "strongly coupled" to the input through the state transition mapping. A system has a reachable state space if *all* the states are coupled to the input in this fashion; taking the "natural mode" viewpoint described in §8, we might say that a system has a reachable state space if and only if each of its natural modes may be "excited" by a suitable choice of input.

In this section, we undertake a similar study of the coupling between the state of time-invariant continuous-time linear system and the system's output. The state and output of an "abstract" linear system are connected by the readout mapping ρ. For the system with state space $\mathbf{R}^n$ constructed as in Example 6.5 from the equations

$$\begin{aligned} \dot{x}(t) &= Ax(t) + Bu(t), \quad t \geq t_o \\ x(t_o) &= x^\circ \\ y(t) &= Cx(t) + Du(t), \end{aligned} \tag{I}$$

we have the relation

$$y(t) = Ce^{(t-t_o)A}x^\circ + \int_{t_o}^{t} Ce^{(t-\tau)A}Bu(\tau)d\tau + Du(t),$$

which holds for $t \geq t_o$. The current state and current input to the system determine the current output directly and completely; we shall attempt in the pages that follow to answer the question, "How much information about the state of a linear system is contained in the output record of the system?"

15.1 Definition: Let there be given, as in Definition 6.1, a real, time-invariant, m-input, p-output continuous-time state space linear system with input function space $\mathbf{U}$, n-dimensional state space X, and readout mapping ρ. A state $x^\circ \in X$ is said to be an *unobservable state* if and only if for every $u \in \mathbf{U}$, we have for every $t \geqslant 0$

$$\rho(t, \phi(t, 0, x^\circ, u), u(t)) = \rho(t, \phi(t, 0, 0, u), u(t)) .$$

Moreover, the system is said to be *observable*, or to have an *observable state space*, if and only if the only $x^\circ \in X$ which is an unobservable state is $x^\circ = 0$. □

15.2 Definition: Let A and C be real matrices having respective sizes $(n \times n)$ and $(p \times n)$. A vector $x^\circ \in \mathbf{R}^n$ is said to be *unobservable with respect to* (A, C) if and only if, for every $t \geqslant 0$, we have

$$Ce^{tA} x^\circ = 0 .$$

Moreover, (A, C) is said to be an *observable pair* if and only if the only $x^\circ \in \mathbf{R}^n$ which is unobservable with respect to (A, C) is $x^\circ = 0$. □

Definition 15.1 merits further elaboration. A state $x^\circ \in X$ of a time-invariant continuous-time system is unobservable precisely when it is *indistinguishable* as an initial condition at time zero from the initial condition $x^\circ = 0$. By saying "x° is indistinguishable from zero," we mean that *every choice of input to the system, for $t \geqslant 0$, gives rise to the same output irrespective of whether the system was started in state x° or 0 at time 0.* Since there is no mention of input functions in Definition 15.2, it is not immediately evident that unobservability of a vector x° with respect to (A, C) is equivalent to unobservability of x° as a state of the linear system with state space $\mathbf{R}^n$ defined by (I). To see that this equivalence does in fact hold, observe that if x° is unobservable with respect to (A, C), then the output of the system defined by (I) is given for $t \geqslant 0$ by

$$y(t) = \int_0^t Ce^{(t-\tau)A} Bu(\tau) d\tau + Du(t) ,$$

which is the same as the output of the system when $x(0) = 0$, regardless of what input function u is applied.

The above line of reasoning leads directly to the following general assertion, which is the "observability version" of Theorem 13.4.

15.3 Theorem: *Let there be given a real, time-invariant, m-input, p-output continuous-time state space linear system having n-dimensional state space X and satisfying the "realizability" Assumptions 7.3. A state $x° \in X$ is unobservable if and only if, for every basis* $\mathbf{x}$ *for X,* $[x°]_{\mathbf{x}}$ *is unobservable with respect to (A, C), where A and C arise as in Theorem 7.8 in the realization of the system provided by* $\mathbf{x}$*. Consequently, such a system has an observable state space if and only if the matrices (A, C) arising in any realization for the system constitute an observable pair.* □

Thus, as in §13, we sacrifice very little generality if we concentrate our study of observability on pairs (A, C) instead of abstract linear systems. Our goal will be to derive simple criteria involving A and C which determine when some $x° \in \mathbf{R}^n$ is unobservable with respect to (A, C) and thereby determine whether (A, C) is an observable pair.

15.4 Definitions: Let A and C be real matrices having respective sizes $(n \times n)$ and $(p \times n)$.

(a) Given $T > 0$, the *T-Observability Gramian* $M_o(T; A, C)$ of (A, C) is the $(n \times n)$ matrix given by

$$M_o(T; A, C) = \int_0^T e^{tA^T} C^T C e^{tA}\, dt \ .$$

(b) The *observability matrix* $Q_o(A, C)$ is the $(np \times n)$ matrix given by

$$Q_o(A, C) = \begin{bmatrix} C \\ CA \\ CA^2 \\ \cdot \\ \cdot \\ \cdot \\ CA^{n-1} \end{bmatrix} .$$ □

It happens that $x° \in \mathbf{R}^n$ is unobservable with respect to (A, C) if and only if $x°$ is in the nullspace of $Q_o(A, C)$; this nullspace will in turn coincide with the nullspace of $M_o(T; A, C)$ for *every* $T > 0$.

15.5 Theorem: *Let A and C be real matrices having respective sizes $(n \times n)$ and $(p \times n)$. A vector $x^\circ \in \mathbf{R}^n$ is unobservable with respect to (A, C) if and only if x° lies in the nullspace of $Q_o(A, C)$.*

Proof: If x° is unobservable with respect to (A, C), then we have

$$Ce^{tA}x^\circ = 0, \quad t \geqslant 0.$$

Taking derivatives and evaluating at $t = 0$, we obtain $Cx^\circ = CAx^\circ = \ldots = CA^k x^\circ = 0$ for every $k \geqslant 0$. Thus, in particular, $CA^k x^\circ = 0$ for $0 \leqslant k \leqslant n-1$, and $Q_o(A, C)x^\circ = 0$.

Conversely, if $Q_o(A, C)x^\circ = 0$, then $CA^k x^\circ = 0$ for every $k \geqslant 0$ by the Cayley-Hamilton Theorem 9.8. Hence, for every $t \geqslant 0$ we have

$$Ce^{tA}x^\circ = \sum_{k=0}^{\infty} \frac{t^k}{k!} CA^k x^\circ = 0,$$

and x° is unobservable with respect to (A, C). □

15.6 Corollary: *Let A and C be real matrices of respective sizes $(n \times n)$ and $(p \times n)$. Then (A, C) is an observable pair if and only if $Q_o(A, C)$ has (full) rank n.*

Proof: By Theorem 15.5, the pair (A, C) fails to be observable precisely when there exists a nonzero x° in the nullspace of $Q_o(A, C)$. The existence of such an x° coincides with rank deficiency in $Q_o(A, C)$, since the rank of $Q_o(A, C)$ is n minus the dimension of its nullspace by Theorem 1.25(b). □

We now investigate the observability Gramian $M_o(T; A, C)$.

15.7 Observations: *(a) $M_o(T; A, C)$ is nonnegative definite for every $T \geqslant 0$.*

(b) The nullspace of $M_o(T; A, C)$ is the same for every $T > 0$, and coincides with the nullspace of $Q_o(A, C)$.

Proof: (a) $M_o(T; A, C)$ is clearly symmetric; moreover, given $x \in \mathbf{R}^n$ we have

$$x^T M_o(T\,;A\,,C)x = \int_0^T x^T e^{tA^T} C^T C e^{tA} x\,dt$$
$$= \int_0^T \| C e^{tA} x \|^2 dt \geqslant 0\,.$$

(b) Let $T > 0$ be given; suppose that $M_o(T\,;A\,,C)x = 0$. We may deduce from the proof of (a) that

$$C e^{tA} x = 0$$

for every $t \in [0, T]$. Thus x is unobservable with respect to $(A\,, C)$, and therefore lies in the nullspace of $Q_o(A\,, C)$. Conversely, if x is in the nullspace of $Q_o(A\,, C)$, then x is unobservable with respect to $(A\,, C)$ by Theorem 15.5, and therefore

$$C e^{tA} x = 0\,, \qquad 0 \leqslant t < \infty\,.$$

We conclude that for every $T > 0$,

$$M_o(T\,;A\,,C)x = \int_0^T e^{tA^T} C^T C e^{tA} x dt = 0\,.$$

Since $T > 0$ was arbitrary, the proof is complete. □

The proof of Observation 15.7(b) reveals the following observability criterion based on $M_o(T\,;A\,,C)$.

15.8 Theorem: *Let A and C be real matrices having respective sizes $(n \times n)$ and $(p \times n)$. A vector $x^\circ \in \mathbf{R}^n$ is unobservable with respect to $(A\,, C)$ if and only if x° is in the nullspace of $M_o(T\,;A\,,C)$ for some (hence all) $T > 0$. Consequently, $(A\,, C)$ is an observable pair if and only if $M_o(T\,;A\,,C)$ is invertible (hence positive definite) for some (and therefore all) $T > 0$.*

Proof: In the proof of Observation 15.7(b), we showed that the nullspace of $M_o(T\,;A\,,C)$ coincides for every $T > 0$ with the nullspace of $Q_o(A\,, C)$. By Theorem 15.5, this last nullspace is precisely the set of n-vectors unobservable with respect to $(A\,, C)$.

Thus $(A\,, C)$ is an observable pair if and only if $M_o(T\,;A\,,C)$ has a trivial nullspace; since $M_o(T\,;A\,,C)$ is $(n \times n)$, this situation is equivalent to invertibility (and hence positive definiteness, by Observation 15.7(a) and the discussion following Theorem 12.5) of $M_o(T\,;A\,,C)$. □

In §§13-14, we encountered the notion of controllability, which turned out to be equivalent to reachability for continuous-time systems but was weaker than reachability in discrete time. There is a corresponding weakened version of observability; it is called *constructibility*, and is equivalent to observability for continuous-time systems. Constructibility is not very interesting in continuous time; for one thing, its definition renders trivially obvious the equivalence of continuous-time observability and constructibitity.

15.9 Definition: Let there be given a real, time-invariant, m-input, p-output continuous-time state space linear system as in Definition 6.1. The system is said to be *constructible*, or to have a *constructible state space*, if and only if for every unobservable state $x^\circ \in X$, there exists a $T > 0$ such that

$$\phi(T, 0, x^\circ, 0) = 0 . \qquad \square$$

15.10 Definition: Let A and C be real matrices having respective sizes $(n \times n)$ and $(p \times n)$. (A, C) is said to be a *constructible pair* if and only if for every $x^\circ \in \mathbf{R}^n$ unobservable with respect to (A, C), there exists a $T > 0$ such that

$$e^{TA} x^\circ = 0 . \qquad \square$$

Definition 15.9 has the following interpretation: a system is constructible if and only if every unobservable initial state x° "dies out" to zero after a *finite* time when zero input is applied to the system. Thus the unobservable states in a constructible system's state space are irrelevant to the behavior of the system "over the long haul." Observe also that (A, C) is a constructible pair precisely when the system with state space $\mathbf{R}^n$ defined by equations (I) as in Example 6.5 is constructible in the sense of Definition 15.9. The following analogue of Theorem 15.3 is a straightforward consequence of the definitions; its proof is therefore left as an exercise for the reader.

15.11 Theorem: *Let there be given a real time-invariant, m-input, p-output continuous-time state space linear system having n-dimensional state space X and satisfying the "realizability" Assumptions 7.3. The system is constructible if and only if, for every choice of basis* $\mathbf{x}$ *for X, (A, C) is a* *constructible pair, where A and C arise as in Theorem 7.8 in the realization of the system provided by* **x**. □

It remains to show the equivalence of observability and constructibility in the continuous-time setting.

15.12 Theorem: *Let A and C be real matrices having respective sizes $(n \times n)$ and $(p \times n)$. (A, C) is an observable pair if and only if (A, C) is a constructible pair.*

Proof: Suppose (A, C) is an observable pair; then the only $x° \in \mathbf{R}^n$ which is unobservable with respect to (A, C) is $x° = 0$, and (trivially) $e^{TA} x° = 0$ for some (in fact every) $T > 0$, so that (A, C) is a constructible pair.

Conversely, assume that (A, C) is a constructible pair. Then any $x° \in \mathbf{R}^n$ unobservable with respect to (A, C) must satisfy $e^{TA} x° = 0$ for some $T > 0$; since e^{TA} is invertible, the only unobservable $x°$ is 0, and (A, C) is an observable pair. □

The reader might wonder why we have bothered at all with the (apparently useless) definition of constructibility and the almost vacuously true Theorem 15.12; the reason is that we wish to emphasize, once again, that continuous-time systems differ fundamentally from discrete-time systems in that the latter (as opposed to the former) need not have an invertible state transition matrix.

For the sake of completeness, we close with an "abstract" version of Theorem 15.12.

15.13 Theorem: *A real, time-invariant, m-input, p-output continuous-time state space linear system satisfying the "realizability" Assumptions 7.3 has an observable state space if and only if it has a constructible state space.* □

The reader is referred to [Brockett], [Desoer], [Kailath], or [Chen] for a discussion of observability for time-varying continuous-time systems.

16. Observability and Constructibility for Time-Invariant Discrete-Time Systems

In this section, we develop the discrete-time analogues of the concepts of observability and constructibility, which were described in §15 for continuous-time systems. Once again, each result has, along with its "abstract" version, a corresponding statement in terms of matrix pairs (A, C). We shall again be referring to the equations

$$x(k+1) = Ax(k) + Bu(k), \qquad k \geqslant k_o$$

$$x(k_o) = x^\circ \tag{II}$$

$$y(k) = Cx(k) + Du(k),$$

and to the time-invariant system with state space $\mathbf{R}^n$ which they define as in Example 6.6. We begin with a pair of definitions which parallel Definitions 15.1 and 15.2.

16.1 Definition: Let there be given, as in Definition 6.2, a real, time-invariant, m-input, p-output discrete-time state space linear system with input function space $\mathbf{U}$, n-dimensional state space X, and readout mapping ρ. A state $x^\circ \in X$ is said to be an *unobservable state* if and only if for every $u \in \mathbf{U}$ we have, for all $k \geqslant 0$,

$$\rho(k, \phi(k, 0, x^\circ, u), u(k)) = \rho(k, \phi(k, 0, 0, u), u(k)).$$

Moreover, the system is said to be *observable*, or to have an *observable state space*, if and only if the only $x^\circ \in X$ which is an unobservable state is $x^\circ = 0$. □

16.2 Definition: Let A and C be real matrices having respective sizes $(n \times n)$ and $(p \times n)$. A vector $x^\circ \in \mathbf{R}^n$ is said to be *discrete-time unobservable with respect to* (A, C) if and only if, for every $k \geqslant 0$,

$$CA^k x^\circ = 0.$$

Moreover, (A, C) is said to be a *discrete-time observable pair* if and only if the only $x^\circ \in \mathbf{R}^n$ which is discrete-time unobservable with respect to (A, C) is $x^\circ = 0$. □

Definition 16.1 and 16.2 have the same intuitive meaning with regard to discrete-time systems as Definitions 15.1 and

15.2 have with regard to continuous-time systems. The idea is that a state x° of a discrete-time system is unobservable precisely when 0 and x°, when used as initial states at time zero, give rise to the same future output regardless of what input is applied after time zero. Once again, inputs are not mentioned in Definition 16.2; still, we may conclude that an n-vector x° is discrete-time unobservable with respect to (A, C) if and only if it is an unobservable state (in the sense of Definition 16.1) of the discrete-time system with state space $\mathbf{R}^n$ defined by equations (II) via Example 6.6. To see this, observe that if x° is discrete-time unobservable with respect to (A, C), then for each $k \geqslant k_o = 0$ the output $y(k)$ of the system defined by (II) is given by

$$y(k) = \sum_{l=0}^{k-1} A^{(k-l-1)} B u(l) + D u(k) ,$$

which is the same for all input functions u as the output of the system when the initial state is $x(0) = 0$.

We have immediately the discrete-time version of Theorem 15.3.

16.3 Theorem: *Let there be given a real, time-invariant, m-input, p-output discrete-time state space linear system having n-dimensional state space X. A state $x^\circ \in X$ is unobservable if and only if, for every choice of basis* $\mathbf{x}$ *for X, $[x^\circ]_{\mathbf{x}}$ is unobservable with respect to (A, C), where A and C arise as in Theorem 7.1 in the realization of the system provided by* $\mathbf{x}$. *Consequently, such a system has an observable state space if and only if the matrices (A, C) arising in any realization for the system constitute an observable pair.* □

In analogy with Definitions 15.4, we have

16.4 Definitions: Let A and C be real matrices having respective sizes $(n \times n)$ and $(p \times n)$.

(a) Given $K > 0$, define the *K-observability Gramian* $M_o(K; A, C)$ of (A, C) as the $(n \times n)$ matrix given by

$$M_o(K; A, C) = \sum_{l=0}^{K-1} (A^T)^{(K-l-1)} C^T C A^{(K-l-1)} .$$

(b) The *observability matrix* $Q_o(A, C)$ of A and C is (once again) the $(np \times n)$ matrix given by

$$Q_o(A, C) = \begin{bmatrix} C \\ CA \\ CA^2 \\ \cdot \\ \cdot \\ \cdot \\ CA^{n-1} \end{bmatrix}. \qquad \square$$

It will again be the case that $x^\circ \in \mathbf{R}^n$ is discrete-time unobservable with respect to (A, C) if and only if x° is in the nullspace of $Q_o(A, C)$; this nullspace will in turn coincide with the nullspace of $M_o(K; A, C)$ for every $K \geqslant n$.

16.5 Theorem: *Let A and C be real matrices having respective sizes $(n \times n)$ and $(p \times n)$. A vector $x^\circ \in \mathbf{R}^n$ is discrete-time unobservable with respect to (A, C) if and only if x° lies in the nullspace of $Q_o(A, C)$.*

Proof: If x° is discrete-time unobservable with respect to (A, C), we have that $CA^k x^\circ = 0$ for *every* $k \geqslant 0$; in particular, $CA^k x^\circ = 0$ when $0 \leqslant k \leqslant n-1$, implying that x° is in the nullspace of $Q_o(A, C)$.

Conversely, if $Q_o(A, C)x^\circ = 0$, then $CA^k x^\circ = 0$ for *all* $k \geqslant 0$ (not just $0 \leqslant k \leqslant n-1$) by the Cayley-Hamilton Theorem 9.8; we conclude that x° is discrete-time unobservable with respect to (A, C). $\square$

16.6 Corollary: *Let A and C be real matrices having respective sizes $(n \times n)$ and $(p \times n)$. Then (A, C) is a discrete-time observable pair if and only if $Q_o(A, C)$ has (full) rank n.*

Proof: By Theorem 16.5, the pair (A, C) fails to be discrete-time observable precisely when there is a nonzero n-vector x° in the nullspace of $Q_o(A, C)$. The existence of such an x° is equivalent to rank deficiency in $Q_o(A, C)$, since the rank of $Q_o(A, C)$ is n minus the dimension of its nullspace by Theorem 1.25(b). $\square$

Our next goal is to establish the connection between properties of the observability Gramian and the foregoing results.

16.7 Observations: *Let A and C be real matrices having respective sizes $(n \times n)$ and $(p \times n)$.*

(a) $M_o(K\ ; A, C)$ is nonnegative definite for every $K \geqslant 0$.

(b) The nullspace of $M_o(K\ ; A, C)$ is the same for every $K \geqslant n$, and, for such K, equals the nullspace of $Q_o(A, C)$.

Proof: (a) $M_o(K\ ; A, C)$ is clearly symmetric; moreover, given $x \in \mathbf{R}^n$ we have

$$x^T M_o(K\ ; A, C)x = \sum_{l=0}^{K-1} x^T (A^T)^{(K-l-1)} C^T C A^{(K-l-1)} x$$

$$= \sum_{l=0}^{K-1} \| C A^{(K-l-1)} x \|^2 \geqslant 0 .$$

(b) Let $K \geqslant n$ be given; suppose $M_o(K\ ; A, C)x = 0$. We may deduce from the proof of (a) that

$$CA^k x = 0$$

at least for $0 \leqslant k \leqslant n-1$; thus $Q_o(A, C)x = 0$, and x is in the nullspace of $Q_o(A, C)$. Conversely, if $Q_o(A, C)x = 0$, then $CA^k x = 0$ for *every* $k \geqslant 0$ by the Cayley-Hamilton Theorem 9.8; thus

$$M_o(K\ ; A, C)x = \sum_{l=0}^{K-1} (A^T)^{(K-l-1)} C^T C A^{(K-l-1)} x = 0 .$$

Since $K \geqslant n$ was arbitrary, the proof is complete. □

We may now state and prove an observability criterion which involves $M_o(K\ ; A, C)$.

16.8 Theorem: *Let A and C be real matrices having respective sizes $(n \times n)$ and $(p \times n)$. A vector $x^\circ \in \mathbf{R}^n$ is discrete-time unobservable with respect to (A, C) if and only if x° lies in the nullspace of $M_o(K\ ; A, C)$ for some (hence all) $K \geqslant n$. Consequently, (A, C) is a discrete-time observable pair if and only if $M_o(K\ ; A, C)$ is invertible (and hence positive definite) for some (and consequently all) $K \geqslant n$.*

Proof: A vector x° is discrete-time unobservable with respect to (A, C) precisely when it is in the nullspace of $Q_o(A, C)$, by Theorem 16.5; this nullspace, by Observation 16.7(b), coincides with the nullspace of $M_o(K; A, C)$ for every $K \geqslant n$. The first part of the theorem follows. As for the second part, (A, C) is a discrete-time observable pair if and only if $Q_o(A, C)$ and therefore $M_o(K; A, C)$ for every $K \geqslant n$ have trivial nullspaces; since $M_o(K; A, C)$ is $(n \times n)$, we conclude that (A, C) is discrete-time observable if and only if $M_o(K; A, C)$ is invertible (hence positive definite, by Observation 16.7(a) and the discussion following Theorem 12.5) for some (and consequently all) $K \geqslant n$. □

We discuss next the concept of discrete-time constructibility, which is not (as it is in the case of continuous-time systems) trivially equivalent to discrete-time observability.

16.9 Definition: Let there be given a real, time-invariant, m-input, p-output discrete-time state space linear system as in Definition 6.2. The system is said to be *constructible*, or to have a *constructible state space*, if and only if for every unobservable state $x^\circ \in X$ there exists a $K > 0$ such that

$$\phi(K, 0, x^\circ, 0) = 0 \,. \qquad \square$$

16.10 Definition: Let A and C be real matrices having respective sizes $(n \times n)$ and $(p \times n)$. (A, C) is said to be a *discrete-time constructible pair* if and only if for every $x^\circ \in \mathbf{R}^n$ which is discrete-time unobservable with respect to (A, C), there exists a $K > 0$ such that

$$A^K x^\circ = 0 \,. \qquad \square$$

As in continuous time, a system is constructible exactly when every unobservable initial state x° "dies out" after a finite time when the zero input is applied to the system. The long-term behavior of a constructible system is therefore unaffected by its unobservable states. Observe that (A, C) is a discrete-time constructible pair if and only if the system with state space $\mathbf{R}^n$ constructed using equations (II) via Example 6.6 is a constructible system in the sense of Definition 16.9. We leave it to the reader to prove the following result, which corresponds with Theorem 16.3.

16.11 Theorem: *Let there be given a real, time-invariant, m-input, p-output discrete-time state space linear system having n-dimensional state space X. The system is constructible if and only if, for every choice of basis $\mathbf{x}$ for X, (A, C) is a discrete-time constructible pair, where A and C arise as in Theorem 7.1 in the realization of the system provided by $\mathbf{x}$.* □

We shall soon demonstrate that any observable discrete-time system (respectively, discrete-time observable pair (A, C)) is a constructible system (respectively, a discrete-time constructible pair). The converse, however, is not true; the key, once again, is that the condition $A^K x^\circ = 0$ does *not* imply that $x^\circ = 0$ as does the condition $e^{TA} x^\circ = 0$ in Definition 15.10. Once again, the state transition matrix $\Phi(k, l) = A^{k-l}$ of a discrete-time system is not necessarily invertible. Consider the example

$$A = \begin{bmatrix} 0 & 0 \\ 1 & 0 \end{bmatrix}; \quad C = [1 \quad 0] .$$

The pair (A, C) is not discrete-time observable; the observability matrix is

$$Q_o(A, C) = \begin{bmatrix} 1 & 0 \\ 0 & 0 \end{bmatrix},$$

which does not have (full) rank 2. However, (A, C) *is* a discrete-time constructible pair; the nullspace of $Q_o(A, C)$, which by Theorem 16.5 is the set of 2-vectors x° which are discrete-time unobservable with respect to (A, C), is the set of all $w \in \mathbf{R}^2$ satisfying $w_1 = 0$. Any such w, however, satisfies $Aw = 0$; thus every unobservable $w \in \mathbf{R}^2$ is annihilated by A, and (A, C) is a discrete-time constructible pair.

16.12 Theorem: *Let A and C be real matrices having respective sizes $(n \times n)$ and $(p \times n)$. If (A, C) is a discrete-time observable pair, then (A, C) is a discrete-time constructible pair.*

Proof: The result is essentially obvious, since observability implies that $x^\circ = 0$ is the only n-vector which is discrete-time unobservable with respect to (A, C), and this particular x° trivially satisfies $A^K x^\circ = 0$ for every $K > 0$. □

The "abstract" result which corresponds with Theorem 16.12 is

16.13 Theorem: *A real, time-invariant, m-input, p-output discrete-time state space linear system is constructible if it has an observable state space.* □

17. The Canonical Structure Theorem

The development in the preceding four sections suggests that the state space of an arbitrary m-input, p-output state space linear system is composed of several kinds of states; some states, it would seem, have more influence on the input-output behavior of the system than others. It is the purpose of the present section to formulate this idea precisely. The principal result, which has become known as the *Canonical Structure Theorem*, was first stated and proved by Kalman in the well-known paper [Kalman]; the proof presented here rests heavily on the vector sum and direct sum constructions (c.f. §5) along with the reachability and observability results of §§13-16.

Before proceeding with the formal exposition, it is worthwhile to consider again the intuitive content of the notions of reachability and observability. A given state space linear system may, as we have often said, be regarded as a mathematical model for a "real-world" process consisting of a "box" which, given an "initial time" t_o or k_o, takes input functions $u \in \mathbf{U}$ to output functions $y \in \mathbf{Y}_{t_o}$ or $y \in \mathbf{Y}_{k_o}$; the *state space* X of such a system models the family of *internal situations* which may obtain inside the box. In this manner, the state space may be thought of as containing all possible *initializations for* and all possible *internal results of* finite time-horizon experiments on the "box."

Suppose that some $x^\circ \in X$ is an unobservable state; as such, it constitutes an initial condition which cannot be distinguished from the zero initial condition by means of input-output experiments on the box. Such an x° might, however, be a reachable state; in that case, we could imagine an experiment in which some input function $u \in \mathbf{U}$ applied to the system starting at time t_{-1} "steers" the state of the system to x° at some time $t_o > t_{-1}$. Observing the system's output $y(t)$ for $t \geqslant t_o$ would *not* reveal any evidence that the given input u was applied before t_o. This situation might actually be *beneficial* in some contexts; for example, it could be that the input applied before t_o is a *disturbance* whose long-term effect on the output we might wish to minimize. On the other hand, consider the following example.

17.1 Example: A certain single-input, single-output continuous-time state space linear system with state space $\mathbf{R}^2$ has a realization (A, B, C, D) given by

$$A = \begin{bmatrix} 1 & 0 \\ 0 & 0 \end{bmatrix}; \quad B = \begin{bmatrix} 1 \\ 1 \end{bmatrix};$$

$$C = [0 \;\; 1]; \quad D = 0$$

Suppose that

$$x(0) = \begin{bmatrix} \epsilon \\ 0 \end{bmatrix};$$

Then even if *zero input* is applied after time $t_o = 0$, the first component of the vector function $t \to x(t)$ blows up as $t \to \infty$. This "unstable" behavior, however, even though it might reflect internal damage to the process modeled by the system, *goes entirely undetected at the output*, since $y(t) \equiv 0$ for $t \geqslant 0$. □

On the other hand, an unobservable state which is *not* reachable is in some sense *completely irrelevant* as far as the input-output behavior of the system is concerned. We can easily conjure up a physical device which might be modeled suitably as a state space linear system with such "irrelevant" states in its state space; think of a box which contains two separate electrical circuits, only one of which has terminals accessible to someone outside the box. If the circuit which is isolated from the outside world is "powered" in some fashion (e.g. by a battery inside the box), then its "state," which has evident physical significance as part of the situation inside the box, might evolve in a complicated way which an experimenter is entirely unable to detect or control.

The preceding paragraphs should alert the reader to the fact that, with regard to input-output experiments on a state space linear system, it is generally true that *some states in the system's state space are, at least to an extent, irrelevant;* the actual *degree of irrelevancy* of such states might not, however, be the same for all of them. We are prompted to seek some sort of "decomposition" of the state space X of a state space linear system which illuminates these features. First, after making a necessary definition, we shall prove two technical lemmas.

17.2 Definition: Let A be a real $(n \times n)$ matrix. A subspace W of $\mathbf{C}^n$ or $\mathbf{R}^n$ is said to be *invariant under* A if and only if whenever $x \in W$, we have $Ax \in W$. □

We have encountered invariant subspaces before, although we haven't until now formally defined them. It is simple to verify that the eigenspaces $E(\lambda_i)$ corresponding to the various eigenvalues λ_i of A are subspaces of $\mathbf{C}^n$ (or $\mathbf{R}^n$, if λ_i is real) which are invariant under A. Similarly, if $\lambda = \mu_0 + i\omega_0$ is a *nonreal* eigenvalue of A having corresponding eigenvector $z_0 \in \mathbf{C}^n$, the subspace of $\mathbf{R}^n$ spanned by $\mathrm{Re}\{z_0\}$ and $\mathrm{Im}\{z_0\}$ (c.f. Example 8.4 and the discussion preceding Lemma 8.7)) is a two-dimensional subspace of $\mathbf{R}^n$ which is invariant under A. Moreover, Proposition 9.2 reveals that the *generalized* eigenspaces $F(\lambda_i)$ corresponding with the various eigenvalues of a not necessarily semi-simple $(n \times n)$ matrix A are subspaces of $\mathbf{C}^n$ which are invariant under A.

We leave it to the reader to verify that if W_1 and W_2 are two subspaces of $\mathbf{R}^n$ (or $\mathbf{C}^n$) which are invariant under a real $(n \times n)$ matrix A, then both $W_1 \cap W_2$ and $W_1 + W_2$ are subspaces of $\mathbf{R}^n$ (or $\mathbf{C}^n$) which are invariant under A. Here, as in §4, $W_1 + W_2$ denotes the *vector sum* of W_1 and W_2; that is, $W_1 + W_2$ is the smallest subspace of $\mathbf{R}^n$ (or $\mathbf{C}^n$) which contains *both* W_1 and W_2.

If W is a k-dimensional subspace of $\mathbf{R}^n$ invariant under A, and if $\mathbf{w}$ is a basis for W, then we may extend the basis $\mathbf{w}$ to a basis $\mathbf{x}$ for $\mathbf{R}^n$ with respect to which the matrix of the linear transformation ${}_AT : \mathbf{R}^n \to \mathbf{R}^n$ has a special form; we have actually used this technique quite a bit already, most notably in §§8-11. More precisely, we may prove the following assertion.

17.3 Lemma: *Suppose that* $\mathbf{x} = \{x^1, x^2, \ldots, x^n\}$ *is a basis for* $\mathbf{R}^n$, *and that some ordered subset* $\mathbf{y} \subset \mathbf{x}$ *is a basis for a* k*-dimensional subspace* W *of* $\mathbf{R}^n$ *which is invariant under* *the* $(n \times n)$ *matrix* A. *Let* P *be the matrix whose* j*th column is* x^j, $1 \leqslant j \leqslant n$, *and let* $\mathbf{k}$ *be the set of indices* j *such that* $x^j \in \mathbf{y}$. *Then if* $j \in \mathbf{k}$, $[P^{-1}AP]_{ij} = 0$ *if* i *is not in* $\mathbf{k}$. *In particular, if* $\mathbf{k} = \{l, \ldots, l+k\}$ *for some* $l \leqslant n-k$, *then* $P^{-1}AP$ *takes the form*

$$\begin{bmatrix} A_{11} & 0 & A_{13} \\ A_{21} & A_{22} & A_{23} \\ A_{31} & 0 & A_{33} \end{bmatrix},$$

where A_{22} *is* $(k \times k)$ *and the other blocks are sized accordingly.*

Proof: Observe first of all (see also the proof of Proposition 9.5) that $P^{-1}AP$ is the matrix with respect to the basis $\mathbf{x}$ of the linear mapping ${}_AT : \mathbf{R}^n \to \mathbf{R}^n$; this conclusion follows from the discussion preceding Definition 4.22. Now, since W is invariant under A, ${}_AT(x^j)$, for $j \in \mathbf{k}$, must be a linear combination of the x^j which involves *only* the $x^j \in \mathbf{y}$; the conclusions of Lemma 17.3 then follow readily from Definition 4.19 of matrices of linear mappings with respect to bases. □

In the present context, we wish to complete a similar maneuver for the A matrix appearing in a realization (A, B, C, D) for a state space linear system; to that end, we make the following observation.

17.4 Lemma: *Let* A, B, *and* C *be real matrices having respective sizes* $(n \times n)$, $(n \times m)$, *and* $(p \times n)$. *Then the following two subspaces of* $\mathbf{R}^n$ *are invariant under* A:

(a) $W_r(A, B)$, *the set of all* $x^1 \in \mathbf{R}^n$ *which are reachable with respect to the pair* (A, B); *and*

(b) $W_u(A, C)$, *the set of all* $x_o \in \mathbf{R}^n$ *which are unobservable with respect to the pair* (A, C).

Proof: By Theorem 13.7, $x^1 \in \mathbf{R}^n$ is reachable with respect to the pair (A, B) if and only if x^1 lies in the range of the *reachability matrix*

$$Q_r(A, B) = [\, B \quad AB \quad \ldots \quad A^{n-1}B \,];$$

since a reachable x^1 may therefore be expressed as a linear combination of the columns of $Q_r(A, B)$, Ax^1 may be expressed as a linear combination of the columns of $AQ_r(A, B)$. But the columns of $AQ_r(A, B)$ are merely the columns of $A^k B$, where k runs from 1 to n; most of these columns are columns of $Q_r(A, B)$; the columns of $A^n B$ are *linear combinations* of the columns of $Q_r(A, B)$ by the Cayley-Hamilton Theorem 9.8. Hence Ax^1 lies in the range of $Q_r(A, B)$, and result (a) follows.

As for (b), $x_o \in \mathbf{R}^n$ is unobservable with respect to (A, C) if and only if x_o lies in the nullspace of $Q_o(A, C)$, by Theorem 15.5, where

$$Q_o(A, C) = \begin{bmatrix} C \\ CA \\ \cdot \\ \cdot \\ \cdot \\ CA^{n-1} \end{bmatrix};$$

that is, $x_o \in W_u(A, C)$ precisely when

$$Cx_o = CAx_o = \ldots = CA^{n-1}x_o = 0.$$

By the Cayley-Hamilton Theorem, once again, we conclude that in this case $CA^k x_o = 0$ for *every* $k \geqslant 0$, from which we deduce readily that Ax_o also lies in the nullspace of $Q_o(A, C)$, and is hence unobservable with respect to (A, C); the assertion in (b) follows immediately. □

Armed with Lemmas 17.3 and 17.4, we now state and prove a *concrete* form of the Canonical Structure Theorem; the *abstract* versions are easy consequences. As in the proof of Lemma 17.4, if A, B, and C are real matrices having respective sizes $(n \times n)$, $(n \times m)$, and $(p \times n)$, denote by $W_r(A, B)$ the subspace of $\mathbf{R}^n$ consisting of the set of all $x^1 \in \mathbf{R}^n$ which are reachable with respect to the pair (A, B); let $k_r(A, B)$ be the dimension of $W_r(A, B)$. Similarly, let $W_u(A, C)$ be the subspace of $\mathbf{R}^n$ of x^o which are *unobservable* with respect to (A, C), and let $k_u(A, C)$ be its dimension. Finally, let $k_{ru}(A, B, C)$ be the dimension of the subspace $W_r(A, B) \cap W_u(A, C)$, which evidently contains all the "states" in $\mathbf{R}^n$ which are reachable with respect to (A, B) but are unobservable with respect to (A, C).

17.5 Theorem: (Version I of the Canonical Structure Theorem) *Let A, B, and C be real matrices having respective sizes $(n \times n)$, $(n \times m)$, and $(p \times n)$. There exists an invertible $(n \times n)$ matrix P such that*

(a) the matrices $\hat{A} = P^{-1}AP$, $\hat{B} = P^{-1}B$, and $\hat{C} = CP$ take the special form

$$\hat{A} = P^{-1}AP = \begin{bmatrix} A_{11} & 0 & A_{13} & 0 \\ A_{21} & A_{22} & A_{23} & A_{24} \\ 0 & 0 & A_{33} & 0 \\ 0 & 0 & A_{43} & A_{44} \end{bmatrix}; \quad \hat{B} = P^{-1}B = \begin{bmatrix} B_1 \\ B_2 \\ 0 \\ 0 \end{bmatrix};$$

and

$$\hat{C} = CP = [\, C_1 \;\; 0 \;\; C_3 \;\; 0 \,].$$

The submatrices above are of the following sizes, with the notation we introduced previously:

submatrix	dimension
A_{22}	$(k_{ru} \times k_{ru})$
A_{11}	$((k_r - k_{ru}) \times (k_r - k_{ru}))$
A_{44}	$((k_u - k_{ru}) \times (k_u - k_{ru}))$
A_{33}	$((n - k_r - k_u + k_{ru}) \times (n - k_r - k_u + k_{ru}))$
B_2	$(k_{ru} \times m)$
B_1	$((k_r - k_{ru}) \times m)$
C_1	$(p \times (k_r - k_{ru}))$
C_3	$(p \times (n - k_r - k_u + k_{ru}))$

The rest of the submatrices are partitioned conformally.

(b) The pair

$$(A_r, B_r) = \left(\begin{bmatrix} A_{11} & 0 \\ A_{21} & A_{22} \end{bmatrix}, \begin{bmatrix} B_1 \\ B_2 \end{bmatrix} \right)$$

is a reachable pair, and

$$(A_o, C_o) = \left(\begin{bmatrix} A_{11} & A_{13} \\ 0 & A_{33} \end{bmatrix}, [\, C_1 \;\; C_3 \,] \right)$$

is an observable pair.

(c) The matrix triple (A_{11}, B_1, C_1) is such that (A_{11}, B_1) is a reachable pair and (A_{11}, C_1) is an observable pair.

Proof: We construct the matrix P by choosing a special basis $\mathbf{x}$ for $\mathbf{R}^n$ as follows. Let $x^1, x^2, \ldots x^{k_r}$ form a basis for $W_r(A, B)$ with the property that the last k_{ru} x^j's, for $j \leqslant k_r$, form a basis for $W_r(A, B) \cap W_u(A, C)$. Meanwhile, choose the *last* $k_u - k_{ru}$ vectors in the basis $\mathbf{x}$ so that *these* vectors, when taken together with $x^{k_r - k_{ru} + 1}, \ldots, x^{k_r}$, form a basis for $W_u(A, C)$. Finally, choose the remaining x^j's (there are $n - k_r - k_u + k_{ru}$ of them) to complete the basis $\mathbf{x}$ for $\mathbf{R}^n$.

Let P be the $(n \times n)$ matrix whose jth column is x^j. Then Lemma 17.3 implies that the four lower leftmost blocks of $\hat{A} = P^{-1}AP$ contain only zeroes, since $\{x^1, \ldots, x^{k_r}\}$ is a basis for $W_r(A, B)$, which is invariant under A by Lemma 17.4. Similarly, the (1, 2) block of $P^{-1}AP$ must also contain zeroes, since the last k_{ru} of the first k_r x^j's form a basis for the subspace $W_r(A, B) \cap W_u(A, C)$, which is also invariant under A. If we let $\mathbf{y}$ be the basis for $W_u(A, B)$ consisting of the last $k_u - k_{ru}$ x^j's together with the last k_{ru} of the first k_r x^j's, then Lemma 17.3, along with invariance under A of $W_u(A, C)$, implies that the (1, 4) and (3, 4) blocks of $\hat{A}$ must contain only zeroes, as well.

Observe next that the range of B is *contained* in $W_r(A, B)$; this is an easy consequence of Theorems 13.7 and 14.5. Since $B = P\hat{B}$, and since the first k_r columns of P form a basis for $W_r(A, B)$, it follows that the the rows of $\hat{B}$ indexed by i for $i > k_r$ must contain only zeroes. Similarly, the nullspace of C *contains* $W_u(A, C)$ by Theorems 15.5 and 16.5. Hence the jth column of $\hat{C} = CP$ must be zero if j indexes a column of P which is a basis vector for $W_u(A, C)$; these columns, however, are precisely the last $k_u - k_{ru}$ columns of P along with the last k_{ru} of the first k_r columns of P. The prescribed form of $\hat{C} = CP$ follows, and the proof of (a) is complete.

As for (b) it is easy to see that $Q_r(\hat{A}, \hat{B}) = P^{-1}Q_r(A, B)$ and $Q_o(\hat{A}, \hat{C}) = Q_o(A, C)P$. Moreover, a simple calculation shows that $Q_r(\hat{A}, \hat{B})$ is the same as $Q_r(A_r, B_r)$ with $n - k_r$ rows of zeroes tacked on at the bottom, and that $Q_o(\hat{A}, \hat{C})$ has k_u columns of zeroes in two bands (of widths k_{ru} and $k_u - k_{ru}$) interspersed between two bands of columns which contain precisely the columns of $Q_o(A_o, C_o)$. Since k_r is the rank of $Q_r(A, B)$, and since P is invertible, it follows that the k_r rows of $Q_r(A_r, B_r)$ must be linearly independent; thus

$Q_r(A_r, B_r)$ has (full) rank k_r, and hence (A_r, B_r) is a reachable pair by Theorem 13.7.

Similarly, the nullspace of $Q_o(\hat{A}, \hat{C})$ must have dimension k_u; hence its $n-k_u$ columns which are not constrained to be zero must be linearly independent. We conclude that the $n-k_u$ columns of $Q_o(A_o, C_o)$ are linearly independent, and that (A_o, C_o) is consequently an observable pair.

The proof of (c) is straightforward given the proof of (b) and the structure of $\hat{A}$, $\hat{B}$, and $\hat{C}$. Merely observe that the first $k_r - k_{ru}$ rows of $Q_o(A_{11}, B_1)$ are *exactly the same* as the first $k_r - k_{ru}$ rows of $Q_r(A_r, B_r)$; these rows are linearly independent because of (b). Likewise, the first $k_r - k_{ru}$ columns of $Q_o(A_{11}, C_1)$ are the same as the first $k_r - k_{ru}$ columns of $Q_o(A_o, C_o)$, which are in turn linearly independent by (b). Thus (A_{11}, B_1) and (A_{11}, C_1) are reachable and observable pairs, respectively. □

It is worth attempting at this point to see how Theorem 17.5 illuminates certain structural features of the time invariant state space linear systems with state space $\mathbf{R}^n$ defined via Examples 6.5 and 6.6 through Equations (I) and (II) of §13. Theorem 17.5 enables us to "separate" the state space of each system into four pieces; the first piece is reachable and, in some sense, "observable;" the second piece is reachable and unobservable; the third piece is "not reachable but observable;" and the last piece is unobservable and, in some sense, "not reachable."

It should be emphasized that "observability" of n-vectors is *not* a well-defined concept; nonetheless, the special forms of the matrices $\hat{A}$, $\hat{B}$, and $\hat{C}$ in Theorem 17.5 provide a useful illustration of the ways in which the reachability of a pair (A, B) is linked with the unobservability of a pair (A, C) through the matrix A and its invariant subspace structure. We remark that if (A, B) is a reachable pair to begin with, then $k_r = n$, and consequently $\hat{A} = A_r$. Similarly, if (A, C) is an observable pair, then $\hat{A} = A_o$. It might be said that the "least degenerate of all possible situations" holds when (A, B) is reachable *and* (A, C) is observable; in this case, $\hat{A} = A_{11} = A$, $\hat{B} = B_1 = B$, and $\hat{C} = C_1 = C$.

17.6 Exercise: Prove the final assertion in the preceding paragraph. □

Our ultimate objective will be to construct, for a given *abstract* time-invariant state space linear system, a realization having *the same special form* as $\hat{A}$, $\hat{B}$, and $\hat{C}$ for its A, B, and C-matrices. This form exhibits, at least in some sense, the way in which states of the system are "built out of components" which possess different reachability and observability properties. A glance at the material in §§7-8 should make it clear that finding a *special realization* for a state space linear system is equivalent to finding a *special basis* for the system's state space X.

Accordingly, let there be given an m-input, p-output time-invariant continuous- or discrete-time state space linear system with n-dimensional state space X. Referring back to Definitions 13.2 and 14.1, we denote by X_r the set of reachable states $x^1 \in X$. Observe that X_r is a *subspace* of X; this last fact should be an evident consequence of the development in §§13-14. Indeed, suppose for the moment that the given system is a continuous-time system; let $\mathbf{x}$ be a basis for X, and let (A, B, C, D) be the corresponding realization for the system.

Theorem 13.4 says that a state $x^1 \in X$ is reachable if and only if $[x^1]_{\mathbf{x}} \in \mathbf{R}^n$ is reachable with respect to the pair (A, B). The set $W_r(A, B)$ of all n-vectors reachable with respect to (A, B) is, by Theorem 13.7, a subspace of $\mathbf{R}^n$, and X_r is merely the inverse image of $W_r(A, B)$ under the isomorphism $x \to [x]_{\mathbf{x}}$ from X to $\mathbf{R}^n$; hence, X_r is a subspace of X. The discrete-time argument is identical.

We may show in exactly the same fashion that the subset X_u of X containing the set of $x^\circ \in X$ of *unobservable* states is a *subspace* of X; in this case, the results of §§15-16 provide the necessary information.

17.7 Theorem: (Version II of the Canonical Structure Theorem) *Let there be given an m-input, p-output time-invariant continuous- or discrete-time state space linear system having n-dimensional state space X. Let X_r and X_u denote, respectively, the reachable and unobservable subspaces of X, and let X_{ru} denote $X_r \cap X_u$. Let k_r, k_u, and k_{ru} denote the dimensions, respectively, of X_r, X_u, and X_{ru}. Then the system possesses a realization of the form $(\hat{A}, \hat{B}, \hat{C}, D)$, where $\hat{A}$, $\hat{B}$, and $\hat{C}$ satisfy the conclusions of Theorem 17.5.*

Proof: Choose a basis $\mathbf{x} = \{x^1, \ldots, x^n\}$ for X (compare with the Proof of Theorem 17.5) as follows:

$$\mathbf{x}_1 = \{x^1, \ldots, x^{k_r}\}$$

forms a basis for X_r;

$$\mathbf{x}_2 = \{x^{k_r - k_{ru} + 1}, \ldots, x^{k_r}\}$$

forms a basis for X_{ru};

$$\mathbf{x}_4 = \{x^{n - k_u + k_{ru} + 1}, \ldots, x^n\},$$

together with $\mathbf{x}_2$, forms a basis for X_u; and

$$\mathbf{x}_3 = \{x^{k_r + 1}, \ldots, x^{n - k_u + k_{ru}}\}$$

completes the set to a basis $\mathbf{x}$ for X.

We claim that the matrices A, B, and C appearing in the realization of the system with respect to the basis $\mathbf{x}$ for X are of the form of $\hat{A}$, $\hat{B}$, and $\hat{C}$ in Theorem 17.5. This is easy to see in view of our earlier observation that the isomorphism $x \to [x]_{\mathbf{x}}$ from X to $\mathbf{R}^n$ maps X_r, X_u, and X_{ru} respectively onto $W_r(A, B)$, $W_u(A, C)$, and $W_{ru}(A, B, C)$; hence the bases $\mathbf{x}_1 \cup \mathbf{x}_2$, $\mathbf{x}_2 \cup \mathbf{x}_4$, and $\mathbf{x}_2$ for these subspaces map onto bases for their respective $\mathbf{R}^n$ counterparts. The prescribed forms for A, B, and C are immediate consequences of the construction employed in the proof of Theorem 17.5(a); the assertions in Theorem 17.5(b) and (c) follow similarly. □

Note that the subspaces X_r, X_u, and X_{ru} are *well-defined subspaces of* X which depend *only* on properties of the system; the subspace $X_3 = \text{span}\{\mathbf{x}_3\}$, however, is *not* determined by system properties. Of course, none of the bases $\mathbf{x}_1$, $\mathbf{x}_2$, $\mathbf{x}_3$, or $\mathbf{x}_4$ is *canonical* in any sense, either. We close by re-stating Theorem 17.7 in terms of the *direct sum* construction (c.f. Definition 4.29).

17.8 Theorem: (Version III of the Canonical Structure Theorem) *With notation as in Theorem 17.7, the state space X may be decomposed into the vector sum of four mutually disjoint subspaces X_1, X_2, X_3, and X_4; the reachable subspace X_r is the vector sum of X_1 and X_2; the unobservable subspace X_u is the vector sum of X_2 and X_4; and $X_r \cap X_u$ is precisely X_2. Hence, by Theorem 4.31, X is isomorphic to the direct sum of four mutually disjoint subspaces X_1, X_2, X_3,* *and* X_4, *of which only* X_2 *is determined uniquely by the system.* □

17.9 Exercise: Prove Theorem 17.8. □

PART III
INPUT-OUTPUT LINEAR SYSTEMS

18. Formal Definitions and General Properties

In this section, we introduce formally the concept of an *input-output linear system*. Like the state space systems of Part II, input-output systems serve as models for real-world processes which may be viewed as "boxes" which take *input functions* to *output functions* in some linear manner. The differences between input-output linear systems and state space linear systems, however, are both philosophical and substantial. Processes which are modeled appropriately with state space linear systems have associated with them some notion of an *internal situation* whose *initial setting* might influence the outcome of an input-output "experiment" on the process. On the other hand, input-output linear systems may be used as models for processes having associated with them *no well-defined concept of what constitutes an internal situation*; the *only* experiments one may perform on such a *black box*-type process are of the input-output variety. Proceeding in the spirit of §6, we propose the following definitions.

18.1 Definition: A *real m-input, p-output continuous-time input-output linear system* consists of the following:

(a) A vector space U of *input functions* $u : \mathbf{R} \to \mathbf{R}^m$;

(b) For each $t_0 \in \mathbf{R}$, a vector space Y_{t_0} of *output functions* $y : [t_0, \infty) \to \mathbf{R}^p$; and

(c) For each $t_0 \in \mathbf{R}$, a mapping $\sigma_{t_0} : \mathbf{U} \to Y_{t_0}$ which satisfies the following conditions:

(i) *linearity:* σ_{t_0} is a *linear* mapping for every $t_0 \in \mathbf{R}$; and

(ii) *causality:* If $t_1 \geqslant t_0$, and if u and u' are input functions in U which satisfy $u(\tau) = u'(\tau)$ for every $\tau \in [t_o, t_1]$, then $y(t) = y'(t)$ for every $t \in [t_o, t_1]$, where $y \in \mathbf{Y}_{t_o}$ is $\sigma_{t_o}(u)$ and $y' \in \mathbf{Y}_{t_o}$ is $\sigma_{t_o}(u')$. □

18.2 Definition: A *real m-input, p-output discrete-time input-output linear system* consists of the following:

(a) A vector space U of *input functions* $u : \mathrm{Z} \to \mathbf{R}^m$;

(b) For each $k_o \in \mathrm{Z}$, a vector space $\mathbf{Y}_{k_o}$ of *output functions* $y : \{k_o, k_o + 1, k_o + 2, \ldots\} \to \mathbf{R}^p$; and

(c) For each $k_o \in \mathrm{Z}$, a mapping $\sigma_{k_o} : \mathrm{U} \to \mathrm{Y}_{k_o}$ which satisfies the following conditions:

(i) *linearity:* σ_{k_o} is a *linear* mapping for every $k_o \in \mathbf{R}$; and

(ii) *causality:* If $k_1 \geqslant k_o$, and if u and u' are input functions in U which satisfy $u(l) = u'(l)$ for every l with $k_o \leqslant l \leqslant k_1$, then $y(k) = y'(k)$ for every k, $k_o \leqslant k \leqslant k_1$, where $y \in \mathbf{Y}_{k_o}$ is $\sigma_{k_o}(u)$ and $y' \in \mathbf{Y}_{k_o}$ is $\sigma_{k_o}(u')$. □

The input functions $u \in \mathrm{U}$ may be visualized as the outputs of *signal generators;* the output function $\sigma_{t_o}(u)$ or $\sigma_{k_o}(u)$ is the "output record" arising from the experiment which entails "hooking up" the signal generator corresponding with u to the system at time t_o or k_o. In this way, the set of input-output mappings σ_{t_o} or σ_{k_o} may be thought of as a cataloguing of all possible input-output experiments on the system in question.

There are several elementary deductions which one can make about input-output linear systems based solely on Definitions 18.1 and 18.2. The first is a direct consequence of linearity, namely that

$$\sigma_{t_o}(0_{\mathrm{U}})(t) \equiv 0, \quad t \geqslant t_o,$$

and that

$$\sigma_{k_o}(0_{\mathrm{U}})(k) \equiv 0, \quad k \geqslant k_o,$$

where 0_{U} denotes the identically zero input function in either continuous or discrete time.

The second easy conclusion which may be drawn is that whenever $u \in \mathrm{U}$, $\sigma_{t_o}(u)$ (respectively, $\sigma_{k_o}(u)$) for every t_o (respectively, k_o) *is the same for all* $t \geqslant t_o$ *(respectively,* $k \geqslant k_o$*)* as the output function which arises when the zero input is applied starting at some time $t_{-1} \leqslant t_o$ (or $k_{-1} \leqslant k_o$) until time t_o (or k_o), at which point the application of u is begun. The significance of this last comment is that our definitions contain within them the implicit assumption that input-output experiments on the systems in question are begun when the system is in a *state of rest*, whatever that might mean in a given context.

More formally, in continuous time the last remark may be rephrased as follows: if $t_{-1} \leqslant t_o$, then

$$\sigma_{t_{-1}}(\mathbf{1}_{t_o} u)(t) = \sigma_{t_o}(u)(t), \qquad t \geqslant t_o,$$

where $\mathbf{1}_s$ is the shifted *unit step function*

$$\mathbf{1}_s(t) = \begin{cases} 1 & t \geqslant s \\ 0 & t < s \,. \end{cases}$$

The discrete time result is similar.

As in §7, we would like to be able to connect the foregoing abstract definitions with the concrete world of numbers and equations. In the case of discrete-time systems, essentially no additional restrictions need be imposed in order for us to be able to anchor our ideas in this fashion. The continuous-time case requires a bit more in the way of assumptions.

Once again as in §7, we assume that the input function space U for an m-input, p-output discrete-time input-output linear system contains *all* real discrete-time m-vector functions $u : \mathrm{Z} \to \mathbf{R}^m$. Also following §7, for each $w \in \mathbf{R}^m$ and each $l \in \mathrm{Z}$, we define $u_{l,w} \in \mathrm{U}$ by

$$u_{l,w}(k) = \begin{cases} w & k = l \\ 0 & k \neq l \,. \end{cases}$$

Given $k_o \in \mathrm{Z}$, define $\hat{W}_{k_o}(k, l)$, for each $l \in \mathrm{Z}$ and $k \geqslant k_o$, as the matrix with respect to standard bases of the linear mapping

$$w \to \sigma_{k_o}(u_{l,w})(k)$$

from $\mathbf{R}^m$ into $\mathbf{R}^p$.

The reader should check that for every $w \in \mathbf{R}^m$, and for every $k \geqslant k_o$ and $l \in \mathrm{Z}$, $\hat{W}_{k_o}(k, l)w$ represents the value of the output of the system at time k given that the input function applied to the system is a *vector pulse* whose value at time l is w and whose value at every other time is zero. This intuitive view of $\hat{W}_{k_o}(k, l)w$ gives the results of Lemma 18.3 below a natural interpretation.

18.3 Lemma: *Given $k_o \in Z$, for each $k \geqslant k_o$ let $\hat{W}_{k_o}(k, l)$ be defined as above. Then*

(a) $\hat{W}_{k_o}(k, l) = 0$ if $l < k_o$ or if $l > k$; and

(b) if $k_o \leqslant q \leqslant k$, then $\hat{W}_{k_o}(k, l) = \hat{W}_q(k, l)$ for each l with $q \leqslant l \leqslant k$.

Proof: Given $u \in U$, observe that the linearity property 18.2(c)(i) implies that for every $k \geqslant k_o$,

$$\sigma_{k_o}(u)(k) = \sum_{l=-\infty}^{\infty} \hat{W}_{k_o}(k, l)u(l).$$

Suppose now that $u(l) = 0$ if $k_o \leqslant l \leqslant k$. Then the causality property 18.2(c)(ii) implies that $\sigma_{k_o}(u)(k) = 0$; thus

$$0 = \sum_{l=-\infty}^{k_o-1} \hat{W}_{k_o}(k, l)u(l) + \sum_{k+1}^{\infty} \hat{W}_{k_o}(k, l)u(l);$$

since $u(l)$ is arbitrary for $l < k_o$ $l > k$, it follows that $\hat{W}_{k_o}(k, l) = 0$ for every $l < k_o$ and $l > k$, which is (a).

As for (b), if $k_o \leqslant q \leqslant k$, and if $u(l) = 0$ for $k_o \leqslant l < q$, then our second remark following Definition 18.2 implies that

$$\sigma_{k_o}(u)(k) = \sigma_q(u)(k)$$

for every $k \geqslant q$. But this means that

$$\sum_{l=q}^{k} \hat{W}_{k_o}(k, l)u(l) = \sum_{l=q}^{k} \hat{W}_q(k, l)u(l);$$

since $u(l)$ is arbitrary for $q \leqslant l \leqslant k$, it follows that $\hat{W}_{k_o}(k, l) = \hat{W}_q(k, l)$ when l is such that $q \leqslant l \leqslant k$. Since q was arbitrary, the result follows. □

The intuitive interpretation of the various parts of Lemma 18.3 should be apparent. Item (a) says that if the "signal generator" puts out a "vector pulse" w which occurs at a time l

before the signal generator is hooked up to the system, then the zero output results. Similarly, Lemma 18.3(b) says that if the signal generator puts out a pulse *after* time k, then that pulse has no influence on the values of the output *before* time k. Finally, Lemma 18.3(c) asserts that the system's output is the same regardless of *when* a pulse generator is connected to the system, provided that the connection is made *before* the pulse occurs.

We may conclude from Lemma 18.3 that given an m-input, p-output discrete-time input-output linear system, there exists a real $(p \times m)$ matrix-valued function

$$(k, l) \rightarrow W(k, l),$$

whose domain is the set of all pairs (k, l) such that $k \geqslant l$, which determines the input-output behavior of the system in the sense that for every $k_o \in \mathrm{Z}$ and for every $u \in \mathrm{U}$, we have

$$\sigma_{k_o}(u)(k) = \sum_{l = k_o}^{k} W(k, l) u(l), \quad k \geqslant k_o .$$

(See the comments following Lemma 7.6 for the state space analogue.) The matrix function $(k, l) \rightarrow W(k, l)$ has a special name, which is suggestive of the role it plays in the analysis of discrete-time input-output linear systems.

18.4 Definition: The matrix function $(k, l) \rightarrow W(k, l)$, $k \geqslant l$ defined as above is called the *weighting pattern* of the given discrete-time system. □

The weighting pattern of a discrete-time input-output linear system is a specification of how the various input values $\{u(l): k_o \leqslant l \leqslant k\}$ contribute to the value of the output at time $k \geqslant k_o$ when the input $u \in \mathrm{U}$ is applied starting at time k_o. It is evident that *knowing the weighting pattern* of a discrete-time system is equivalent to *knowing* σ_{k_o} *for all* $k_o \in \mathrm{Z}$. The weighting pattern is therefore a *complete specification* of the input-output behavior of the system.

In order to prove the continuous-time version of Lemma 18.3 and, subsequently, to define weighting patterns for continuous-time input-output systems, we must make additional assumptions about the input-output mappings σ_{t_o}. Assumptions 18.5 resemble Assumptions 7.3, which enabled us to conclude that *realizations* existed for continuous-time state space linear systems. Assumptions 18.5 pertain to a real, m-input, p-output continuous-time input-output linear system as in Definition 18.1.

18.5 Assumptions:

(a) The input function space U contains at least all of the continuous functions $U : \mathbf{R} \to \mathbf{R}^m$.

(b) For every $t_o \in \mathbf{R}$, the mapping $u \to \sigma_{t_o}(u)$ is given by *an integral with continuous kernel* plus a certain *finite number of evaluations;* that is, for every $t_o \in \mathbf{R}$, there exists for each $t \geqslant t_o$

(i) a real $(p \times m)$ matrix-valued function $\tau \to \hat{W}_{t_o}(t, \tau)$, which is continuous in τ at least on $[t_o, t]$,

(ii) a finite number $N(t_o ; t)$ of times $t_k(t_o ; t)$, $1 \leqslant k \leqslant N(t_o ; t)$, and

(iii) $N(t_o ; t)$ real *nonzero* $(p \times m)$ matrices $D_k(t_o ; t)$, $1 \leqslant k \leqslant N(t_o ; t)$, such that for every $u \in \mathbf{U}$, $\sigma_{t_o}(u)(t)$ is given for $t \geqslant t_o$ by

$$\int_{-\infty}^{\infty} \hat{W}_{t_o}(t, \tau) u(\tau) d\tau + \sum_{k=1}^{N(t_o ; t)} D_k(t_o ; t) u(t_k(t_o ; t)). \qquad \square$$

Assumption 18.5(b) is somewhat unconventional, since it allows for the possibility of input-output systems which contain a *pure delay* component; such systems, however, have become more and more important during recent years as models for real-world processes arising in applications. As the reader might already have guessed, our next step will be to prove the continuous-time counterpart of Lemma 18.3, which will result in a considerable simplification of the notation introduced in Assumptions 18.5.

18.6 Lemma: *With notation as in Assumption 18.5, the following conditions hold for every $t_o \in \mathbf{R}$ and for every $t \geqslant t_o$:*

(a) $\hat{W}_{t_o}(t, \tau) = 0$ if $\tau < t_o$ or if $\tau > t$;

(b) If $s \in [t_o, t]$, then $\hat{W}_{t_o}(t, \tau) = \hat{W}_s(t, \tau)$ for all $\tau \in [s, t]$;

(c) *Each* $t_k(t_o\,;t)$ *lies in* $[t_o\,,t]$; *moreover, if* $s \in [t_o\,,t]$, *then* $\{t_k(s\,;t)\}$ *is a subset of* $\{t_k(t_o\,;t)\}$. *In fact,* $N(s\,;t)$ *is equal to the number of* k*-values for which* $t_k(t_o\,;t) \in [s\,,t]$, *and if* $t_k(t_o\,;t) = t_l(s\,;t)$ *for some* k *and* l, *then* $D_k(t_o\,;t) = D_l(s\,;t)$.

Proof: Suppose that $u \in \mathrm{U}$ is identically zero on $[t_o\,,t)$; then the causality condition 18.1(c)(ii) says that $\sigma_{t_o}(u)(t) = 0$. In the notation of Assumptions 18.5, we have

$$0 = \int_{-\infty}^{t_o} \hat{W}_{t_o}(t\,,\tau)u(\tau)d\tau + \int_{t}^{\infty} \hat{W}_{t_o}(t\,,\tau)u(\tau)d\tau$$

$$+ \sum D_k(t_o\,;t)u(t_k(t_o\,;t))\,,$$

where the sum is taken over

$$\{k : t_k(t_o\,;t) \in (-\infty, t_o) \cup (t\,,\infty)\}\,.$$

Since u may be an arbitrary continuous function on $(-\infty, t_o)$ and $(t\,,\infty)$, we conclude (c.f. Fact 7.4) that $\hat{W}_{t_o}(t\,,\tau) = 0$ if $\tau < t_o$ or $\tau > t$; this is (a). Moreover, the first part of (c) follows because of the nonzeroness assumption on the $\{D_k\}$.

Now suppose that $s \in [t_o\,,t]$, and that $u(\tau) \equiv 0$ on $[t_o\,,s)$. The causality property again implies that

$$\sigma_{t_o}(u)(t) = \sigma_s(u)(t)\,,$$

which reads, in the notation of Assumptions 18.5,

$$\int_{s}^{t} \hat{W}_{t_o}(t\,,\tau)u(\tau)d\tau + \sum D_k(t_o\,;t)u(t_k(t_o\,;t))$$

$$= \int_{s}^{t} \hat{W}_s(t\,,\tau)u(\tau)d\tau + \sum D_k(s\,;t)u(t_k(s\,;t))\,.$$

The sum on the left-hand side is taken over $\{k : t_k(t_o\,;t) \in [s\,,t]\}$, and the sum on the right-hand side is over $1 \leqslant k \leqslant N(s\,;t)$.

We may choose u so that u is zero at each $t_k(t_o\,;t)$ and at each $t_k(s\,;t)$; then Fact 7.4 implies that $\hat{W}_{t_o}(t\,,\tau) = \hat{W}_s(t\,,\tau)$ for all $\tau \in [s\,,t]$, which is result (b); canceling the integrals, we obtain

$$\sum D_k(t_o\,;t)u(t_k(t_o\,;t)) = \sum D_k(s\,;t)u(t_k(s\,;t))\,.$$

Since u may be an arbitrary continuous function on $[s\,,t]$, it is evident that the $t_k(t_o\,;t) \in [s\,,t]$ and the $t_k(s\,;t)$ must match up, and that the $D_k(t_o\,;t)$ must match up with the $D_k(s\,;t)$ in exactly the same way. This proves the rest of (c). □

The result of Lemma 18.6 may be restated in a nice way which parallels the restatement of Lemma 18.3 for discrete-time systems. Given an m-input, p-output continuous-time input-output linear system satisfying Assumptions 18.5, we have shown that for every $t \in \mathbf{R}$ there exists a real $(p \times m)$ matrix function $\tau \to W(t,\tau)$ which is continuous in τ on $(-\infty, t]$, along with a *countable* sequence of "times" $\{t_k(t)\}$ and real, nonzero $(p \times m)$ matrices $D_k(t)$, such that for every $t_o \in \mathbf{R}$,

$$\sigma_{t_o}(u)(t) = \int_{t_o}^{t} W(t,\tau)u(\tau)d\tau + \sum_{[t_o,t]} D_k(t)u(t_k(t)) ,$$

where the summation notation means "sum over all k such that $t_k(t) \in [t_o, t]$."

18.7 Exercise: Justify the restatement of Lemma 18.6 given in the preceding paragraph. In particular, why is the total number of t_k's countable? Also, why, exactly, is one correct in suppressing the various t_o dependencies? □

Lemma 18.6 and the remark which follows it give us most of the necessary ingredients for a suitable definition of the *weighting pattern* of a continuous-time input-output linear system which satisfies Assumptions 18.5. Nonetheless, the straight evaluation or *delay* terms corresponding with the $D_k(t)$'s make it difficult to arrive at a "clean" definition of continuous-time weighting pattern which parallels Definition 18.4 for discrete-time systems. Such a parallel definition will turn out to be extremely useful in §19 when we discuss an *operational calculus* (based on Laplace transforms) for time-invariant continuous-time input-output systems.

In order to achieve the parallel definition, we need to introduce a construction which is probably more or less familiar to most readers and which, while it is extremely useful, seems unpleasantly non-rigorous. The central motivating idea is that we would like to be able to express in terms of *integrals* essentially all linear mappings which take functions to numbers.

Let $\mathbf{F}$ denote the set of all functions $f : \mathbf{R} \to \mathbf{R}$; $\mathbf{F}$ is evidently a real vector space, and perhaps the "simplest" *linear functionals* (c.f. Definition 5.1(b)) on $\mathbf{F}$ are the mappings $\Delta_T : \mathbf{F} \to \mathbf{R}$ given by $f \to f(T)$; each of these mappings takes a function $f \in \mathbf{F}$ to its value at some fixed $T \in \mathbf{R}$. One can think of the value of $\Delta_T(f)$, at least for *continuous* $f \in \mathbf{F}$, as being given by the limit as $\epsilon \to 0$ of the *average value* of f over the interval $[T-\epsilon, T+\epsilon]$:

$$\Delta_T(f) = \lim_{\epsilon \to 0} \frac{1}{2\epsilon} \int_{T-\epsilon}^{T+\epsilon} f(\tau)d\tau \ .$$

Let $p_\epsilon(t)$, $t \in \mathbf{R}$, be given by

$$p_\epsilon(t) = \begin{cases} 1 & |t| \leqslant \epsilon \\ 0 & |t| > \epsilon \ ; \end{cases}$$

there is, of course, no function from $\mathbf{R}$ to $\mathbf{R}$ which is equal to

$$\lim_{\epsilon \to 0} \frac{1}{2\epsilon} p_\epsilon(t) \ ,$$

since this limit is 0 for $t \neq 0$ and ∞ for $t = 0$. Nonetheless, it turns out to be useful to pretend that there *is* such a limiting "function" $t \to \delta(t)$, since then the formula for $\Delta_T(f)$, $f \in \mathbf{F}$, at least after a sloppy interchange of limit and integration, reduces to

$$\Delta_T(f) = \int_{-\infty}^{\infty} \delta(T-\tau) f(\tau)d\tau \ . \tag{*}$$

The preceding paragraph illustrates, at least in a primitive way, the evolution of the notion of the so called *Dirac delta function* or *unit impulse* $t \to \delta(t)$. There are, of course, sophisticated functional analytic ways of defining the delta function in terms of Lebesgue measure and integration; the reader is referred to [Rudin] for a treatment along those lines. We won't be so picky in what follows; we'll act as if there *were* a function $t \to \delta(t)$ which satisfied equation (*) above along with every $f \in \mathbf{F}$. It should be emphasized that this pretense, at least in the present context, is merely a vehicle for *notational simplification* and renders palatable certain otherwise unintelligible formulas.

18.8 Definition: Given an m-input, p-output continuous-time input-output linear system satisfying Assumptions 18.5, let $(t, \tau) \to W(t, \tau)$, for $t \geqslant \tau$, $\{D_k(t)\}$, for $t \in \mathbf{R}$, and $\{t_k(t)\}$, for $t \in \mathbf{R}$, be defined as in Lemma 18.6 and the discussion which follows. The *weighting pattern* of the system is defined as the $(p \times m)$ matrix "function"

$$\tilde{W}(t, \tau) = W(t, \tau) + \sum_k D_k(t)\delta(\tau - t_k(t)),$$

whose domain is $\{(t, \tau) \in \mathbf{R}^2 : t \geqslant \tau\}$. □

The usefulness of Definition 18.8 should be apparent. The input-output behavior of a continuous-time input-output system may now be described in terms of its weighting pattern as follows: for every $u \in \mathrm{U}$ and for every $t_o \in \mathrm{R}$,

$$\sigma_{t_o}(u)(t) = \int_{t_o}^{t} \tilde{W}(t, \tau)u(\tau)d\tau .$$

Implicit in the last equation is the convention that the "region of integration" for a definite integral *includes* the end points of the interval of integration; this convention ensures that if $t_k(t) = t_o$ or t for some k, then none of the "evaluation components" of the response $\sigma_{t_o}(u)(t)$ is obliterated by the integral. We may formalize this convention by *defining*

$$\int_{t_o}^{t} f(\tau)d\tau = \lim_{\epsilon,\, \delta \downarrow 0} \int_{t_o - \epsilon}^{t+\delta} f(\tau)d\tau .$$

As in the case of discrete-time systems, it should be emphasized that *knowing the weighting pattern* of a continuous-time input-output linear system is *equivalent* to knowing everything about the input-output behavior of the system.

Before giving some examples of input-output linear systems, we discuss the notions of *time-invariance* and *steady-state response*. Each of these items may be explained either in terms of the abstract definitions 18.1 and 18.2 or in terms of weighting patterns. As in §6, if $u : \mathbf{R} \to \mathbf{R}^m$ is a given m-vector function and $s \in \mathbf{R}$ is given, we define ${}_s u : \mathbf{R} \to \mathbf{R}^m$ by

$${}_s u(t) = u(t-s), \quad t \in \mathbf{R};$$

similarly, if $u : \mathrm{Z} \to \mathbf{R}^m$, and $j \in \mathrm{Z}$, then

$${}_j u(k) = u(k-j), \quad k \in \mathrm{Z}.$$

18.9 Definition: An m-input, p-output continuous-time linear system is said to be *time-invariant* if and only if

(a) for every $s \in \mathbf{R}$, we have

(i) whenever $u \in \mathbf{U}$, ${}_s u \in \mathbf{U}$, as well;

(ii) if $y \in \mathbf{Y}_{t_0}$, then the function $t \to y(t-s)$, defined on $[t_0 + s, \infty)$, is in $\mathbf{Y}_{t_0+s}$; and

(b) for every $t, s, t_0 \in \mathbf{R}$, with $t \geqslant t_0$, and for every $u \in \mathbf{U}$:

$$\sigma_{t_0+s}({}_s u)(t+s) = \sigma_{t_0}(u)(t). \qquad \square$$

18.10 Definition: An m-input, p-output discrete-time input-output linear system is said to be *time-invariant* if and only if

(a) for every $j \in \mathbf{Z}$, we have

(i) whenever $u \in \mathbf{U}$, ${}_j u \in \mathbf{U}$, as well;

(ii) if $y \in \mathbf{Y}_{k_0}$, then the function $k \to y(k-j)$, defined on $\{k_0 + j, k_0 + j + 1, \ldots\}$, is in $\mathbf{Y}_{k_0+j}$; and

(b) for every $k, j, k_0 \in \mathbf{Z}$, with $k \geqslant k_0$, and for every $u \in \mathbf{U}$:

$$\sigma_{k_0+j}({}_j u)(k+j) = \sigma_{k_0}(u)(k). \qquad \square$$

The intuitive content of Definitions 18.9 and 18.10 should be clear. Say we're given a continuous-time system; applying the input $u \in \mathbf{U}$ starting at time $t_0 \in \mathbf{R}$ gives output $t \to \sigma_{t_0}(u)(t)$, $t \geqslant t_0$; performing the "same input-output experiment starting at another time" entails applying the input ${}_s u$ starting at time $t_0 + s$. If the system is time-invariant, the two experiments should yield "the same results;" hence the output function arising from the second experiment, which is defined on $[t_0 + s, \infty)$, should be the s-shifted version of the output function arising from the first experiment.

There is an easy way to characterize time-invariance of an input-output linear system in terms of its *weighting pattern*. Simply stated, an input-output linear system is time invariant if and only if its weighting pattern is invariant with respect to an equal shift in both of its time arguments. More specifically, we have the following result.

18.11 Theorem: *(a) An m-input, p-output discrete-time input-output linear system is time-invariant if and only if it* *satisfies the assumptions in Definition 18.10(a) and its weighting pattern satisfies*

$$W(k, l) = W(k + j, l + j)$$

for every $k \geq l \in \mathrm{Z}$ *and for every* $j \in \mathrm{Z}$.

(*b*) *An m-input, p-output continuous-time input-output linear system satisfying Assumptions 18.5 is time-invariant if and only if it satisfies the assumptions in Definition 18.9(a) and its weighting pattern satisfies*

$$\tilde{W}(t, \tau) = \tilde{W}(t + s, \tau + s)$$

for every $t \geq \tau \in \mathrm{R}$ *and for every* $s \in \mathrm{R}$.

Proof: Since the systems in question satisfy the necessary assumptions with regard to the "richness" of U, Y_{t_o}, and Y_{k_o}, we merely need to show that, under these assumptions, conditions 18.9(b) and 18.10(b) are equivalent to the properties of the respective weighting patterns which are given in the theorem statement.

For the discrete-time case, condition 18.10(b) and the definition of the weighting pattern say that for every $u \in \mathrm{U}$, $j, k_o \in \mathrm{Z}$, and $k \geq k_o$, we have

$$\sum_{l = k_o + j}^{k + j} W(k + j, l) u(l - j) = \sum_{l = k_o}^{k} W(k, l) u(l) .$$

Now let $q = l - j$ in the first sum and use the fact that $u \in \mathrm{U}$ may be chosen arbitrarily from among *all* m-vector functions $u : \mathrm{Z} \to \mathrm{R}^m$ to conclude from the equation

$$\sum_{q = k_o}^{k} W(k + j, q + j) u(q) = \sum_{l = k_o}^{k} W(k, l) u(l)$$

that $W(k, l) = W(k + j, l + j)$ for every $k \geq l \in \mathrm{Z}$ and $j \in \mathrm{Z}$. The argument just given may be reversed to show that time-invariance is a consequence of the shift-invariance condition on the weighting patterns.

The proof of the continuous-time result is similar, although there are impulses floating about which make things a bit trickier. First, assume that the given system is time-invariant; then for every $u \in \mathrm{U}$ and for every $t_o \in \mathrm{R}$, $t \geq t_o$, and $s \in \mathrm{R}$, we may rewrite the stipulation of Definition 18.9(b) in terms of the system's weighting pattern as follows:

$$\int_{t_o+s}^{t+s} W(t+s,\tau)u(\tau-s)d\tau + \sum_{[t_o+s,\,t+s]} D_k(t+s)u(t_k(t+s))$$

$$= \int_{t_o}^{t} W(t,\tau)u(\tau)d\tau + \sum_{[t_o,\,t]} D_k(t)u(t_k(t)) . \qquad (**)$$

Now change variables in the first integral and conclude that the following equation holds for any continuous $u \in \mathbf{U}$ which *vanishes* at each $t_k(t)$ and each $t_k(t+s)$:

$$\int_{t_o}^{t} W(t,\tau)u(\tau)d\tau = \int_{t_o}^{t} W(t+s,\xi+s)u(\xi)d\xi .$$

Fact 7.4 then implies that $W(t,\tau) = W(t+s,\tau+s)$ except possibly at the countable set of τ values $\{t_k(t), t_k(t+s)\}$. The continuity of $\tau \to W(t,\tau)$ on $(-\infty, t]$, however, enables us to conclude that $W(t,\tau) = W(t+s,\tau+s)$ for all $\tau \leqslant t$.

Having eliminated the integral terms from equation (**), we are left with

$$\sum_{[t_o,\,t]} D_k(t)u(t_k(t)) = \sum_{[t_o+s,\,t+s]} D_k(t+s)u(t_k(t+s)) ;$$

since this equation must hold for *all* $u \in \mathbf{U}$, it is simple to show that

$$\{t_k(t+s) \in [t_o+s, t+s]\} = \{t_k(t)+s : t_k(t) \in [t_o, t]\} ,$$

and that $D_k(t+s) = D_k(t)$ when k is such that $t_k(t+s) = t_k(t)+s$. It then follows readily that

$$\tilde{W}(t,\tau) = \tilde{W}(t+s,\tau+s)$$

for every $t \in \mathbf{R}$, $\tau \leqslant t$, and $s \in \mathbf{R}$.

Conversely, the shift-invariance condition on $\tilde{W}$ implies by way of equation (**) that the system is time-invariant in the sense of Definition 18.9, and the proof of Theorem 18.11 is complete. □

The continuous-time part of Theorem 18.11 merits closer examination. Suppose that we are given a continuous-time input-output system which satisfies Assumptions 18.5, and which therefore possesses a weighting pattern. For each $t \in \mathbf{R}$, renumber the set of $t_k(t)$ in *decreasing* order; that is, index the $t_k(t)$'s so that $t_0(t) > t_1(t) > t_2(t) > \ldots$. It then follows directly from the proof of Theorem 18.11 that if the system is time-invariant, then for every $s \in \mathbf{R}$,

$$t_k(t+s) = t_k(t) + s \ , \quad k \in \mathrm{Z} \ .$$

The last equation enables us to define a set of *delays* T_k, $k \in \mathrm{Z}$, by way of $t_k(t) = t - T_k$; to see this, plug in $t = 0$, and define $T_k = -t_k(0)$, $k \in \mathrm{Z}$.

Observe that $T_k \geqslant 0$ for every $k \in \mathrm{Z}$; observe also that the proof of Theorem 18.11 implies that each $D_k(t)$ is equal to a constant $(p \times m)$ matrix D_k independent of t. The shift-invariance condition on the "continuous" part $W(t, \tau)$ of the weighting pattern implies that $W(t, \tau) = W(t-\tau, 0)$ for every $t \in \mathrm{R}$ and for every $\tau \leqslant t$. Hence, an m-input, p-output continuous-time input-output linear system satisfying 18.9(a) is time-invariant if and only if its weighting pattern is given for every $t \geqslant \tau$ by

$$\tilde{W}(t, \tau) = W(t-\tau, 0) + \sum_{k=0}^{\infty} D_k \, \delta(t-\tau-T_k) \, ;$$

that is,

$$\sigma_{t_o}(u)(t) = \int_{t_o}^{t} W(t-\tau, 0) u(\tau) d\tau + \sum D_k u(t-T_k)$$

for every $u \in \mathrm{U}$, $t_o \in \mathrm{R}$, and $t \geqslant t_o$. The last sum is taken over $\{k : t - T_k \geqslant t_o\}$.

18.12 Exercise: Flesh out the argument leading to the last equation. □

We turn now to a consideration of the *steady-state* properties of input-output linear systems. Imagine encountering a "black box" which has had the same "signal generator" hooked up to its input for a very long time. If suitable "stability" requirements are satisfied, all the "transient" behavior of the process will have died out, and the output will reflect the manner in which the properties of the box determine the system's *long-term* response to the input signal u. Naturally, one might expect that such *long-term* response to an input function would be well-defined for *some* input functions, but not for others.

18.13 Definition: Let there be given a real, m-input, p-output continuous-time input-output linear system as in Definition 18.1. If $u \in \mathrm{U}$, then we say that the *steady-state* *response of the system to the input u* is defined if and only if for every $t \in \mathbf{R}$, the limit

$$\sigma(u)(t) = \lim_{t_o \to -\infty} \sigma_{t_o}(u)(t)$$

exists, in which case the *steady-state response of the system to the input u* is given by the function $t \to \sigma(u)(t)$, $t \in \mathbf{R}$. □

18.14 Definition: Let there be given a real, m-input, p-output discrete-time input-output linear system as in Definition 18.2. If $u \in \mathbf{U}$, then we say that the *steady-state response of the system to the input u* is defined if and only if for every $k \in \mathbf{Z}$, the limit

$$\sigma(u)(k) = \lim_{k_o \to -\infty} \sigma_{k_o}(u)(k)$$

exists, in which case the *steady-state response of the system to the input u* is given by the function $k \to \sigma(u)(k)$, $k \in \mathbf{Z}$. □

There is a simple type of input function to which the steady-state response of an input-output linear system is always defined. Let $\mathbf{U}$ be the input function space of such a system; suppose that $u \in \mathbf{U}$ is such that $u(t) = 0$ for all $t < \hat{t}$, or $u(k) = 0$ for all $k < \hat{k}$ in discrete time, where $\hat{t}$ or $\hat{k}$ is given. It is an elementary consequence of Definitions 18.13 and 18.14 that the steady-state response of the given system to the input u is well-defined, and agrees with $\sigma_{\hat{t}}(u)$ on $[\hat{t}, \infty)$ or $\sigma_{\hat{k}}(u)$ on the set of times $k \geq \hat{k}$. Thus, the steady-state response of an input-output linear system to such a "right-sided" input function is always well-defined.

18.15 Example: Consider the "real-world process" defined by the following electrical circuit.

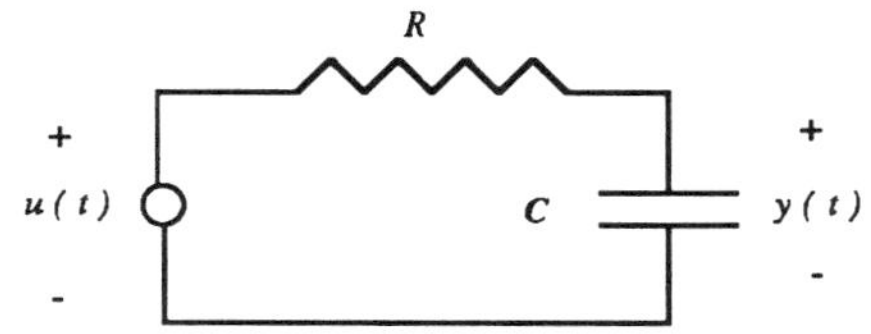

Figure 18.1 A real-world process

This process may be modeled by a single-input, single-output time-invariant continuous-time input-output linear system with input function space **U** consisting of all the continuous functions $u : \mathbf{R} \to \mathbf{R}$, output function space $\mathbf{Y}_{t_0}$ consisting of all continuous functions $y : [t_0, \infty) \to \mathbf{R}$, and input-output mappings

$$\sigma_{t_0}(u)(t) = \frac{1}{RC}\int_{t_0}^{t} e^{-\frac{t-\tau}{RC}} u(\tau)d\tau \ .$$

Observe how the implicit "state of rest at t_0" assumption is built into the definition of the σ_{t_0}. Evidently, the steady-state response of the system to $u \in \mathbf{U}$ exists if and only if

$$\int_{-\infty}^{t} e^{-\frac{t-\tau}{RC}} u(\tau)d\tau$$

exists for every $t \in \mathbf{R}$; hence, in particular, the steady state response to an input function of the form $t \to e^{\alpha t}$ is defined if and only if $\alpha > -RC^{-1}$. □

In anticipation of the development in the two sections which follow, we now define the notion of *impulse response* for time-invariant input-output systems.

18.16 Definition: The *impulse response* of an m-input, p-output time-invariant continuous- or discrete-time input-output linear system which possesses a weighting pattern is defined as follows:

(a) For a continuous-time system, the impulse response is the $(p \times m)$ matrix function $t \to H(t)$ which is zero for $t < 0$ and is given for $t \geqslant 0$ by

$$\begin{aligned} H(t) &= \tilde{W}(t, 0) \\ &= W(t, 0) + \sum D_k \delta(t - T_k), \quad t \geqslant 0 . \end{aligned}$$

(b) For a discrete-time system, the impulse response is the $(p \times m)$ discrete-time matrix function $k \to H(k)$ given by

$$H(k) = \begin{cases} W(k, 0) & k \geqslant 0 \\ 0 & k < 0 . \end{cases}$$ □

Suppose, then, that a given m-input, p-output time-invariant input-output linear system possesses a weighting pattern, and that the steady-state response $y = \sigma(u)$ of the system to the input $u \in \mathbf{U}$ is defined. In terms of the system's impulse response H, $y(t)$ is given for $t \in \mathbf{R}$ by

$$y(t) = \int_{-\infty}^{t} H(t-\tau)u(\tau)d\tau$$

for a continuous-time system, and by

$$y(k) = \sum_{l=-\infty}^{k} H(k-l)u(l), \qquad k \in \mathbf{Z}$$

for a discrete-time system.

The impulse response of a time-invariant input-output linear system is so named because it *is*, essentially, the *response of the system to a unit impulse input*. Such a characterization makes rigorous sense for discrete-time systems, but in continuous time gives rise to the usual analytical hassles which come up whenever impulses are present.

18.17 Exercise: Prove the statement in the preceding paragraph; that is, show that the impulse response of a time-invariant discrete-time input-output linear system is simply $\sigma(\delta)$, where $\delta \in \mathbf{U}$ is given by the "discrete-time impulse"

$$\delta(k) = \begin{cases} 1 & k = 0 \\ 0 & k \neq 0, \end{cases}$$

and that the impulse response of a continuous-time system is the output of the system when the "input" is the Dirac delta function. □

We close this section by showing how a given state space linear system defines in a natural way an associated input-output linear system.

18.18 Definition: Let there be given a real, m-input, p-output continuous-time state space linear system as in Definition 6.1. Define its *associated input-output linear system* as the input-output system with input function space U, output function spaces $\mathbf{Y}_{t_0}$, and input-output mappings $\sigma_{t_0} : \mathbf{U} \rightarrow \mathbf{Y}_{k_0}$ defined in terms of the readout mapping ρ and the state transition mapping ϕ as follows:

$$\sigma_{t_o}(u)(t) = \rho(t, \phi(t, t_o, 0, u), u(t)); \quad u \in \mathbf{U}, t \geqslant t_o . \square$$

18.19 Definition: Let there be given a real, m-input, p-output discrete-time state space linear system as in Definition 6.2. Define its *associated input-output linear system* as the input-output linear system with input function space U, output function spaces $\mathbf{Y}_{k_o}$, and input-output mappings

$$\sigma_{k_o}(u)(k) = \rho(k, \phi(k, k_o, 0, u), u(k)); \quad u \in \mathbf{U}, k \geqslant k_o . \square$$

Definitions 18.18 and 18.19 will be of considerable importance later on, when we discuss *realization theory* for input-output linear systems. Input-output realization theory lies at the heart of many "real-world" problems which involve the *synthesis* of processes which have prescribed input-output behavior. The reader should note that these definitions may be restated conveniently in terms of the *overall response functions* (c.f. Definitions 6.3 and 6.4) of the state space systems involved.

18.20 Exercise: Show that the input-output linear system associated with a finite-dimensional continuous-time state space linear system via Definition 18.18 has only *one* "$t_k(t)$-term" in its weighting pattern; explain the significance of this fact in view of Example 6.13. □

18.21 Exercise: Show that if a continuous- or discrete-time state space linear system is *time-invariant* (c.f. Definitions 6.10 and 6.11), then its associated input-output linear system is also time-invariant. □

19. Frequency Responses and Transfer Functions of Time-Invariant Continuous-Time Systems

In the preceding section, we laid a foundation for the *time domain* analysis of input-output linear systems. In the present section and the one which follows, we focus our attention on a collection of ideas and techniques which comprise the so called *frequency domain* approach to the analysis of time-invariant input-output systems. Frequency domain techniques have a long and colorful history; they are rooted mathematically in the functional analytic methods of theoretical physics which have developed over the last century or so, but their contribution to the genesis of the "systems approach" to engineering problems, particularly electrical ones, cannot be overestimated. One of their principal selling points is that they appeal strongly to *empirical intuition;* the meaning of this last comment should be made clear in what follows.

The frequency domain approach to time-invariant input-output linear systems depends heavily on the application of various types of *transform analysis*. It might be said that all transform analysis stems from the pioneering work of Fourier on trigonometric series; since Fourier's time, the transform which bears his name has become one of the cornerstones of modern mathematical physics, particularly quantum mechanics. During the last half century, the *Laplace transform* has played a similar role in the theory of time-invariant continuous-time input-output linear systems. There is a vast mathematical literature on all manner of transforms; readers who have a basic acquaintance with Lebesgue integration and complex analysis are referred to [Dym and McKean] for a very readable account of Fourier and Laplace transforms along with Fourier series. As it happens, the *analytical* aspects of transform theory are, in some sense, less important in the analysis of input-output linear systems than the *algebraic* aspects. Accordingly, we shall not spend a great deal of time proving analytical facts about transforms; we shall strive, however, to be as accurate as possible and to give suitable references when analytical statements are necessary.

It is arguable that the basic tenet of Fourier analysis is that a large class of time functions $t \to f(t), t \in \mathbf{R}$, may be expressed as or approximated by a *linear combination of sinusoids*. The meaning of the phrase "linear combination" depends on the context; in general, the linear combinations arising in Fourier analysis are *infinite*, and may take the form either of infinite sums or of integrals. Given the premise that the family of time functions which may be approximated by "linear combinations of sinusoids" is very *large*, it would seem reasonable to expect that the response of an input-output linear system to a broad class of input functions would be determined, in some sense, by the system's response to sinusoidal input functions; because of linearity, the response of the system to a linear combination of such "simple" input functions would need to be the *same linear combination* of the responses of the system to the individual sinusoidal components.

Consider, then, a real, m-input, p-output time-invariant continuous-time input-output linear system whose input function space $\mathbf{U}$ contains at least the sinusoidal input functions $t \to w \cos \omega_0 t$ and $t \to w \sin \omega_0 t$, $t \in \mathbf{R}$ for every $w \in \mathbf{R}^m$. It is convenient to pretend that the system admits *complex* input functions of the form $t \to u(t) = u_1(t) + iu_2(t)$, $t \in \mathbf{R}$, $u_j \in \mathbf{U}$, with the response to such an input function given by $\sigma_{t_0}(u) = \sigma_{t_0}(u_1) + i\,\sigma_{t_0}(u_2)$ for every $t_0 \in \mathbf{R}$.

Given $\omega_0 \in \mathbf{R}$, suppose that the *steady-state responses* (c.f. Definition 18.13) of the system to the inputs $t \to w \cos \omega_0 t$ and $t \to w \sin \omega_0 t$ are defined for every $w \in \mathbf{R}^m$. With notation as in Definition 18.16 and the discussion preceding Exercise 18.12, let the impulse response of the system be the $(p \times m)$ matrix function $t \to H(t)$ which is zero for $t < 0$ and is given for $t \geqslant 0$ by

$$\begin{aligned} H(t) &= \tilde{W}(t, 0) \\ &= W(t, 0) + \sum_{k=0}^{\infty} D_k \delta(t - T_k). \end{aligned}$$

Once again, $\{T_k : k \geqslant 0\}$ is an increasing sequence of *nonnegative delays*. Consider the steady-state response $t \to y(t)$ of the system to the input function $t \to \mathbf{w}^j e^{i\omega_0 t}$, where $\mathbf{w}^j$ is the jth standard basis vector for $\mathbf{R}^m$; the function $\mathbf{y}$ is given by

$$\begin{aligned} y(t) &= \int_{-\infty}^{\infty} \tilde{W}(t, \tau) \mathbf{w}^j e^{i\omega_0 \tau} d\tau \\ &= \int_{-\infty}^{\infty} H(t - \tau) \mathbf{w}^j e^{i\omega_0 \tau} d\tau. \end{aligned}$$

A simple change of variable in the last integral gives the following formula for the kth component of the output function y:

$$[y(t)]_i = [\hat{H}(\omega_0)]_{kj} e^{i\omega_0 t}, \quad t \in \mathbf{R}, \quad 1 \leqslant k \leqslant p, \quad (*)$$

where the $(p \times m)$ matrix function $\omega \to \hat{H}(\omega)$ defined by the last equation is given at $\omega \in \mathbf{R}$ by

$$\hat{H}(\omega) = \int_{-\infty}^{\infty} H(\tau) e^{-i\omega\tau} d\tau.$$

Note that in the last integral, τ "really" only runs from 0 to ∞, since $H(\tau) = 0$ for $\tau < 0$.

The function $\omega \to \hat{H}(\omega)$, $\omega \in \mathbf{R}$, which is well-defined by our assumption about the existence of steady-state responses to sinusoidal inputs, is called the *frequency response* of the system; before formalizing the definition of frequency response, it is worthwhile to examine more closely the intuitive content of equation (*). It says, essentially, that if the *input* to the system is a sinusoid $t \to we^{i\omega_0 t}$ of a certain frequency ω_0, then the *output* of the system is also a sinusoid which has *the same frequency* ω_0. Moreover, the factor by which the input sinusoid is multiplied in order to yield the output sinusoid is the value $\hat{H}(\omega_0)$ of the frequency response of the system at the given frequency.

Thus the frequency response, as one might expect, measures the way in which the system's *amplification* or *attenuation* of a sinusoidal input function depends on the input function's frequency. As it happens, the frequency response of the system determines a good deal more; we shall see, in the light of Fourier analysis, that it serves to specify how the various *frequency components* of a (possibly non-sinusoidal) input function contribute to the corresponding components of the output function.

Before proceeding further, we present the following definition, of which many variants exist.

19.1 Definition: A function $f : \mathbf{R} \to \mathbf{R}$ is said to be *Fourier transformable* if and only if the following integral exists in some sense of the meaning of the word "integral:"

$$\hat{f}(\omega) = \int_{-\infty}^{\infty} f(t) e^{-i\omega t}\, dt\,, \qquad \omega \in \mathbf{R}; \tag{FT}$$

in this case, the complex-valued function $\omega \to \hat{f}(\omega)$, $\omega \in \mathbf{R}$, is called the *Fourier transform* of the function f. □

Naturally, a vector- or matrix-valued function is said to be Fourier transformable if and only if each of its component functions is, in which case the Fourier transform is defined component-wise. The "senses of the meaning of integral" to which we are alluding in Definition 19.1 are many; it behooves us to mention a few of them.

If f is a continuous function which vanishes outside of a bounded interval, then f is Fourier transformable, and the integral defining its Fourier transform $\hat{f}(\omega)$ may be interpreted as the "ordinary Riemann integral." If f is absolutely integrable in the sense of Lebesgue, then the above integral may be viewed as a standard Lebesgue integral which defines $\hat{f}(\omega)$ for *every* $\omega \in \mathrm{R}$; moreover, $\omega \to \hat{f}(\omega)$ will be a bounded, continuous complex-valued function. If f is *square* integrable in the sense of Lebesgue, then equation (FT) may be understood in the following *mean square* sense: there exists an essentially unique complex-valued function $\omega \to \hat{f}(\omega)$ for which

$$\lim_{T \to \infty} \int_{-\infty}^{\infty} |\hat{f}(\omega) - \int_{-T}^{T} f(t)e^{-i\omega t}\, dt\, |^2 d\omega = 0\, .$$

For a much more thorough discussion of these technical "details," which have a rich mathematical heritage, see [Dym and McKean].

A standard and quite useful convention is to regard the *unit impulse* $t \to \delta(t)$ as being Fourier transformable, with Fourier transform given by $\hat{\delta}(\omega) \equiv 1$ for every $\omega \in \mathrm{R}$. This convention makes sense in view of the discussion in §18 of the impulse along with the equation (FT) above.

Definition 19.1 fails to illuminate the way in which the Fourier transformability of some function f enables one to express f as a "linear combination of sinusoids;" suffice it to say that under a wide range of circumstances, one may write

$$f(t) = \frac{1}{2\pi} \int_{-\infty}^{\infty} \hat{f}(\omega)e^{i\omega t}\, d\omega\, , \quad t \in \mathrm{R}\, , \qquad \text{(FTI)}$$

where the so-called "inversion integral" which gives f in terms of $\hat{f}$ may be interpreted in as many ways as the integral in equation (FT). A point that should be emphasized is that *regardless* of the validity of equation (FTI), *knowledge of* $\omega \to \hat{f}(\omega)$ *determines a Fourier transformable function* f *in an essentially unique way*. Once again, the reader is referred to [Dym and McKean].

The formula (FTI) expresses f as a "linear combination of sinusoids" which is weighted by $\hat{f}$. Thus $\hat{f}(\omega)$ for ω near some ω_o reveals, at least in some sense, the "frequency content" of f at frequencies near ω_o. In the context of input-output linear systems, we may formalize the notion of frequency response in terms of Fourier transforms as follows.

19.2 Definition: Given an m-input, p-output time-invariant continuous-time input-output linear system with $(p \times m)$ impulse response $t \to H(t)$ defined as in §18, we say that the *system possesses a frequency response* if and only if H is Fourier transformable; in this case, the *frequency response* of the system is the Fourier transform $\omega \to \hat{H}(\omega)$ of H. □

In the notation of Definition 18.16(a), keeping in mind the convention that the unit impulse has a constant Fourier transform, we have the formula

$$\hat{H}(\omega) = \int_{-\infty}^{\infty} W(t, 0)e^{-i\omega t}\,dt + \sum_{k=0}^{\infty} D_k e^{-i\omega T_k}, \quad \omega \in \mathbf{R},$$

for the frequency response of an m-input, p-output time-invariant continuous-time input-output linear system with Fourier transformable impulse response $t \to H(t)$.

A bit of thought should convince the reader that if the steady-state response of a system to every sinusoidal input is well-defined (c.f. the discussion at the start of this section), then the system possesses a frequency response; it is somewhat less obvious that the converse *is not true*. Cases in which the converse fails, however, are somewhat pathological, and never come up in the finite-dimensional theory on which we shall be focusing the bulk of our attention. In order to understand how the frequency response of an input-output linear system reflects the way in which the system "acts" on the frequency components of input functions which are not necessarily sinusoidal, we need the following basic fact; for a rigorous treatment of a more general version, see [Dym and McKean].

19.3 Fact: *Suppose that* $f_1 : \mathbf{R} \to \mathbf{R}$ *and* $f_2 : \mathbf{R} \to \mathbf{R}$ *are both Fourier transformable. Suppose also that the convolution* $f : \mathbf{R} \to \mathbf{R}$ *of* f_1 *and* f_2, *which is defined by the formula*

$$f(t) = \int_{-\infty}^{\infty} f_1(\tau) f_2(t-\tau)\,d\tau, \quad t \in \mathbf{R},$$

exists and is Fourier transformable. Then the Fourier transform of f *is the product of the Fourier transforms of* f_1 *and* f_2; *i.e.,* $\hat{f}(\omega) = \hat{f}_1(\omega)\hat{f}_2(\omega)$, $\omega \in \mathbf{R}$.

Proof (sketch): The Fourier transform of f is given at $\omega \in \mathbf{R}$ by

$$\hat{f}(\omega) = \int_{-\infty}^{\infty} [\int_{-\infty}^{\infty} f_1(\tau) f_2(t-\tau) d\tau] e^{-i\omega t} dt \ ;$$

Assuming that interchange of the order of integration is permissible, we have

$$\begin{aligned}\hat{f}(\omega) &= \int_{-\infty}^{\infty} f_1(\tau) [\int_{-\infty}^{\infty} f_2(t-\tau) e^{-i\omega(t-\tau)} dt] e^{-i\omega\tau} d\tau \\ &= \hat{f}_2(\omega) \int_{-\infty}^{\infty} f_1(\tau) e^{-i\omega\tau} d\tau \\ &= \hat{f}_1(\omega) \hat{f}_2(\omega) \ .\end{aligned}$$

[We remark that the interchange of order of integration is permissible if both f_1 and f_2 are absolutely integrable (in which case, by the way, f is also absolutely integrable); the interchange is also justified in certain other cases provided that the convolution f is absolutely integrable.] □

Suppose, then, that we are given an m-input, p-output time-invariant continuous-time input-output linear system with a Fourier transformable impulse response H. Suppose that the steady-state response of the system to the input function $u \in U$ is defined; all this means is that the convolution

$$\begin{aligned} y(t) &= \int_{-\infty}^{\infty} H(t-\tau) u(\tau) d\tau \\ &= \int_{-\infty}^{\infty} H(\tau) u(t-\tau) d\tau \end{aligned}$$

is well-defined; the second equality follows from a simple change of variable in the integral. If both u and y are *Fourier transformable*, then by Fact 19.3 we have $\hat{y}(\omega) = \hat{H}(\omega)\hat{u}(\omega)$ for every $\omega \in R$; that is to say, the (p-vector) Fourier transform of y is simply the (m-vector) Fourier transform of u multiplied on the left by the ($p \times m$) frequency response $\hat{H}$.

Thus, if the "frequency content" of u is concentrated in some "frequency band" over which the magnitude of $\hat{H}$ is, say, very *small*, then the output y will be very small, as well; the system, in this case, acts as a *filter* which "filters out" signals such as u.

19.4 Example: Consider the electrical network of Example 18.15. We view the network as defining a single-input, single-output time-invariant continuous-time input-output linear system with input function space U containing all the continuous functions $u : \mathbf{R} \to \mathbf{R}$ and output functions spaces $\mathbf{Y}_{t_0}$ equal to the set of all functions $y : \mathbf{R} \to \mathbf{R}$. The input-output mappings σ_{t_0} are given by

$$\sigma_{t_0}(u)(t) = \frac{1}{RC}\int_{t_0}^{t} e^{-\frac{1}{RC}(t-\tau)} u(\tau)d\tau\,, \quad t \geqslant t_0\,.$$

The impulse response of the system is given by

$$H(t) = \frac{1}{RC} e^{-\frac{t}{RC}} 1(t)\,, \quad t \in \mathbf{R}\,;$$

As usual, $t \to 1(t)$ denotes the *unit step function* which is zero for $t < 0$ and 1 for $t \geqslant 0$. The frequency response of the system is therefore

$$\hat{H}(\omega) = \frac{RC^{-1}}{i\omega + RC^{-1}}\,, \quad \omega \in \mathbf{R}\,.$$

Observe that $\hat{H}(\omega)$ decays fairly rapidly as $\omega \to \pm\infty$; thus the system serves to *attenuate* high-frequency inputs much more radically than it does low-frequency inputs. The system is perhaps the simplest example of what an engineer might call a *low-pass filter*. □

We alluded earlier to the fact that frequency domain methods appeal to one's *empirical intuition*. Perhaps some justification for this assertion is in order, since it is probably the strongest motivating force behind the development of frequency domain techniques over the last half century or so. The essential idea is that one ought to be able to learn much about a system by performing *input-output experiments* on it. By "exciting" the system with a sufficiently wide class of readily available inputs, a good deal of information about the overall input-output behavior should be obtainable.

Sinusoids are arguably the simplest of "interesting" input functions which are readily available in a variety of contexts. Good approximations to the frequency response of an input-output linear system may be found by measuring (approximately, of course) its steady-state response to a large number of sinusoids having different frequencies. Once the frequency response has been approximated, an estimate of the system's impulse response may be made. Once the impulse response is "known," so is everything else about the system.

The reader might wonder at this point what sorts of systems do *not* possess frequency responses. Evidently, if the impulse response $t \to H(t)$ of an m-input, p-output time-invariant continuous-time input-output linear system decays rapidly enough as $t \to \infty$, then the system *will* possess a frequency response, since H will be Fourier transformable. Intuitively, then, a system which doesn't have a frequency response will be one whose impulse response $t \to H(t)$ *grows*, or, at the very least, *fails to decay sufficiently fast* as $t \to \infty$.

Given such a system, suppose the steady state response y of the system to one of its input functions u is defined; then y is given at each $t \in \mathbf{R}$ by

$$y(t) = \int_{-\infty}^{\infty} H(t-\tau)u(\tau)d\tau .$$

If $H(t-\tau)$ *grows* as $\tau \to -\infty$, then the system's output at each time $t \in \mathbf{R}$ "depends more strongly" on the distant "past" of the input function u than it does on the values of u at times "close" to t. Such a situation is somewhat unsettling, at least at face value; nonetheless, many "real-world processes" may be modeled by such systems. Fundamental examples of such processes are *active* electrical networks which serve as *amplifiers* in many applications. In Example 19.4, if a *negative* value is assumed for the resistor R, then such an active network is obtained. In that case, the impulse response $H(t)$ *grows exponentially* as $t \to \infty$; the steady-state response of the system to a great many input signals is, however, defined.

19.5 Exercise: Show that the steady-state response of the system in Example 19.4 to input function u is well-defined whenever u is given by $u(t) = e^{\alpha t}$, $t \in \mathbf{R}$, where α is any real number bigger than $-(RC)^{-1}$, *regardless* of the signs of R and C. □

Exercise 19.5 illustrates the important fact that the steady-state response of an input-output linear system to a large class of inputs may be well-defined *even if the system does not possess a frequency response*. An input function of the form $t \to e^{s_0 t}$, where $s_0 = \sigma_0 + i\omega_0$ is a complex number, may be regarded as a *generalized sinusoidal input*; the steady-state response of the system in Example 19.5 to such an input function is well-defined provided that σ_0 is large enough. If

$RC < 0$, then σ_o must, in particular, be *positive* in order that $t \to e^{s_o t}$ be an "admissible steady-state input function."

Suppose, more generally, that $t \to H(t)$ is the impulse response of an m-input, p-output time-invariant continuous-time input-output linear system. Suppose also that there exists some real σ_a such that $e^{-\sigma_a t} H(t) \to 0$ as $t \to \infty$. Given $w \in \mathbf{R}^m$, consider the input signal $t \to we^{s_o t}$ where $\sigma_o = \mathrm{Re}\{s_o\} > \sigma_a$. The constraint on σ_o guarantees that the steady-state response $t \to y(t)$ of the system to this input function is well-defined; a quick calculation shows that y is given by

$$y(t) = [\int_{-\infty}^{\infty} H(\tau)e^{-s_o \tau} d\tau] we^{s_o t}, \quad t \in \mathbf{R}.$$

The quantity in brackets is a *constant* which depends only on s_o and on the system's impulse response; in view of the foregoing discussion, it resembles the "frequency response of the system at the *complex frequency* s_o."

We are prompted to make the following definition.

19.6 Definition: Let $t \to H(t)$, $t \in \mathbf{R}$, be the impulse response of an m-input, p-output time-invariant continuous-time input-output linear system. If there exists some $\sigma \in \mathbf{R}$ such that $e^{-\sigma t} H(t) \to 0$ as $t \to \infty$, we say that the system *possesses a transfer function*. In this case, let σ_a be given by

$$\sigma_a = \inf_{\sigma \in \mathbf{R}} \{\sigma : \lim_{t \to \infty} e^{-\sigma t} H(t) = 0\};$$

The *transfer function* of the system is then defined to be the complex-valued function $s \to G(s)$ having domain $\{s \in \mathbf{C} : \mathrm{Re}\{s\} > \sigma_a\}$ and given on that domain by

$$G(s) = \int_{-\infty}^{\infty} H(t)e^{-st} dt . \qquad \square$$

Note that the integral defining $G(s)$ in the last equation is well-defined whenever $\mathrm{Re}\{s\} > \sigma_a$ because $H(t) = 0$ for $t < 0$. The transfer function of a time-invariant continuous-time input-output linear system may be viewed as a generalized version of the system's frequency response; indeed, if $\sigma_a < 0$ in Definition 19.6, then the system possesses a frequency response $\hat{H}$, and $\hat{H}(\omega) = G(i\omega)$ for every $\omega \in \mathbf{R}$. Observe that

$\sigma_a = -\infty$ is, indeed, a possibility; in particular, $\sigma_a = -\infty$ whenever there exists $t_1 \in \mathbf{R}$ such that $H(t) = 0$ for every $t > t_1$. A system for which such a t_1 exists is often, for obvious reasons, said to have *finite memory*. The transfer function generalizes the frequency response in precisely the same way that the *Laplace transform*, which we define next, generalizes the Fourier transform.

19.7 Definition: Let $f : \mathbf{R} \to \mathbf{R}$ be a function; suppose that there exists some $\sigma \in \mathbf{R}$ for which $e^{-\sigma t} f(t) \to 0$ as $t \to \pm\infty$. For such an f, define $\sigma_a(f)$ and $\sigma_b(f)$, respectively, as the infimum and supremum of the set

$$\{\sigma \in \mathbf{R} : \lim_{|t| \to \infty} e^{-\sigma t} f(t) = 0\} .$$

If $\sigma_a(f) < \sigma_b(f)$, then f is said to be *Laplace transformable*. The *Laplace transform* of f is then defined to be the complex valued function $s \to \tilde{f}(s)$ having domain

$$\{s \in \mathbf{C} : \sigma_a(f) < \mathrm{Re}\{s\} < \sigma_b(f)\}$$

and given on that domain by

$$\tilde{f}(s) = \int_{-\infty}^{\infty} f(t) e^{-st} dt . \qquad \square$$

Implicit in Definition 19.7 is the assumption that f is "reasonable" enough to be integrated in some sense, at least over bounded intervals. The domain

$$\{s \in \mathbf{C} : \sigma_a(f) < \mathrm{Re}\{s\} < \sigma_b(f)\}$$

is called the *region of convergence* for the Laplace transform of f; we shall employ the notation

$$ROC(f) = \{s \in \mathbf{C} : \sigma_a(f) < \mathrm{Re}\{s\} < \sigma_b(f)\} .$$

The choice of the term *region of convergence* is based on the fact that the integral defining $\tilde{f}(s)$ "converges" precisely for those $s \in \mathbf{C}$ lying in $ROC(f)$. The idea is that $e^{-st} f(t) \to 0$ exponentially as $t \to \pm\infty$ if $s \in ROC(f)$. Note that $ROC(f)$ is an infinite vertical rectangular *strip* in the complex plane. $ROC(f)$ is an entire *half-plane* if one of $\sigma_a(f)$ or $\sigma_b(f)$ is infinite, and is all of $\mathbf{C}$ if both are infinite.

In fact, if $f : \mathbf{R} \to \mathbf{R}$ is Laplace transformable, then it is clear that $\sigma_b(f) = \infty$ whenever $f(t) = 0$ for all t less than some $t_o \in \mathbf{R}$; a similar assertion holds for $\sigma_b(f)$. The reader should check, however, that f need not vanish outside a bounded interval in order that $ROC(f)$ be *the entire complex plane*. Consider, for example, $f:\mathbf{R} \to \mathbf{R}$ defined by $f(t) = e^{-t^2}, t \in \mathbf{R}$.

If f in Definition 19.7 is a vector- or matrix-valued function, then its Laplace transform, if it exists, is taken in a component-wise fashion. Given such an $f:\mathbf{R} \to \mathbf{R}^n$, observe that any $\sigma \in \mathbf{R}$ for which $e^{-\sigma t} f(t) \to 0$ as $t \to \pm\infty$ lies in *all* of the $ROC(f_i)$, $1 \leqslant i \leqslant n$. Hence, f is Laplace transformable precisely when there is a *nonempty intersection* between the regions of convergence of the Laplace transforms of all of f's component functions.

It is useful, just as in the case of the Fourier transform, to adopt the convention that the unit impulse $t \to \delta(t)$ is Laplace transformable, and has Laplace transform $\tilde{\delta}(s) \equiv 1$, $s \in \mathbf{C}$. The reader may easily verify that an m-input, p-output time-invariant continuous-time input-output linear system possesses a transfer function if and only if its impulse response is Laplace transformable; in this case, the transfer function is the Laplace transform of the impulse response, and is given (in the notation of Definition 18.16) by

$$\begin{aligned} G(s) &= \int_{-\infty}^{\infty} W(t,0)e^{-st}\,dt + \sum_{k=0}^{\infty} D_k e^{-sT_k} \\ &= \int_0^{\infty} H(t)e^{-st}\,dt\,, \end{aligned}$$

with region of convergence given by

$$ROC(H) = \{s \in \mathbf{C} : \mathrm{Re}\{s\} > \sigma_a(H)\}\,.$$

Implicit in the specification of $ROC(H)$ is that $\sigma_b(H) = \infty$, in the notation of Definition 19.7. This is true because $H(t) = 0$ for $t < 0$.

There is a vast mathematical literature on the Laplace transform and its relationship with the Fourier transform; the reader is referred, once again, to [Dym and McKean] along with [Rudin], and also to the classic monograph [Paley and Wiener]. We summarize here without proof some of the many interesting analytical features of Laplace transforms.

19.8 Facts: *Let $f : \mathbf{R} \to \mathbf{R}$ be Laplace transformable; let $\sigma_a(f)$, $\sigma_b(f)$, and ROC(f) be defined as above. Then*

(a) $s \to \tilde{f}(s)$, $s \in$ ROC(f), is an analytic function on its domain of definition.

(b) If $\sigma_a(f) < 0 < \sigma_b(f)$, then f is Fourier transformable; its Fourier transform $\hat{f}$ is given at each $\omega \in \mathbf{R}$ by $\hat{f}(\omega) = \tilde{f}(i\omega)$.

(c) Let $\tilde{\sigma} \in \mathbf{R}$ be such that $\sigma_a(f) < \tilde{\sigma} < \sigma_b(f)$; let $\tilde{\Omega}$ be the upwardly directed vertical contour $\mathrm{Re}\{s\} = \tilde{\sigma}$ in the complex plane. Then f is given at $t \in \mathbf{R}$ by the contour integral

$$f(t) = \frac{1}{2\pi i} \int_{\Omega} \tilde{f}(s) e^{st}\, ds \ . \qquad \square$$

The integral in Fact 19.8(c) resembles (and, under the appropriate assumptions, *reduces* to) the "inversion integral" which gives a Fourier transformable f in terms of its Fourier transform $\hat{f}$; compare equation (FTI) and the discussion following Definition 19.1. In addition, as one might expect, there is an operational rule for Laplace transforms which parallels the property of Fourier transforms which is described in Fact 19.3.

19.9 Fact: *Let $f_1 : \mathbf{R} \to \mathbf{R}$ and $f_2 : \mathbf{R} \to \mathbf{R}$ be Laplace transformable functions. Suppose that the regions of convergence ROC(f_1) and ROC(f_2) have a nonempty intersection. Then the convolution f of f_1 and f_2, given by*

$$f(t) = \int_{-\infty}^{\infty} f_1(\tau) f_2(t-\tau)\, d\tau , \qquad t \in \mathbf{R} ,$$

is well-defined and Laplace transformable. Moreover, ROC(f) contains the intersection ROC$(f_1) \cap$ ROC(f_2), and $\tilde{f}(s)$ is given for each s in the intersection by the product $\tilde{f}_1(s)\tilde{f}_2(s)$.

Proof: Define σ_a and σ_b by

$$\sigma_a = \sup\{\sigma_a(f_1), \sigma_a(f_2)\}$$
$$\sigma_b = \inf\{\sigma_b(f_1), \sigma_b(f_2)\} ;$$

the assumption that ROC(f_1) overlaps ROC(f_2) guarantees that $\sigma_a < \sigma_b$. For each $\sigma \in (\sigma_a, \sigma_b)$, the quantities $f_1(\tau)e^{-\sigma\tau}$ and $f_2(t-\tau)e^{-\sigma(t-\tau)}$ decay, for each $t \in \mathbf{R}$, exponentially as $\tau \to \pm\infty$. In addition, for each $\tau \in \mathbf{R}$, the function $t \to e^{-\sigma t} f_1(\tau) f_2(t-\tau)$ decays exponentially as $t \to \pm\infty$, and hence may be integrated. We obtain, after a simple change of variable,

$$\int_{-\infty}^{\infty} e^{-\sigma t} f_1(\tau) f_2(t-\tau) dt = e^{-\sigma\tau} f_1(\tau) \tilde{f}_2(\sigma).$$

This result may in turn be integrated with respect to τ, since $\sigma \in ROC(f_2)$:

$$\int_{-\infty}^{\infty} [\int_{-\infty}^{\infty} e^{-\sigma t} f_1(\tau) f_2(t-\tau) dt] d\tau = \tilde{f}_1(\sigma) \tilde{f}_2(\sigma).$$

The same argument works if σ is replaced by $s = \sigma + i\omega$, for any real ω. The result is

$$\tilde{f}_1(s) \tilde{f}_2(s) = \int_{-\infty}^{\infty} [\int_{-\infty}^{\infty} e^{-st} f_1(\tau) f_2(t-\tau) dt] d\tau.$$

Observe that the inner integrand *also* decays exponentially as $\tau \to \pm\infty$; hence it is integrable with respect to τ, and the order of integration may be interchanged. This yields

$$\tilde{f}_1(s) \tilde{f}_2(s) = \int_{-\infty}^{\infty} e^{-st} [\int_{-\infty}^{\infty} f_1(\tau) f_2(t-\tau) d\tau] dt ,$$

which shows simultaneously that f is well-defined and that $e^{-st} f(t) \to 0$ as $t \to \pm\infty$ for all s in $ROC(f_1) \cap ROC(f_2)$. □

Fact 19.9 provides us with a valuable operational tool; this tool is the by-product of an "extension" of the simple relation (c.f. the discussion following Fact 19.3) between the Fourier transforms of the input and output of a linear system possessing a frequency response.

Suppose that $u \in U$ is an input function for a given m-input, p-output time-invariant continuous-time input-output linear system. Suppose also that u is Laplace transformable, and that the region of convergence $ROC(u)$ for u's Laplace transform overlaps the region of convergence $ROC(H)$ for the system's transfer function $s \to G(s)$, $s \in ROC(H)$. Fact 19.9 implies that the convolution

$$y(t) = \int_{-\infty}^{\infty} H(t-\tau) u(\tau) d\tau$$

is well-defined and Laplace transformable, with Laplace transform given at least on $ROC(H) \cap ROC(u) \subset ROC(y)$ by

$$\tilde{y}(s) = G(s)\tilde{u}(s).$$

The convolution y, however, is simply the *steady-state response* of the system to the input u. We have therefore arrived at a simple test, stated in terms of regions of convergence, which determines whether the steady-state response of a given system to a certain input does indeed exist.

20. *Frequency Responses and Transfer Functions of Time-Invariant Discrete-Time Systems*

Frequency domain techniques for time-invariant discrete-time input-output linear systems are of fairly recent vintage when compared with the corresponding continuous-time methods. Their development has taken place largely over the last fifty years, and has accelerated considerably with the advent of the computer age. Nonetheless, as is often the practice in applied mathematics, it is possible to present an overview of discrete-time frequency response techniques which is in essence *completely parallel* to the development in §19. Such an exposition would, however, obscure the history of the subject unnecessarily. As a result, we shall try consistently to accompany the definitions and theorems in this section with comments which should alert the reader to certain fundamental similarities and differences between continuous- and discrete-time frequency domain analysis; we shall also attempt to describe, at least briefly, how the two theories *complement* each other in the "real world" of systems engineering.

It might be said that there are two distinct types of discrete-time functions, namely, those which are *intrinsically* discrete-time functions, and those which arise from the *sampling* of continuous-time functions. Prototypical examples of the first kind of discrete-time functions are bank balances, stock market closing averages, population statistics, and other "measurable quantities" which do *not* evolve in continuous time. It is difficult to establish for such discrete-time functions a "frequency domain formalism" which is as rich or as intuitively appealing as that for continuous-time functions. On the other hand, discrete-time functions which are merely sampled versions of continuous-time functions have "frequency domain properties" which reflect in many ways the corresponding properties of the continuous-time functions from which they were obtained. Historically, discrete-time frequency domain analysis has centered largely on the study of this second kind of discrete-time function; nonetheless, the mathematical formalism makes sense in either context.

In order to understand the importance of the foregoing comments, consider the words *frequency content* and *period*; how do they apply to continuous- and discrete-time functions? If $\omega_0 \in \mathbf{R}$ is very large, then the function $t \to e^{i\omega_0 t}$, $t \in \mathbf{R}$, is evidently a "high frequency" periodic function; its period is $2\pi\omega_0^{-1}$. On the other hand, the discrete-time complex exponential function $k \to e^{i\Omega k}$, $k \in \mathrm{Z}$ might *not* be "periodic in k;" it *will* be periodic if Ω_0 is a rational multiple of 2π, but in that case $e^{i\Omega_0 k} = e^{i(\Omega_0 + N2\pi)k}$ for every $k \in \mathrm{Z}$ and every integer N.

Thus a high Ω_0 will *not* necessarily guarantee a "high frequency" discrete-time complex exponential. Observe, however, that if a rational number ρ is close to being *irrational* (that is, if the "lowest denominator" for ρ is very high), then the "period" of the periodic discrete-time complex function $k \to e^{i2\pi\rho k}$, $k \in \mathrm{Z}$ is also very high, and its "frequency" is therefore low. Moreover, one could argue that the *shortest* "period" possible for a discrete-time function is 1; is it to be concluded that a discrete-time function has "frequency" at most 2π? As might be expected, the answer is, "yes and no;" before proceeding further, we make a definition which parallels Definition 19.1 for continuous-time functions.

20.1 Definition: A function $f : \mathrm{Z} \to \mathbf{R}$ is said to be *discrete-time Fourier transformable* if and only if for every $\Omega \in \mathbf{R}$ the following sum exists in some sense of the meaning of the word "sum:"

$$\hat{f}(\Omega) = \sum_{k=-\infty}^{\infty} f(k) e^{-i\Omega k} \; ; \qquad \text{(DTFT)}$$

in this case, the complex-valued function $\Omega \to \hat{f}(\Omega)$, $\Omega \in \mathbf{R}$, is called the *discrete-time Fourier transform* of the function f. □

As usual, a vector- or matrix-valued discrete-time function is discrete-time Fourier transformable if and only if each of its component functions has that property, in which case the discrete-time Fourier transform is taken componentwise. The "senses of the meaning of *sum*" bear a striking resemblance to the "senses of the meaning of integral" which accompany Definition 19.1. The easiest case occurs when $k \to f(k)$ *vanishes* except for finitely many values of k; then $\Omega \to \hat{f}(\Omega), \Omega \in \mathbf{R}$, is evidently well-defined and is a *trigonometric polynomial* in Ω; its period is 2π if $f(1)$ or $f(-1)$ is nonzero, and is $2\pi k_o^{-1}$ if k_o is the greatest common divisor of the set of all integers k for which $f(k) \neq 0$.

If f is *absolutely summable* in the sense that

$$\sum_{k=-\infty}^{\infty} |f(k)|$$

exists, then $\hat{f}$ is a bounded, continuous, periodic function of Ω whose value at each Ω is given by the sum in (DTFT); its period is also $2\pi k_o^{-1}$, with k_o defined as above. If f is *square summable*, i.e., if

$$\sum_{k=-\infty}^{\infty} |f(k)|^2$$

exists, then $\hat{f}$ exists in the sense of the *mean square;* that is, there exists a periodic $\hat{f}:\mathbf{R} \to \mathbf{R}$, whose period is an integer divisor of 2π, such that

$$\lim_{K \to \infty} \int_{-\pi}^{\pi} |\hat{f}(\Omega) - \sum_{k=-K}^{K} f(k)e^{-i\Omega k}|^2 d\Omega = 0 .$$

Thus, the discrete-time Fourier transform $\Omega \to \hat{x}(\Omega)$, $\Omega \in \mathbf{R}$, of a discrete-time Fourier transformable function $k \to x(k)$, $k \in \mathbf{Z}$, is *periodic* in Ω with period at most 1. As in the case of the continuous-time Fourier transform (c.f. Equation (FTI) of §19), we have an "inversion formula" for the discrete-time Fourier transform:

$$x(k) = \frac{1}{2\pi}\int_{-\pi}^{\pi} \hat{x}(\Omega)e^{i\Omega k} d\Omega \, ; \qquad \text{(DTFTI)}$$

this relation holds under a wide variety of circumstances if the integral is interpreted correctly. If $\hat{x}$ is piecewise continuous and has only finitely many discontinuities on $[-\pi, \pi]$, then equation (DTFT) holds pointwise at every point of continuity of $\hat{x}$; the reader is referred once again to [Dym and McKean]. It is difficult to see how (DTFTI) provides a connection between $\hat{x}$ and the *frequency content* of the function x. Are we to conclude that all the information about the frequency content of the function x is contained in the "frequency band" $[-\pi, \pi] \subset \mathbf{R}$?

We may obtain an illuminating *partial* answer to the last question by considering the situation where $k \to x(k)$ arises from the *sampling* of a continuous-time function $t \to f(t), t \in \mathbf{R}$; accordingly, suppose that

$$x(k) = f(kT), \quad k \in \mathbf{Z},$$

for some $T > 0$. Suppose that f is a band limited function; that is, the Fourier transform $\omega \to \hat{f}(\omega)$ of f *vanishes* outside of some bounded interval $[-\omega_m, \omega_m] \subset \mathbf{R}$. It will turn out that if T is small enough, then $\hat{x}$ determines $\hat{f}$ *completely*. More precisely, we have the following classic result, which is known as the *Sampling Theorem*.

20.2 Theorem: (The Sampling Theorem) *Suppose that $f: \mathbf{R} \to \mathbf{R}$ is Fourier transformable, and that $\hat{f}(\omega) = 0$ for every ω with $|\omega| > \omega_m$, where $\omega_m > 0$. Suppose also that $\omega \to \hat{f}(\omega)$ is at least piecewise continuous on $[-\omega_m, \omega_m]$. Let T be such that $0 < T < \frac{\pi}{\omega_m}$. Let $x: \mathbf{Z} \to \mathbf{R}$ be defined by*

$$x(k) = f(kT), \quad k \in \mathbf{Z}.$$

Then x is discrete-time Fourier transformable; moreover,

$$\hat{f}(\omega) = \begin{cases} T\hat{x}(\omega T) & -\omega_m \leqslant \omega \leqslant \omega_m \\ 0 & |\omega| > \omega_m ; \end{cases}$$

or, equivalently,

$$\hat{x}(\Omega) = \frac{1}{T}\hat{f}\left(\frac{\Omega}{T}\right), \quad -\pi < \Omega < \pi.$$

Proof: It may be shown (see, for example, [Dym and McKean]) that the bandlimitedness assumption implies that f and $\hat{f}$ satisfy equation (FTI) of §19; hence,

$$f(t) = \int_{-\omega_m}^{\omega_m} \hat{f}(\omega) e^{i\omega t} d\omega, \quad t \in \mathbf{R}.$$

Thus $x(k) = f(kT)$ is given for $k \in \mathbf{Z}$ by

$$x(k) = \int_{-\omega_m}^{\omega_m} \hat{f}(\omega) e^{i\omega kT} d\omega.$$

Set $\Omega = \omega T$, change variables in the last integral, and obtain, for every $k \in Z$,

$$x(k) = \int_{-\omega_m T}^{\omega_m T} \frac{1}{T} \hat{f}\left(\frac{\Omega}{T}\right) e^{i\Omega k}\, d\Omega$$

$$= \int_{-\pi}^{\pi} \frac{1}{T} \hat{f}\left(\frac{\Omega}{T}\right) e^{i\Omega k}\, d\Omega ,$$

where the last equality follows from the facts that $\omega_m T < \pi$ and that $\hat{f}(\omega) = 0$ for $|\omega| > \omega_m$. The last equation is the same as Equation (DTFTI) above with $\hat{x}(\Omega)$ replaced by

$$\frac{1}{T} \hat{f}\left(\frac{\Omega}{T}\right).$$

The result follows from the piecewise continuity of $\hat{f}$, along with the comments accompanying Equation (DTFTI). □

Thus the discrete-time Fourier transform of a rapidly enough sampled version x of a bandlimited continuous-time function f is simply the periodic extension of a *re-scaled* version of the Fourier transform of f. The "high frequency" components of f are reflected in the behavior near $\Omega = \pm\pi$ of $\hat{x}$. The minimum (angular) frequency of sampling which satisfies the hypothesis of Theorem 20.2 is called the *Nyquist rate* for the function f. Thus, if ω_m is the infimum of all values of ω_1 such that $\hat{f}(\omega) = 0$ for $|\omega| > \omega_1$, then then the Nyquist rate for f is $2\omega_m$. The assumption that f is *bandlimited* is what makes Theorem 20.2 "work;" it may appear to be a strong assumption, but is essentially true of any f which can be measured "in practice."

Having established at least a *weak* connection between the discrete-time Fourier transform and *frequency content*, we now consider the discrete-time versions of *frequency response* and *transfer function*. The following definition parallels Definition 19.2 almost exactly, and has essentially the same motivation behind it.

20.3 Definition: Given an m-input, p-output time-invariant discrete-time input-output linear system with $(p \times m)$ impulse response $k \to H(k)$ as defined in §18, we say that *the system possesses a frequency response* if and only if H is discrete-time Fourier transformable; in this case, the *frequency response* of the system is the discrete-time Fourier transform

$\Omega \to \hat{H}(\Omega)$ of the impulse response H. □

In the notation of Definition 18.16 and the discussion which follows, if a time-invariant discrete-time input-output linear system possesses a frequency response, it is given by

$$\hat{H}(\Omega) = \sum_{k=-\infty}^{\infty} H(k)e^{-i\Omega k}$$
$$= \sum_{k=0}^{\infty} W(k,0)e^{-i\Omega k} .$$

It is once again convenient to pretend that the systems under consideration admit *complex* input functions u; the output functions $\sigma_{k_o}(u)$ corresponding with such complex input functions are defined for each $k_o \in \mathbf{Z}$ via

$$\sigma_{k_o}(u) = \sigma_{k_o}(\mathrm{Re}\{u\}) + i\,\sigma_{k_o}(\mathrm{Im}\{u\}) .$$

Under this convention, it is easily seen (c.f. the discussion following Definition 18.16) that if $\Omega_o \in \mathbf{R}$ and $w \in \mathbf{R}^m$ are such that the *steady-state response* $k \to y(k)$ of the system to the input $k \to u(k) = we^{i\Omega_o k}$, $k \in \mathbf{Z}$, is defined, then

$$y(k) = \hat{H}(\Omega_o)we^{i\Omega_o k} , \quad k \in \mathbf{Z} .$$

Thus, just as in the case of continuous-time systems, if the steady-state response of a discrete-time system to a given *sinusoidal* input is defined, then that response is a *sinusoid of the same frequency*.

Again in analogy with continuous-time systems, we may prove a result which is similar to Fact 19.3 and which enables one to interpret the frequency response of a discrete-time system as an indicator of how the system *acts* on the frequency components of Fourier transformable inputs.

20.4 Fact: *Suppose that* $x_1 : \mathbf{Z} \to \mathbf{R}$ *and* $x_2 : \mathbf{Z} \to \mathbf{R}$ *are both discrete-time Fourier transformable. Suppose also that the discrete-time convolution* $x : \mathbf{Z} \to \mathbf{R}$ *of* x_1 *and* x_2, *which is defined by the formula*

$$x(k) = \sum_{l=-\infty}^{\infty} x_1(l)x_2(k-l) , \quad k \in \mathbf{Z} ,$$

exists and is discrete-time Fourier transformable. Then the discrete-time Fourier transform of x *is the product of the discrete-time Fourier transforms of* x_1 *and* x_2; *i.e.*,

$$\hat{x}(\Omega) = \hat{x}_1(\Omega)\hat{x}_2(\Omega), \quad \Omega \in \mathbf{R}.$$

Proof (sketch): The discrete-time Fourier transform of x is given at $\Omega \in \mathbf{R}$ by

$$\hat{x}(\Omega) = \sum_{k=-\infty}^{\infty} \left[\sum_{l=-\infty}^{\infty} x_1(l)x_2(k-l) \right] e^{-i\Omega k}.$$

Now perform an interchange in the order of summation; we obtain

$$\begin{aligned} \hat{x}(\Omega) &= \sum_{l=-\infty}^{\infty} x_1(l)e^{-i\Omega l} \left[\sum_{k=-\infty}^{\infty} x_2(k-l)e^{-i\Omega(k-l)} \right] \\ &= \hat{x}_2(\Omega) \sum_{l=-\infty}^{\infty} x_1(l)e^{-i\Omega l} \\ &= \hat{x}_1(\Omega)\hat{x}_2(\Omega). \end{aligned}$$

Of course, we haven't justified the interchange in order of summation; it is permissible if both x_1 and x_2 are absolutely summable, in which case f is absolutely summable, as well; the interchange is also justified under a wide range of other circumstances. See [Dym and McKean] for a thorough discussion.□

Suppose, then, that a given m-input, p-output time-invariant discrete-time input-output linear system possesses a frequency response; that is, its impulse response $k \to H(k)$ is discrete-time Fourier transformable. Let $\Omega \to \hat{H}(\Omega)$ be the system's frequency response; let $u \in \mathbf{U}$ be an input function to which the system has a well-defined steady-state response $y : \mathbf{Z} \to \mathbf{R}^p$. Then for every $k \in \mathbf{Z}$, we have

$$\begin{aligned} y(k) &= \sum_{l=-\infty}^{\infty} H(k-l)u(l) \\ &= \sum_{l=-\infty}^{k} H(k-l)u(l); \end{aligned}$$

the last equality holds because $H(k-l) = 0$ for $k < l$. If y is discrete-time Fourier transformable, then, by Fact 20.4, $\hat{y}(\Omega) = \hat{H}(\Omega)\hat{u}(\Omega)$ for each $\Omega \in \mathbf{R}$.

As in the case of continuous-time systems, those discrete-time input-output systems which do *not* possess frequency responses are those whose impulse response functions $k \to H(k)$

either grow or do not decay sufficiently fast as $k \to \infty$. Some very innocuous real-world processes may be modeled effectively using such systems; consider the following example.

20.5 Example: A certain bank pays an annual twelve percent interest, which is compounded and credited monthly, on its savings accounts. One of its depositors, whose name is Terry, is paid $u(k)$ dollars during the kth month by his employer, which is Cornell University. Assume that the zeroth month was March, 1939, the month of Terry's birth. Let $y(k)$ denote the balance in Terry's savings account at the end of the kth month. Assuming that he deposits ninety-five percent of his pay immediately upon receiving it each month, we have

$$y(k) = 1.01y(k-1) + .95u(k), \quad k > 0.$$

We have assumed that $u(k) = 0$ for $k < 312$, and $y(0) = 0$. Alternatively, assuming that $y(0) = 0$, for each $k > 0$

$$y(k) = \sum_{l=0}^{k-1} .95(1.01)^{k-l-1}u(l+1).$$

We claim that the above real-world process may be modeled as a single-input, single-output time-invariant discrete-time input-output linear system with input function space $\mathbf{U}$ consisting of *all* $u : \mathbf{Z} \to \mathbf{R}$ and output function spaces $\mathbf{Y}_{k_o}$ consisting of *all* $y : \{k_o, k_o+1, k_o+2, \ldots\} \to \mathbf{R}$. The impulse response of the proposed system is given by

$$H(k) = \begin{cases} .95(1.01)^k & k \geqslant 0 \\ 0 & k < 0. \end{cases}$$

We leave it to the reader to check that the balance $y(k)$ in Terry's account at the end of the kth month (for $k > 0$) is, in the notation of §18, simply $\sigma_0(u)(k)$, where $u(k)$ is defined to be zero if $k \leqslant 0$. Note that $H(k)$ *grows* as $k \to \infty$. □

As in the case of continuous-time systems, we have a notion of *transfer function* which, at least in some sense, serves as a "generalized frequency response" for time-invariant discrete-time input-output linear systems which do not necessarily possess frequency responses. Moreover, there is a discrete-time transform, the *z-transform*, which generalizes the discrete-time Fourier transform in the same way that the Laplace transform generalizes the (continuous-time) Fourier transform.

Referring to the discussion which precedes Definition 19.6, suppose that we are given a time-invariant discrete-time system, and that its impulse response is the $(p \times m)$ matrix function $k \to H(k)$. Suppose also that there is some positive R_a such that $R_a^{-k} H(k) \to 0$ as $k \to \infty$. Given $w \in \mathbf{R}^m$ and $\Omega_o \in \mathbf{R}$, consider the input function $k \to w z_o^{\,k}$, where $z_o = R_o e^{i\Omega_o}$ and $R_o > R_a$. The constraint on R_o guarantees that the steady-state response $y : \mathbf{Z} \to \mathbf{R}^p$ of the system to the input u is well-defined; in fact, as may be computed easily by means of the formulas in the discussion following Definition 18.16, y is given by

$$y(k) = [\sum_{l=0}^{\infty} H(l) z_o^{-l}] w z_o^k , \qquad k \in \mathbf{Z} .$$

The sum converges since its terms decay *exponentially* as $l \to \infty$; its value is a *constant* which depends only on R_o, Ω_o, and the system's impulse response; in view of Definition 20.3, the sum resembles the "frequency response" of the system at the "complex frequency" z_o.

The foregoing discussion motivates the following definitions, which parallel Definitions 19.6 and 19.7.

20.6 Definition: Let $k \to H(k)$, $k \in \mathbf{Z}$, be the impulse response of an m-input, p-output time-invariant discrete-time input-output linear system. If there exists some $R > 0$ such that $R^{-k} H(k) \to 0$ as $k \to \infty$, we say that the system *possesses a transfer function*. In this case, define R_a by

$$R_a = \inf_{R > 0} \{ R : \lim_{k \to \infty} R^{-k} H(k) = 0 \} .$$

The *transfer function* of the system is then defined to be the complex-valued function $z \to G(z)$ having domain $\{ z \in \mathbf{C} : |z| > R_a \}$ and given on that domain by

$$G(z) = \sum_{k=0}^{\infty} H(k) z^{-k} . \qquad \square$$

20.7 Definition: Let $f : \mathbf{Z} \to \mathbf{R}$ be a discrete-time function; suppose that there exists some $R > 0$ such that $R^{-k} f(k) \to 0$ as $k \to \pm\infty$. For such an f, define $R_a(f)$ and $R_b(f)$, respectively, as the infimum and supremum of the set

$$\{R > 0: \lim_{|k| \to \infty} R^{-k} f(k) = 0\} .$$

If $R_a(f) < R_b(f)$, then f is said to be *z-transformable.* The *z-transform* of f is then defined to be the complex-valued function $z \to \tilde{f}(z)$ having domain

$$\{z \in \mathbf{C}: R_a(f) < |z| < R_b(f)\} .$$

The above domain is called the *region of convergence* for the z-transform $\tilde{f}$ of f, and is denoted by $ROC(f)$. For $z \in ROC(f)$, $\tilde{f}$ is given by

$$\tilde{f}(z) = \sum_{k=-\infty}^{\infty} f(k) z^{-k} . \qquad \square$$

Observe that the region of convergence $ROC(f)$ for a z-transformable f is a possibly unbounded *open annulus* in the complex plane; if it is bounded, its boundary consists of two circles centered at zero. Note that by our definition, $ROC(f)$ never includes $0 \in \mathbf{C}$; thus the largest possible region of convergence for the z-transform of a discrete-time function is $\{z \in \mathbf{C}: z \neq 0\}$.

If f in Definition 20.7 is a vector- or matrix-valued discrete-time function, then f is said to be z-transformable if and only if all of its component functions are z-transformable *and* their regions of convergence have a nonempty intersection; in this case, the z-transform of f is taken componentwise, and has region of convergence given by that overlap. For such an f, $R_a(f)$ and $R_b(f)$ are easily seen to be, respectively, the maximum and minimum of the corresponding quantities for the component functions of f.

A little thought should convince the reader that a given m-input, p-output time-invariant discrete-time input-output linear system possesses a transfer function if and only if its impulse response $k \to H(k)$ is z-transformable. In this case, the system's transfer function is the z-transform of its impulse response; following the notation of Definition 18.15, we have the following two formulas for the transfer function; the formulas are equivalent because $H(k)$ is equal to zero for $k < 0$ and is equal to $W(k, 0)$ for $k \geqslant 0$.

$$G(z) = \sum_{k=-\infty}^{\infty} H(k) z^{-k}$$

$$= \sum_{k=0}^{\infty} W(k, 0) z^{-k}$$

The domain of definition of $z \to G(z)$, namely $ROC(H)$, is given by $|z| > R_a(H)$, since $H(k) \equiv 0$ for $k < 0$ implies that $R_b(H) = \infty$.

There is a complex analytic theory for the z-transform which roughly parallels the corresponding theory for the Laplace transform; the theory of the Laplace transform is, in some sense, *more difficult* because of the fact that the contour integrals which appear in various formulas (c.f. the inversion formula in Fact 19.8(c)) are always taken over unbounded contours. The reader is referred to [Dym and McKean] or [Rudin] for proofs of the following analytical properties of the z-transform. Observe that these results bear a striking resemblance to Facts 19.8; in a manner of speaking, Facts 20.8 may be obtained from Facts 19.8 by replacing σ_a and σ_b by R_a and R_b, s with z, real parts with magnitudes, and infinite rectangular vertical strips in the complex plane with circular annuli in the complex plane.

20.8 Facts: *Let* $f : \mathbf{Z} \to \mathbf{R}$ *be* z*-transformable; let* $R_a(f)$, $R_b(f)$, $ROC(f)$, *and* $\tilde{f} : ROC(f) \to \mathbf{C}$ *be defined as above. Then*

(*a*) $z \to \tilde{f}(z)$ *is an analytic function on* $ROC(f)$.

(*b*) *If* $R_a(f) < 1 < R_b(f)$, *then* f *is discrete-time Fourier transformable; its discrete-time Fourier transform* $\hat{f}$ *is given at each* $\Omega \in \mathbf{R}$ *by* $\hat{f}(\Omega) = \tilde{f}(e^{i\Omega})$.

(*c*) *Let* $\tilde{R} > 0$ *be such that* $R_a(f) < \tilde{R} < R_b(f)$; *let* $\tilde{C}$ *be the circular contour* $|z| = \tilde{R}$ *in the complex plane, directed counterclockwise. Then* f *is given at each* $k \in \mathbf{Z}$ *by*

$$f(k) = \frac{1}{2\pi i} \int_{\tilde{C}} \tilde{f}(z) z^{-k-1} dz \ . \qquad \square$$

The integral in Fact 20.8(c) reduces to the "inversion integral" for the discrete-time Fourier transform (cf. equation (DTFTI) above) when the hypothesis of Fact 20.8(b) holds; Fact 20.8(c) is easily proved using Fact 20.8(a) and the Cauchy integral formula from complex analysis. Our last major result concerning z-transforms is an operational rule which generalizes the corresponding rule for discrete-time Fourier transforms (c.f. Fact 20.4).

20.9 Fact: *Let $f_1 : \mathbf{Z} \to \mathbf{R}$ and $f_2 : \mathbf{Z} \to \mathbf{R}$ be z-transformable discrete-time functions. Suppose that the regions of convergence $ROC(f_1)$ and $ROC(f_2)$ have a nonempty intersection. Then the discrete-time convolution f of f_1 and f_2, given by*

$$f(k) = \sum_{l=-\infty}^{\infty} f_1(l) f_2(k-l), \quad k \in \mathbf{Z},$$

is well-defined and is z-transformable. Moreover, $ROC(f)$ contains the intersection $ROC(f_1) \cap ROC(f_2)$. Furthermore, $\tilde{f}(z)$ is given for each $z \in ROC(f)$ by the product $\tilde{f}_1(z)\tilde{f}_2(z)$.

Proof: Define R_a and R_b by

$$\begin{aligned} R_a &= \max\{R_a(f_1), R_a(f_2)\} \\ R_b &= \inf\{R_b(f_1), R_b(f_2)\}\ ; \end{aligned}$$

the assumption that $ROC(f_1)$ overlaps $ROC(f_2)$ guarantees that $R_a < R_b$. For each $R \in (R_a, R_b)$, the quantity $R^{-l} f_1(l)$ decays exponentially as $l \to \pm\infty$; similarly, for each $k \in \mathbf{Z}$, $R^l f_2(k-l)$ decays exponentially as $l \to \pm\infty$. In addition, for each $l \in \mathbf{Z}$, the function $k \to R^{-k} f_1(l) f_2(k-l)$ decays exponentially as $k \to \pm\infty$, and hence may be summed over k. We obtain, after a simple change of summation index,

$$\sum_{k=-\infty}^{\infty} R^{-k} f_1(l) f_2(k-l) = R^{-l} f_1(l) \tilde{f}_2(R).$$

This result may in turn be summed over l, since $R \in ROC(f_1)$:

$$\sum_{l=-\infty}^{\infty} \left[\sum_{k=-\infty}^{\infty} R^{-k} f_1(l) f_2(k-l) \right] = \tilde{f}_1(R) \tilde{f}_2(R).$$

The same argument works if R is replaced by $z = Re^{i\Omega}$ for any real Ω. The result is

$$\tilde{f}_1(z)\tilde{f}_2(z) = \sum_{l=-\infty}^{\infty} \left[\sum_{k=-\infty}^{\infty} z^{-k} f_1(l) f_2(k-l) \right].$$

Observe that the inner summands *also* decay exponentially as $l \to \pm\infty$; hence the order of summation may be interchanged. This yields

$$\tilde{f}_1(z)\tilde{f}_2(z) = \sum_{k=-\infty}^{\infty} z^{-k} [\sum_{l=-\infty}^{\infty} f_1(l) f_2(k-l)],$$

which shows simultaneously that f is well-defined and that $z^{-k} f(k) \to 0$ as $k \to \pm\infty$ for every z in the intersection of the two regions $ROC(f_1)$ and $ROC(f_2)$. □

Fact 20.9 may be applied to the analysis of time-invariant discrete-time input-output linear systems in the same fashion as Fact 19.9 may be used in the continuous-time setting. Indeed, suppose that we are given an m-input, p-output time-invariant discrete-time input-output linear system with input function space U; suppose that the system's impulse response is $k \to H(k)$, $k \in \mathbb{Z}$, and that the system possesses a transfer function $\tilde{H}(z)$, $z \in ROC(H)$. Suppose also that $u \in$ U is z-transformable, and that $ROC(u)$ overlaps $ROC(H)$. We may then conclude from Fact 20.9 that the steady-state response y of the system to the input u exists; indeed, y is simply the convolution of H with u:

$$y(k) = \sum_{l=-\infty}^{\infty} H(k-l)u(l), \quad k \in \mathbb{Z}.$$

Moreover, also by Fact 20.9, y is z-transformable; its z-transform $\tilde{y}$ is defined at least for every z which lies in the overlap $ROC(H) \cap ROC(u)$, and is given for such z by

$$\tilde{y}(z) = \tilde{H}(z)\tilde{u}(z).$$

The last equation is evidently a generalization of the formula (c.f. the discussion following Fact 20.4) which relates the discrete-time Fourier transforms of the input and output of a time-invariant discrete-time system through the system's frequency response.

21. *Realizations and McMillan Degree*

In this section, we examine in detail some important connections between the theory of state space linear systems and the theory of input-output linear systems. These connections comprise an elegant and aesthetically appealing segment of linear system theory, and their development, which has occurred largely over the last thirty-five years, has been carried out painstakingly (and, some might argue, *ad nauseum*) by many of the more widely respected researchers in the field. From the standpoint of building intuition about mathematical modeling, there is much to be gained from investigating the ways in which state space systems give rise to input-output systems, and vice versa. Furthermore, many fundamental engineering questions involving the *analysis* and *synthesis* of various real-world processes may be phrased and answered in terms of the *realization theory* of time-invariant linear systems, which is addressed in the pages which follow.

Let us begin by recalling that in Definitions 18.18 and 18.19, we introduced the notion of the *input-output linear system associated with a given state-space linear system*. The idea was that a state space linear system with input function space U and output function spaces Y_{t_o} in continuous time or Y_{k_o} in discrete time gave rise in a *canonical way* to linear input-output mappings $\sigma_{t_o}:\mathrm{U}\to\mathrm{Y}_{t_o}$ or $\sigma_{k_o}:\mathrm{U}\to\mathrm{Y}_{k_o}$; the output function $\sigma_{t_o}(u)\in\mathrm{Y}_{t_o}$ (or $\sigma_{k_o}(u)\in\mathrm{Y}_{k_o}$) arising from the input function u was simply the output of the state space system given that the input u was applied to the system starting from a *zero initial state* at time t_o (or k_o). More precisely, in the notation of Definitions 6.1 and 6.2, we had

$$\sigma_{t_o}(u)(t)=\rho(t\,,\phi(t\,,t_o\,,0,u),u(t))\,,\qquad t\geqslant t_o\,,$$

in continuous time, and

$$\sigma_{k_o}(u)(k)=\rho(k\,,\phi(k\,,k_o\,,0,u),u(k))\,,\qquad k\geqslant k_o\,,$$

in discrete time. Immediately, one is prompted to determine exactly when an input-output linear system has input-output mappings which arise in this fashion.

21.1 Definition: A given real m-input, p-output continuous- (respectively, discrete-) time input-output linear system which possesses a weighting pattern is said to be *realizable* if and only if there exists a real m-input, p-output continuous- (respectively, discrete-) time state space linear system with which the given input-output system is associated in the above manner. In this case, a *realization* of the given input-output system is any realization of any state space system with which the given system is so associated. □

In Definition 21.1, we are ignoring, albeit implicitly, a somewhat trivial special case; it is worth disposing of that case now so that it won't cause problems later on. Suppose that an m-input, p-output continuous- or discrete-time input-output linear system is *memoryless* in the sense that for any input function $u \in \mathrm{U}$ and for any $t_o \in \mathrm{R}$ (or $k_o \in \mathrm{Z}$), the value $\sigma_{t_o}(u)(t)$ for every $t \geqslant t_o$ (respectively, the value of $\sigma_{k_o}(u)(k)$ for every $k \geqslant k_o$) depends *only* on the value of u at that *particular* time t (respectively, at time k). It is then easy to show using the results and techniques of §18 that the input-output mappings σ_{t_o} (or σ_{k_o}) are given by

$$\sigma_{t_o}(u)(t) = D(t)u(t), \quad t \geqslant t_o,$$

in continuous time, or

$$\sigma_{k_o}(u)(k) = D(k)u(k), \quad k \geqslant k_o,$$

in discrete time, where the D's are $(p \times m)$ continuous- or discrete-time matrix functions. Such memoryless systems are, in some sense, "realizable" by means of "state space linear systems" whose "state spaces" have dimension *zero*. In the remainder of the section, we shall deal *only* with input-output linear systems which are *not* memoryless.

Recall from §18 (in particular from Assumptions 18.5 and Definitions 18.4 and 18.8) that every discrete-time input-output system possesses a weighting pattern, and that "almost every" continuous-time input-output system possesses one, as well; the reader is also referred to §7, and in particular to Definitions 7.2 and 7.9, for a discussion of realizations for state space linear systems. It is useful at this point to review these definitions; in so doing, we shall see that stipulating *realizability* for a given input-output linear system imposes certain elementary constraints on the *weighting pattern* of the system.

Suppose, then, that we are given an m-input, p-output continuous-time state space linear system whose state space X has dimension n. Let U and Y_{t_o}, for $t_o \in \mathrm{R}$, be the input and output function spaces (c.f. Definition 6.1). We require that the system satisfy Assumptions 7.3, so that realizations for the system do, indeed, exist. Let $t \to A(t)$, $t \to B(t)$, $t \to C(t)$, and $t \to D(t)$ be the matrix functions in the realization of the system corresponding to some choice of basis $\mathbf{x}$ for X; these matrix functions have respective sizes $(n \times n)$, $(n \times m)$, $(p \times n)$, and $(p \times m)$. For $\tau \in \mathbf{R}$, let $t \to \Phi(t, \tau)$, be the unique $(n \times n)$ matrix solution to the differential equation

$$\frac{d}{dt}\Phi(t, \tau) = A(t)\Phi(t, \tau), \quad t \in \mathbf{R},$$

which satisfies $\Phi(\tau, \tau) = I_n$. In the notation of §§6-7 (see, in particular, Theorem 7.8), the state transition and readout mappings are given in coordinates with respect to the basis $\mathbf{x}$ for X and the standard bases for $\mathbf{R}^m$ and $\mathbf{R}^p$ by

$$[\phi(t, t_0, x^\circ, u)]_{\mathbf{x}} = \Phi(t, t_0)[x^\circ]_{\mathbf{x}} + \int_{t_0}^{t} \Phi(t, \tau)B(\tau)u(\tau)d\tau$$

and

$$\rho(t, x, u) = C(t)[x]_{\mathbf{x}} + D(t)u(t).$$

Let $y : [t_0, \infty) \to \mathbf{R}^p$ be the output function arising from the input u given that the system was started in state x° at time t_0. Then for $t \geqslant t_0$, we have

$$\begin{aligned} y(t) &= \rho(t, \phi(t, t_0, x^\circ, u), u(t)) \\ &= C(t)[\phi(t, t_0, x^\circ, u)]_{\mathbf{x}} + D(t)u(t) \\ & C(t)\Phi(t, t_0)[x^\circ]_{\mathbf{x}} + \int_{t_0}^{t} C(t)\Phi(t, \tau)B(\tau)u(\tau)d\tau + D(t)u(t). \end{aligned}$$

Consider, now, the associated input-output linear system, as in Definition 18.18. This system has input function space U, output function spaces Y_{t_0}, and input-output mappings $\sigma_{t_0} : \mathrm{U} \to \mathrm{Y}_{t_0}$ given by

$$\begin{aligned} \sigma_{t_0}(u)(t) &= \rho(t, \phi(t, t_0, 0, u), u(t)) \\ &= \int_{t_0}^{t} C(t)\Phi(t, \tau)B(\tau)u(\tau)d\tau + D(t)u(t), \quad t \geqslant t_0. \end{aligned}$$

Introducing the ubiquitous unit impulse once again, we find that

$$\begin{aligned} \sigma_{t_0}(u)(t) &= \int_{t_0}^{t} [C(t)\Phi(t, \tau)B(\tau) + D(t)\delta(t-\tau)]u(\tau)d\tau \\ &= \int_{t_0}^{t} \tilde{W}(t, \tau)u(\tau)d\tau, \end{aligned}$$

where $\tilde{W}(t, \tau)$ is defined to be the quantity in brackets. After referring to Definition 18.8, it becomes evident that the *weighting pattern* of the associated input-output system is precisely

$(t, \tau) \to \tilde{W}(t, \tau)$.

We may therefore rephrase Definition 21.1, at least as it applies to continuous-time systems, in terms of *weighting patterns*.

21.2 Observation: *A given real m-input, p-output continuous-time input-output linear system satisfying Assumptions 18.5 is realizable if and only if there exist an integer $n > 0$ and a quadruple of continuous real matrix-valued functions (A, B, C, D) having respective sizes $(n \times n)$, $(n \times m)$, $(p \times n)$, and $(p \times m)$ such that the weighting pattern of the input-output system is given for each $t \geqslant \tau$ by*

$$\tilde{W}(t, \tau) = C(t)\Phi(t, \tau)B(\tau) + D(t)\delta(t-\tau),$$

where $\Phi(t, \tau)$ is the unique $(n \times n)$ matrix solution to

$$\frac{d}{dt}\Phi(t, \tau) = A(t)\Phi(t, \tau)$$

$$\Phi(\tau, \tau) = I_n . \qquad \square$$

The argument which leads to a discrete-time version of Observation 21.2 is quite similar. Suppose that we are given a real m-input, p-output discrete-time state space linear system whose state space X has dimension n; denote the input and output function spaces by $\mathbf{U}$ and $\mathbf{Y}_{k_o}$, respectively. Let $\mathbf{x}$ be a basis for X, and let $k \to A(k)$, $k \to B(k)$, $k \to C(k)$, and $k \to D(k)$ be the matrix functions in the realization of the system taken with respect to the basis $\mathbf{x}$ for X. Let $\Phi(k, l)$, for $k \geqslant l$, solve

$$\Phi(k+1, l) = A(k)\Phi(k, l)$$

with

$$\Phi(l, l) = I_n .$$

In the notation of §§6-7, we have

$$[\phi(k, k_o, 0, u)]_{\mathbf{x}} = \sum_{l=k_o}^{k-1} \Phi(k, l+1)B(l)u(l),$$

and

$$\rho(k, \phi(k, k_o, 0, u), u(k)) = C(k)[\phi(k, k_o, 0, u)]_{\mathbf{x}} + D(k)u(k).$$

Thus the input-output system associated with the given state-space system then has input-output mappings $\sigma_{k_o} : \mathrm{U} \to \mathrm{Y}_{k_o}$ given by

$$\sigma_{k_o}(u)(k) = \sum_{l = k_o}^{k-1} C(k)\Phi(k, l+1)B(l)u(l) + D(k)u(k), \quad k \geqslant k_o ;$$

the last expression, for each $k \geqslant k_o$, equals

$$\sum_{l = k_o}^{k} [C(k)\Phi(k, l+1)B(l)1(k-l-1) + D(k)\delta(k-l)]u(l) .$$

As usual, $k \to 1(k)$ denotes the discrete-time unit step and $k \to \delta(k)$ denotes the discrete-time impulse.

In the notation and terminology of Definition 18.4, we see that the quantity in brackets must be the *weighting pattern* of the associated discrete-time system. The discrete-time analogue of Observation 21.2 may therefore be stated as follows.

21.3 Observation: *A given real m-input, p-output discrete-time input-output linear system is realizable if and only if there exist an integer $n > 0$ and a quadruple of discrete-time matrix functions (A, B, C, D) having respective sizes $(n \times n)$, $(n \times m)$, $(p \times n)$, and $(p \times m)$ such that the weighting pattern of the input-output system is given for $k \geqslant l$ by*

$$W(k, l) = C(k)\Phi(k, l+1)B(l)1(k-l-1) + D(k)\delta(k-l) ,$$

where

$$\Phi(k+1, l) = A(k)\Phi(k, l) , \quad k \geqslant l$$

$$\Phi(l, l) = I_n . \qquad \square$$

Observations 21.2 and 21.3 give a rather cumbersome criterion on the weighting pattern of a given input-output system which is equivalent to the system's realizability. We shall see in the pages which follow that the situation is somewhat simpler if we specialize to *time-invariant* input-output systems. Nonetheless, it is worth stating two simple results which relate the realizability of an input-output system with a *factorizability* condition on its weighting pattern. As we have often seen to be the case, the continuous- and discrete-time results are *different* precisely because the state transition matrix $\Phi(k, l)$ of a discrete-time state space linear system need not be invertible, whereas $\Phi(t, \tau)$ for a continuous-time state space system is *always* invertible.

21.4 Theorem: *Let* $(t, \tau) \to \tilde{W}(t, \tau)$, *defined for* $t \geqslant \tau$, *be the weighting pattern of a given real* m*-input,* p*-output continuous-time input-output linear system. The system is realizable if and only if there exist an integer* $n > 0$ *and continuous real matrix functions* $t \to F_1(t)$, $t \to F_2(t)$, *and* $t \to D(t)$ *having respective sizes* $(p \times n)$, $(n \times m)$, *and* $(p \times m)$ *such that*

$$\tilde{W}(t, \tau) = F_1(t) F_2(\tau) + D(t)\delta(t - \tau).$$

Proof: Suppose first that the system is realizable. By Observation 21.2, we may write

$$\tilde{W}(t, \tau) = C(t)\Phi(t, \tau)B(\tau) + D(t)\delta(t - \tau)$$

for every $t \geqslant \tau$ and some suitable matrix functions A, B, C, and D. Let A (and hence Φ) be $(n \times n)$; by property (SG) in the proof of Theorem 2.8, we may factor $\Phi(t, \tau)$ into the product $\Phi(t, 0)\Phi(0, \tau)$. Setting $F_1(t) = C(t)\Phi(t, 0)$ and $F_2(\tau) = \Phi(0, \tau)B(\tau)$ proves that $\tilde{W}$ may be written in the prescribed fashion.

Conversely, given a "factorization" of $\tilde{W}$ having the form in the theorem statement, set $A(t) = 0$, $C(t) = F_1(t)$, and $B(t) = F_2(t)$ for every $t \in \mathbf{R}$. Since $A \equiv 0$, $\Phi(t, \tau) = I_n$ for every $t, \tau \in \mathbf{R}$, and we have exhibited $\tilde{W}$ in the form required by Observation 21.2. It follows that the system is realizable. □

The analogous discrete-time result is a sort of "one-way" version of Theorem 21.4. The key to the difference is that the state transition matrix $\Phi(k, l)$ of a discrete-time state space system does *not* necessarily factor into the product of a function of k times a function of l.

21.5 Theorem: *If the weighting pattern* $W(k, l)$ *for a given real* m*-input,* p*-output discrete-time input-output linear system may be written*

$$W(k, l) = F_1(k) F_2(l) 1(k - l - 1) + D(k)\delta(k - l),$$

where F_1 *is* $(p \times n)$, F_2 *is* $(n \times m)$, *and* D *is* $(p \times m)$, *then the system is realizable.*

Proof: Set $C(k) = F_1(k)$, $A(k) = I_n$, and $B(k) = F_2(k)$ for all $k \in \mathbf{Z}$. Since $\Phi(k, l) = I_n$ for all $k \geqslant l$, the result follows immediately from Observation 21.3.□

A simple counterexample to the converse of Theorem 21.5 may be constructed as follows. Let $A(k) = 0$, $B(k) = C(k) = 1$, and $D(k) = 0$ for all $k \in \mathbf{Z}$. Consider the state space linear system with state space $\mathbf{R}^1$ arising as in Example 6.6 from the matrix quadruple (A, B, C, D); since $A(k) \equiv 0$, we have

$$\Phi(k, l) = \begin{cases} 1 & k = l \\ 0 & k \neq l \,. \end{cases}$$

Thus, the weighting pattern of the associated input-output system is given for $k \geqslant l$ by

$$\begin{aligned} W(k, l) &= C(k)\Phi(k, l+1)B(l)1(k-l-1) + D(k)\delta(k-l) \\ &= \delta(k-l-1) \,. \end{aligned}$$

It is easy to show that $\delta(k-l-1)$ may *not* be factored into the product of a function of k and a function of l. Once again, we have encountered a fundamental *difference* between corresponding continuous- and discrete-time results which is due to the noninvertibility of discrete-time state transition matrices.

For the rest of this section, we focus our attention on *time-invariant* input-output linear systems. Realization theory for such systems is much richer than that for time-varying systems. This fact should come as no surprise; as we saw in §§19-20, there is an entire *frequency domain formalism* which applies solely to time-invariant systems. Our exposition is based on the premise that there are many intuitively illuminating ways of viewing the realization problem for input-output linear systems. The fundamental ideas which underlie the interpretation of state space and input-output linear systems as models for *real-world processes* should always be kept in mind.

Specifically, recall that an *input-output* linear system serves as a model for a process consisting of a "black box" which accepts input functions and emits output functions; a *state space* linear system models such a process consisting of a "box" which has something inside of it; associated with this internal apparatus is a well-defined notion of *state*, or *internal situation* for the box. In a sense, then, state space linear systems model physical objects which can be assembled out of "hardware" according to some recipe, or which are already assembled and need to be analyzed. Input-output linear systems, on the other hand, model processes which one might wish to *synthesize;* there might, however, be many "hardware recipes" which one could follow in building a box whose input-output behavior imitates that of the process in question.

Finding a specific *realization* for an input-output linear system is tantamount to *specifying a particular such recipe.* Many of the questions which realization theory seeks to answer are related to the *relative economy* of various such recipes. Naturally, if hardware is expensive, one might wish to synthesize a certain input-output behavior using the *fewest components possible;* as might be expected, the complexity of a given input-output behavior limits the simplicity of any physical process which can be expected to generate it.

Time-invariant realization theory involves a host of algebraic results, many of which have complex analytic overtones. We shall be concerned almost exclusively with the algebra; most of the fundamental manipulations required in the course of proving the central results may be carried out *formally,* without regard for the fact that complex functions are involved. In this spirit, we make the following definition, which will be central in what follows.

21.6 Definition: Let s (or z) represent a complex indeterminate. By a *real rational function of* s *(or of* z*)* we mean a ratio of two polynomials in s (or z) which have *real coefficients.* A rational function is called *proper* if the degree of the numerator polynomial is less than or equal to the degree of the denominator polynomial; it is called *strictly proper* if the degree of the numerator polynomial is *strictly less* than the degree of the denominator polynomial. □

By dividing out the coefficient of the highest power of s or z in the denominator polynomial, we may always assume that a real rational function f takes the form

$$f(s) = \frac{p_0 s^{n_1} + p_1 s^{n_1-1} + \ldots + p_{n_1-1} s + p_{n_1}}{s^{n_2} + q_1 s^{n_2-1} + \ldots + q_{n_2-1} s + q_{n_2}},$$

or

$$f(z) = \frac{p_0 z^{n_1} + p_1 z^{n_1 - 1} + \ldots + p_{n_1 - 1} z + p_{n_1}}{z^{n_2} + q_1 z^{n_2 - 1} + \ldots + q_{n_2 - 1} z + q_{n_2}},$$

where the p_i and q_i are *real numbers*. For our purposes, it suffices that we regard s and z as *formal indeterminates*, rather than as "complex variables." Any rational function $f = pq^{-1}$ has an interpretation as a *complex function* whose domain consists of the open subset of the complex plane which one obtains by removing the roots of the denominator polynomial q; this interpretation will not be of great importance to us in the present section, however.

A *real vector* or *matrix rational function* of s or z is simply a vector or matrix each of whose elements is a rational function. Such a "rational vector" or "rational matrix" is said to be *proper* or *strictly proper* if and only if each of its elements has the corresponding property. As we shall see, transfer functions of realizable time-invariant input-output linear systems, at least roughly speaking, turn out to be proper rational matrix functions. Our first results along these lines are the following basic observations; they hinge on the easily demonstrated assertion (see Exercise 18.21) that the input-output linear system associated with a given *time-invariant* state space linear system is itself time-invariant.

21.7 Observations: (a) *Let there be given a real m-input, p-output time-invariant continuous-time state space linear system satisfying Assumptions 7.3; let $t \to H(t)$ be the impulse response of the associated input-output linear system. Then H is Laplace transformable (i.e., the input-output system possesses a transfer function); moreover, the transfer function $s \to G(s)$ agrees on $ROC(H)$ with a proper real $(p \times m)$ matrix rational function.*

(b) *Let there be given a real m-input, p-output time-invariant discrete-time state space linear system; let $k \to H(k)$ be the impulse response of the associated input-output linear system. Then H is z-transformable (i.e., the input-output system possesses a transfer function); moreover, the transfer function $z \to G(z)$ agrees on $ROC(H)$ with a proper real $(p \times m)$ matrix rational function.*

Proof: (a) Let (A, B, C, D) be any constant realization for the state space system; the discussion at the end of §7 shows that such a realization exists. In the notation of Observation

21.2, the weighting pattern for the associated input-output linear system is given by

$$\tilde{W}(t,\tau) = C\Phi(t,\tau)B + D\delta(t-\tau) .$$

From §2 and the discussion at the end of §7, we know that $\Phi(t,\tau) = e^{(t-\tau)A}$; finally, invoking Definition 18.16(a), we obtain the following formula for the input-output system's impulse response:

$$H(t) = \begin{cases} Ce^{tA}B + D\delta(t) & t \geqslant 0 \\ 0 & t < 0 . \end{cases}$$

We now demonstrate that $t \to H(t)$ is Laplace transformable. From §§8-9, we know that every entry in $H(t)$ is a linear combination of terms of the form $t^k e^{\lambda t} 1(t)$, where $k \geqslant 0$ and λ is an eigenvalue of A. Hence if $\sigma \in \mathbf{R}$ is larger than the real part of any eigenvalue of A, then $e^{-\sigma t}H(t) \to 0$ as $t \to \pm\infty$. We conclude that H is Laplace transformable; moreover, $ROC(H)$ *includes* the part of the complex plane to the right of all the eigenvalues of A. As for the transfer function of the system (i.e., the Laplace transform $s \to G(s)$ of H), at least for every s in the part of $ROC(H)$ which lies to the right of all of A's eigenvalues, we have

$$\begin{aligned} G(s) &= \int_{-\infty}^{\infty} [Ce^{tA}B\,1(t) + D\delta(t)]e^{-st}\,dt \\ &= \int_{0}^{\infty} Ce^{-(sI_n - A)t}Bdt + D . \end{aligned}$$

Observe that the integrand is the derivative with respect to t of $-C(sI_n - A)^{-1}e^{-(sI_n - A)t}B$; since $\mathrm{Re}\{s\}$ is bigger than the real part of any eigenvalue of A, the last quantity vanishes as $t \to \infty$, and we obtain

$$\begin{aligned} G(s) &= -[-C(sI_n - A)^{-1}e^{-(sI_n - A)t}B + D]\,|_{t=0} \\ &= C(sI_n - A)^{-1}B + D . \end{aligned}$$

The final formula exhibits $G(s)$ as a matrix each of whose entries is a ratio of real polynomials in s. To see this, refer to the discussion following Exercise 1.32; the idea is that

$$(sI_n - A)^{-1} = \frac{1}{\det(sI_n - A)}\,\mathrm{adj}(sI_n - A) .$$

The determinant is a polynomial in s of degree n (assuming that A is $(n \times n)$); each entry in the matrix $\mathrm{adj}(sI_n - A)$, being the determinant of a proper submatrix of $(sI_n - A)$, is a polynomial in s of degree $(n-1)$ or less. Hence $C(sI_n - A)^{-1}B$ is a *strictly proper* real $(p \times m)$ rational matrix function of s; adding on D to obtain $G(s)$ makes $G(s)$ at worst a *proper* real $(p \times m)$ rational matrix function of s, and the proof of (a) is complete modulo the assertion that if $s \in ROC(H)$ and $\mathrm{Re}\{s\}$ is *not* greater than the real part of every eigenvalue of A, then $G(s)$ is *still* given by the formula above.

This last assertion follows from the fact that if $ROC(H)$ contains some eigenvalue λ_0 of A, then, by definition of $ROC(H)$, none of the terms in e^{tA} which involves $e^{\lambda_0 t}$ may appear in $H(t)$. It is easily checked that the integration which leads to the formula for $G(s)$ goes through, in this case, for *all* $s \in ROC(H)$.

As for the discrete-time result (b), let (A, B, C, D) be any constant realization for the state space system; once again, the discussion at the end of §7 (c.f. Exercise 6.12) shows that such a realization exists. Assuming that A is $(n \times n)$, the results of §3 and §7 show that $\Phi(k, l) = A^{(k-l)}$ when $k > l$ and I_n when $k = l$. The weighting pattern of the associated input-output linear system, following Observation 21.3, is therefore given by

$$W(k, l) = CA^{(k-l-1)}B\,1(k-l-1) + D\,\delta(k-l), \quad k \geqslant l,$$

where A^0 is understood to be I_n. From Definition 18.16(b), we obtain the following formula for the impulse response:

$$H(k) = CA^{(k-1)}B\,1(k-1) + D\,\delta(k).$$

We now demonstrate the z-transformability of H. Again by the results of §9, the entries of $H(k)$, for $k > 0$, are linear combinations of terms of the form

$$\binom{k-1}{l}\lambda^l,$$

where $0 \leqslant l \leqslant k-1$ and λ is an eigenvalue of A. It follows that if the magnitude of z is greater than the magnitudes of all the eigenvalues of A, then $z^{-k}H(k) \to 0$ as $k \to \pm\infty$. It may be concluded that H is z-transformable, and that $ROC(H)$ *includes* the infinite circular annulus in the complex plane which is centered at 0 and lies *outside* the largest eigenvalue of A.

The transfer function $z \to G(z)$, $z \in ROC(H)$, of the input-output system is the z-transform of H. Its value at any $z \in ROC(H)$ is

$$G(z) = \sum_{k=-\infty}^{\infty} z^{-k} H(k)$$

$$= \sum_{k=1}^{\infty} [z^{-k} CA^{(k-1)}B] + D \ .$$

We now derive a "nice" expression for the infinite sum. Suppose first of all that $z \in ROC(H)$ has magnitude greater than that of any eigenvalue of A. Consider the series

$$\sum_{k=1}^{\infty} z^{-k} A^{k-1} ;$$

we may rewrite this sum as

$$z^{-1} \sum_{k=0}^{\infty} z^{-k} A^{k} \ .$$

The N th partial sum of the series is

$$z^{-1} \sum_{k=0}^{N} z^{-k} A^{k} = z^{-1}(I_n - z^{-1}A)^{-1}(I_n - z^{-(N+1)}A^{(N+1)}) \ .$$

The term $z^{-(N+1)}A^{(N+1)}$ goes to zero as $N \to \infty$ because of the magnitude restriction on z; hence the series sums to $z^{-1}(I_n - z^{-1}A)^{-1}$, which is in turn equal to $(zI_n - A)^{-1}$.

Hence

$$G(z) = C(zI_n - A)^{-1}B + D \ ,$$

at least for those $z \in ROC(H)$ which exceed all the eigenvalues of A in magnitude. An argument similar to that in the proof of (a) shows that the last formula for $G(z)$ holds for *all* $z \in ROC(H)$; an *identical* argument to the one used in the proof of (a) enables us to conclude that G is a proper rational function of z on $ROC(H)$, and the proof of (b) is complete. □

Observation 21.7 implies that if an m-input, p-output time-invariant input-output linear system is realizable, then it *must* possess a transfer function and that the transfer function *must*, on its region of convergence, agree with a proper rational function of s in continuous time or of z in discrete time. What sorts of time-invariant input-output systems, one might ask, are *not* realizable? The easiest examples of such systems are continuous-time systems which contain *pure delay terms* in their impulse responses.

21.8 Example: Consider the single-input, single-output time-invariant continuous-time input-output linear system whose impulse response is

$$H(t) = e^{-t}\,1(t) + \delta(t-3)\ .$$

Its input-output mappings σ_{t_o}, for each $t_o \in \mathbf{R}$, are easily found to be

$$\sigma_{t_o}(u)(t) = \int_{t_o}^{t} e^{-(t-\tau)} u(\tau)d\tau + u(t-3)1(t-3-t_o)\ , \qquad t \geqslant t_o\ .$$

H is certainly Laplace transformable; the transfer function of the system is

$$G(s) = \frac{1}{s+1} + e^{-3s}\ , \qquad \mathrm{Re}\{s\} > -1\ .$$

G evidently does *not* agree on its region of convergence with any rational function of s, and the system is therefore *not* realizable. □

We are now ready to state and prove one of the fundamental results in the realization theory of time-invariant input-output linear systems. It states essentially that the *converse* of Observation 21.7 holds, as well.

21.9 Theorem: *(a) A real m-input, p-output time-invariant continuous-time input-output linear system is realizable if and only if it possesses a transfer function which agrees on its region of convergence with a real proper $(p \times m)$ matrix rational function of s.*

(b) A real m-input, p-output time-invariant discrete-time input-output linear system is realizable if and only if it possesses a transfer function which agrees on its region of convergence with a real proper $(p \times m)$ matrix rational function of z.

Proof: We know from Observation 21.7 that realizability of a system *implies* existence and proper rationality of the system's transfer function in both continuous and discrete time. As for the converse, suppose that a real m-input, p-output time-invariant continuous-time input-output linear system has impulse response $t \to H(t)$, $t \in \mathbf{R}$, and transfer function $s \to G(s)$, $s \in ROC(H)$. Suppose that G agrees on $ROC(H)$ with a real proper $(p \times m)$ matrix rational function of s; we shall denote this matrix of rational functions by $G(s)$, as well. Since G is proper, every element approaches a finite limit as $|s| \to \infty$; define the real $(p \times m)$ matrix D by

$$D = \lim_{|s| \to \infty} G(s).$$

Set $\hat{G}(s) = G(s) - D$; we claim that $\hat{G}$ is a *strictly* proper $(p \times m)$ matrix rational function of s. Let $q(s)$ be the monic lowest common denominator polynomial of the entries in $\hat{G}(s)$; suppose that $q(s)$ has degree n. Since $\hat{G}$ is strictly proper, we may write

$$\hat{G}(s) = \frac{1}{q(s)} R(s),$$

where $R(s)$ is a $(p \times m)$ matrix each of whose entries is a polynomial in s having degree at most $n-1$. We may therefore *expand* $q(s)$ and $R(s)$ as follows:

$$\begin{aligned} q(s) &= s^n + q_1 s^{n-1} + \ldots + q_{n-1} s + q_n \\ R(s) &= R_1 s^{n-1} + R_2 s^{n-2} + \ldots + R_{n-1} s + R_n , \end{aligned}$$

where the q_i are real numbers and the R_i are real $(p \times m)$ matrices. Now define the real matrices A, B, and C in a "blockwise" fashion as follows:

$$A = \begin{bmatrix} 0_m & I_m & \ldots & 0_m \\ \cdot & 0_m & \ldots & \cdot \\ \cdot & 0_m & \ldots & \cdot \\ 0_m & \cdot & \ldots & I_m \\ -q_n I_m & -q_{n-1} I_m & \ldots & -q_1 I_m \end{bmatrix}; \quad B = \begin{bmatrix} 0_m \\ 0_m \\ \cdot \\ \cdot \\ I_m \end{bmatrix};$$

$$C = \begin{bmatrix} R_n & R_{n-1} & \ldots & R_2 & R_1 \end{bmatrix}.$$

The respective sizes of A, B, and C are $(N \times N)$, $(N \times m)$, and $(p \times N)$, where $N = nm$. As usual, 0_m denotes the $(m \times m)$ matrix of zeroes and I_m denotes the $(m \times m)$ identity matrix. It is not too hard to show that $\hat{G}(s) = C(sI_N - A)^{-1}B$. In view of Observation 21.2 and the proof of Observation 21.7, since $G(s) = C(sI_N - A)^{-1}B + D$, the matrix quadruple (A, B, C, D) is a realization for the system, and the proof of (a) is complete. As for the discrete-time result (b), the proof is exactly identical with $k \to H(k)$ substituted for $t \to H(t)$ and z substituted for s. □

Thus, the existence of a realization for a time-invariant input-output linear system is *equivalent* to the existence and proper rationality of its transfer function matrix. Observe that the realizations constructed in the proof of Theorem 21.9 were quadruples (A, B, C, D) of *constant* matrices; any state space system realized by such a quadruple (keeping in mind the results of §7 along with Examples 6.5 and 6.6) must be a *time-invariant* state space system.

This is not to suggest that *every* realization for a time-invariant input-output linear system must be constant; indeed, as Example 21.11 shows, every realizable time-invariant input-output linear system possesses time-varying realizations. Nonetheless, we are more interested, as a rule, in *constant* realizations for time-invariant systems. Theorem 21.9 and the development which precedes it enable us to characterize the set of *all* constant realizations for a realizable time-invariant input-output linear system in the following concise fashion.

21.10 Facts: (a) *Let a given real m-input, p-output time-invariant continuous-time input-output linear system have impulse response $t \to H(t)$ and transfer function $s \to G(s)$, $s \in ROC(H)$. If the system is realizable, the set of all constant realizations for the system is the set of all real matrix quadruples (A, B, C, D) which satisfy the two equivalent conditions*

$$H(t) = Ce^{tA}B\,1(t) + D\,\delta(t), \qquad t \in \mathbf{R}$$
$$G(s) = C(sI_n - A)^{-1}B + D, \qquad s \in ROC(H).$$

(b) *Let a given real m-input, p-output time-invariant discrete-time input-output linear system have impulse response $k \to H(k)$ and transfer function $z \to G(z)$, $z \in ROC(H)$. If the system is realizable, the set of all constant realizations for the system is the set of all real matrix quadruples (A, B, C, D) which satisfy the two equivalent conditions*

$$H(k) = CA^{k-1}B\,1(k-1) + D\,\delta(k), \qquad k \in \mathbf{Z}$$
$$G(z) = C(zI_n - A)^{-1}B + D, \qquad z \in ROC(H).$$

□

Although Facts 21.10 give a convenient description of the family of all constant realizations for a realizable time-invariant input-output linear system, they say essentially nothing about the *size* of that family, or even about the *sizes of the various matrices themselves*. In the proof of Theorem 21.9, we managed to construct, for a realizable m-input, p-output time-invariant system, a realization whose "A-matrix" had size $(nm \times nm)$, where n was the degree of the lowest common denominator polynomial for the (rational) entries in the system's transfer function matrix. It is an interesting and not very difficult exercise to *dualize* that argument in such a way that there results a realization whose "A-matrix" is $(np \times np)$.

Keeping in mind the philosophical discussion at the start of the section, the *size of the A-matrix* in a constant realization for a time-invariant input-output system reflects, in some sense, the *complexity* of a certain state space system whose input-output behavior mimics that of the given input-output system. The following example shows that a realizable system may have realizations whose A-matrices are of *arbitrarily large size;* a large part of the remainder of the section is devoted to discussing those realizations of a system whose A-matrices are as *small* as possible.

21.11 Example: Suppose that a 1-input, 2-output system has transfer function

$$G(s) = \begin{bmatrix} \dfrac{1}{s^2+3s+1} \\ \dfrac{3s}{s^2+3s+1} \end{bmatrix}.$$

We factor $G(s)$ as in the proof of Theorem 21.9:

$$G(s) = \frac{1}{s^2+3s+1}\left\{\begin{bmatrix}1\\0\end{bmatrix} + \begin{bmatrix}0\\3s\end{bmatrix}\right\}.$$

A realization for the system is then given by (A, B, C, D), with $D = 0$ and

$$A = \begin{bmatrix} 0 & 1 \\ -1 & -3 \end{bmatrix}; \quad B = \begin{bmatrix} 0 \\ 1 \end{bmatrix};$$

$$C = \begin{bmatrix} 1 & 0 \\ 0 & 3 \end{bmatrix}.$$

The reader may, however, verify that *another* realization for the system is given by $(\hat{A}, \hat{B}, \hat{C}, 0)$, where

$$\hat{A} = \begin{bmatrix} -3 & 0 & 1 & 0 \\ 0 & -3 & 0 & 1 \\ -1 & 0 & 0 & 0 \\ 0 & -1 & 0 & 0 \end{bmatrix}; \quad \hat{B} = \begin{bmatrix} 0 \\ 3 \\ 1 \\ 0 \end{bmatrix};$$

$$\hat{C} = \begin{bmatrix} 1 & 0 & 0 & 0 \\ 0 & 1 & 0 & 0 \end{bmatrix}.$$

It may be verified that constant realizations for this system whose A-matrices are *arbitrarily large* may be constructed by taking the matrices in, say, $(A, B, C, 0)$ and "augmenting" them as follows:

$$\tilde{A} = \begin{bmatrix} A & 0_{(2\times k)} \\ 0_{(k\times 2)} & I_k \end{bmatrix}; \quad \tilde{B} = \begin{bmatrix} B \\ 0_{(k\times 1)} \end{bmatrix};$$

$$\tilde{C} = \begin{bmatrix} C & 0_{(2\times k)} \end{bmatrix}.$$

In fact, if the I_k in $\tilde{A}$ is replaced by, say tI_k, then a *time-varying* realization for the given system is obtained. The problem of characterizing *all* the realizations for the system, or even all the *constant* ones, would seem to be quite difficult indeed. □

Even though a realization having an arbitrarily *large* A-matrix may be found for a given realizable time-invariant input-output system, there is evidently a *smallest positive integer* n for which there exists a constant realization of the system whose A-matrix is $(n \times n)$.

21.12 Definition(s): A constant realization (A, B, C, D) for a real m-input, p-output time-invariant continuous- or discrete-time input-output linear system, where A is $(n \times n)$, is said to be a *minimal realization* for the system if and only if n is *as small as possible* among all constant realizations of the system. This smallest n is called the *McMillan degree* of the input-output system. □

The McMillan degree is named after Brockway McMillan, whose famous paper [McMillan] is widely acknowledged as containing the first major published work on the realization problem for time-invariant input-output linear systems. Since the weighting pattern, impulse response, and transfer function constitute *equivalent* specifications of an input-output linear system, a realizable system's McMillan degree is often referred to as the McMillan degree of its weighting pattern, of its impulse response, or of its transfer function. We shall often observe these lexical conventions in what follows.

It can be seen from the proof of Theorem 21.9 that the McMillan degree of a realizable m-input, p-output time-invariant input-output linear system is bounded from above by nm, where n is the degree of the lowest common denominator polynomial for the entries in its (necessarily rational) $(p \times m)$ transfer function matrix. A "dual" argument shows that the McMillan degree is bounded from above by np, as well. In §22, we shall explore various other characterizations of a system's McMillan degree in terms of properties of the system's impulse response and transfer function. Suffice it to say that, speaking roughly, the McMillan degree of an input-output system reflects the *complexity* of the *simplest state space linear system* whose input-output behavior mimics that of the given input-output system.

Suppose we are given a minimal realization for some input-output linear system; using the technique in Example 21.11, we may *augment* the matrices and come up with (non-minimal) realizations for the system whose A-matrices are very large.

In checking through Example 21.11, the reader may have observed that the augmentation was done in such a way so that the "new" part of the larger matrix $\tilde{A}$ was *canceled*, in some sense, when the product $\tilde{C}(sI_{n+k}-\tilde{A})^{-1}\tilde{B}$ was taken. If $(\tilde{A}, \tilde{B}, \tilde{C}, \tilde{D})$ is thought of as a realization for a state space system, then $\tilde{A}$ may be interpreted as governing the *internal dynamics*, or *state transition behavior*, of the state space system. The *size* of $\tilde{A}$ reflects the *dimension* of the state space. The fact that part of $(sI_{n+k}-\tilde{A})^{-1}$ is canceled when multiplied on the left by C and on the right by B shows that some of the state space has no effect on the input-output behavior of the state space system.

In view of the intuitive interpretations of reachability and observability which were developed in §§13-17, the following result should come as no surprise. It might justly be called the *Fundamental Theorem of Time-Invariant Input-Output Realization Theory.*

21.13 Theorem: *Let (A, B, C, D) be a constant realization for a given m-input, p-output time-invariant continuous- or discrete-time input-output linear system. (A, B, C, D) is a minimal realization for the system if and only if (A, B) is a reachable pair and (A, C) is an observable pair.*

Proof: Since we shall be using many of the various reachability and observability criteria from §§13-17, a glance at those sections might be in order. First, we show that if (A, B, C, D) is *not* a minimal realization, then either (A, B) is not a reachable pair or (A, C) is not an observable pair.

We begin by proving this result for continuous-time systems. Let A be $(n \times n)$; let $(\hat{A}, \hat{B}, \hat{C}, \hat{D})$ be another realization for the system, and let $\hat{A}$ be $(r \times r)$, where $r < n$. Suppose that $t \to H(t)$ is the impulse response of the system. From Facts 21.10, we deduce that $D = \hat{D}$ and that

$$Ce^{tA}B\,1(t) = H(t) - D\,\delta(t) = \hat{C}e^{t\hat{A}}\hat{B}\,1(t)\,, \qquad t \in \mathbf{R}\,.$$

We now begin to manipulate. It follows easily from the preceding equation that

$$Ce^{\sigma A}e^{(t-\tau)A}B = \hat{C}e^{\sigma\hat{A}}e^{(t-\tau)\hat{A}}\hat{B}\,, \qquad t, \tau, \sigma \in \mathbf{R}\,.$$

Multiply both sides on the left by $e^{\sigma A^T}C^T$ and on the right by $B^T e^{(t-\tau)A^T}$; after integrating both sides from 0 to t over both τ and σ, the left-hand side becomes

$$\Big[\int_0^t e^{\sigma A^T}C^T C e^{\sigma A}\,d\sigma\Big]\Big[\int_0^t e^{(t-\tau)A}BB^T e^{(t-\tau)A^T}\,d\tau\Big]\,,$$

and the right-hand side becomes

$$\Big[\int_0^t e^{\sigma A^T}C^T \hat{C} e^{\sigma\hat{A}}\,d\sigma\Big]\Big[\int_0^t e^{(t-\tau)\hat{A}}\hat{B}B^T e^{(t-\tau)A^T}\,d\tau\Big]\,.$$

The left-hand side is the product of the observability Gramian $M_o(t\,; A, C)$ and the reachability Gramian $M_r(t\,; A, B)$; see Definitions 13.5 and 15.4. By Theorems 13.7 and 15.8, both of these matrices (and hence their product) must be invertible if (A, B) is reachable and (A, C) is observable.

Both of these matrices are $(n \times n)$; the right-hand side of the last equation, however, is the product of an $(n \times r)$ matrix and an $(r \times n)$ matrix; thus the right-hand side must have rank *less than or equal to* r, which is in turn less than n. We conclude that one of the Gramians on the left-hand side is not invertible, from which it follows that either (A, B) is not a reachable pair or (A, C) is not an observable pair.

The proof of the discrete-time result is similar; again, assume that A is $(n \times n)$ and that $\hat{A}$ is $(r \times r)$, with $r < n$. Since both quadruples realize the same discrete-time system, we know from Facts 21.10 that $D = \hat{D}$ and (in particular) that for every $k > n$ and every nonnegative j and $l < k$, we have

$$CA^{j} A^{k-l} B = \hat{C}\hat{A}^{j}\hat{A}^{k-l}\hat{B} .$$

Multiply both sides on the left by $(A^T)^j C^T$ and on the right by $B^T (A^T)^{k-l}$ and sum over both j and l from 1 to k. We are left once again with a product of Gramians on the left-hand side (c.f. Definitions 14.4 and 16.4) and the product of an $(n \times r)$ matrix and an $(r \times n)$ matrix on the right-hand side. From Theorems 14.7 and 16.8, we conclude that at least one of the Gramians is not invertible, implying that either (A, B) is not (discrete-time) reachable or (A, C) is not (discrete-time) observable.

As for the converse, we appeal to Theorem 17.5, which we called Version I of the Canonical Structure Theorem. Suppose that (A, B, C, D) is a constant minimal realization for the given (continuous-time) input-output linear system. By Theorem 17.5, we may find an invertible $(n \times n)$ matrix P such that $(\hat{A}, \hat{B}, \hat{C}, D)$ defined by $\hat{A} = P^{-1}AP$, $\hat{B} = P^{-1}B$, $\hat{C} = CP$, take the special form given in the statement of Theorem 17.5. It is easy to check that

$$C(sI_n - A)^{-1}B = \hat{C}(sI_n - \hat{A})^{-1}\hat{B} ;$$

in fact, in the notation of Theorem 17.5,

$$\hat{C}(sI_n - \hat{A})^{-1}\hat{B} = C_1(sI_{n_{11}} - A_{11})^{-1}B_1 .$$

The last two equations imply that (A_{11}, B_1, C_1, D) constitutes another realization of the system. If A_{11} is of a size $(n_{11} \times n_{11})$ which is *smaller* than that of A, the minimality of the original realization (A, B, C, D) is contradicted; hence n_{11} must equal n, and we must have $\hat{A} = A_{11}$, $\hat{B} = B_1$, and $\hat{C} = C_1$. Theorem 17.5(c) then asserts that $(\hat{A}, \hat{B})$ is a reachable pair and $(\hat{A}, \hat{C})$ is an observable pair.

Now, in the notation of §13 and §15, we have $Q_r(\hat{A}, \hat{B}) = P^{-1}Q_r(A, B)$ and $Q_o(\hat{A}, \hat{C}) = Q_o(A, C)P$; by Theorem 13.7 and Corollary 15.6, $Q_r(\hat{A}, \hat{B})$ and $Q_o(\hat{A}, \hat{B})$ must both have rank n. The same must hold for $Q_r(A, B)$ and $Q_o(A, C)$, since P is invertible; thus by Theorem 13.7 and Corollary 15.6, (A, B) is a reachable pair and (A, C) is an observable pair.

The proof of the discrete-time version of the converse is left as an exercise for the reader. □

Thus, the minimality of a constant realization (A, B, C, D) for a given time-invariant continuous- or discrete-time input-output linear system is *equivalent* to the reachability of (A, B) and observability of (A, C). We turn now to the problem of characterizing the family of *all* minimal realizations for a given system. Example 21.11 demonstrates the existence of many *non-minimal* realizations for any system; as it happens, there exist infinitely many *minimal* realizations, as well. The set of all such realizations, however, has a nice characterization which is reminiscent of the one-to-one correspondence between realizations of a *state space* linear system and choices of *basis* for its state space. We shall spend most of the rest of the present section developing this characterization.

We begin by stating a preliminary result which will lead eventually to a convenient characterization of the McMillan degree of an input-output system in terms of its weighting pattern or transfer function.

21.14 Fact: *Two real matrix quadruples (A, B, C, D) (respective sizes $(n \times n)$, $(n \times m)$, $(p \times n)$, and $(p \times m)$) and $(\hat{A}, \hat{B}, \hat{C}, \hat{D})$ (respective sizes $(\hat{n} \times \hat{n})$, $(\hat{n} \times m)$, $(p \times \hat{n})$, and $(p \times m)$ are realizations of the same real m-input, p-output time-invariant continuous- or discrete-time input-output linear system if and only if $\hat{D} = D$ and $CA^kB = \hat{C}\hat{A}^k\hat{B}$ for every nonnegative $k \in \mathbb{Z}$.*

Proof: By Facts 21.10, the two quadruples realize the same continuous-time system if and only if

$$Ce^{tA}B\,1(t) + D\,\delta(t) = \hat{C}e^{t\hat{A}}\hat{B}\,1(t) + \hat{D}\,\delta(t)\,, \qquad t \in \mathbb{R}\,,$$

and the same discrete-time system if and only if

$$CA^{k-1}B\,1(k-1) + D\,\delta(k) = \hat{C}\hat{A}^{k-1}\hat{B}\,1(k-1) + \hat{D}\,\delta(k)$$

for every $k \in \mathbf{Z}$. The second equation is clearly equivalent to the conditions $\hat{D} = D$ and $CA^k B = \hat{C}\hat{A}^k \hat{B}$ for every $k \geqslant 0$.

If these last conditions hold, then the first equation is true since in that case

$$\begin{aligned} Ce^{tA}B &= \sum_{k=0}^{\infty} \frac{t^k}{k!} CA^k B \\ &= \sum_{k=0}^{\infty} \frac{t^k}{k!} \hat{C}\hat{A}^k \hat{B} \\ &= \hat{C}e^{t\hat{A}}\hat{B}\,, \quad t \geqslant 0\,. \end{aligned}$$

Conversely, if the first equation holds, then taking the limit as $t \downarrow 0$ of the k th derivative of each side gives $CA^k B = \hat{C}\hat{A}^k\hat{B}$, and we're done. □

21.15 Corollary: *If (A, B, C, D) is any constant (not necessarily minimal) realization for a given m-input, p-output time-invariant continuous- or discrete-time input-output linear system, and if A is $(n \times n)$, then for every invertible $(n \times n)$ matrix P, the quadruple $(\hat{A}, \hat{B}, \hat{C}, D)$, where $\hat{A} = P^{-1}AP$, $\hat{B} = P^{-1}B$, and $\hat{C} = CP$ is another realization for the system.*

Proof: It is evident that $\hat{A}^k = P^{-1}A^k P$ for every $k \geqslant 0$; hence $\hat{C}\hat{A}^k\hat{B} = CA^k B$ for every $k \geqslant 0$, and the result follows from Fact 21.14. □

The matrices $R_k = CA^k B$ appearing in Fact 21.14 depend *only on the system* realized by (A, B, C, D) and not on the realization itself. They are often called the *Markov matrices* of the system. For a discrete-time system, they have an evident description in terms of the system's impulse response $k \to H(k)$; in fact, $R_k = H(k+1)$ for every $k \geqslant 0$. The Markov matrices of a continuous-time system may be viewed as coefficients in a power series expansion of the continuous part of the system's impulse response:

$$H(t) = \sum_{k=0}^{\infty} [R_k \frac{t^k}{k!}] + D\,\delta(t)\,, \qquad t \in \mathbf{R}\,.$$

In either case, the Markov matrices have a complex analytic interpretation as coefficients in a Laurent series expansion of the *transfer function* of the system. The idea is that, since the transfer function of a realizable system agrees on its region of convergence with a real *proper* $(p \times m)$ matrix rational function of s (or of z), it may be regarded as defining a function of s (or of z) which is analytic on some neighborhood of infinity in the complex plane. It is easy to check that the Laurent series expansion of $G(s) = C(sI_n - A)^{-1}B + D$ about $s = \infty$ comes out as

$$\begin{aligned} G(s) &= D + s^{-1}CB + s^{-2}CAB + \ldots + s^{-k-1}CA^kB + \ldots \\ &= D + \sum_{k=1}^{\infty} s^{-k} R_{k-1}; \end{aligned}$$

the discrete-time expansion is obtained by replacing s with z in the last equation. Markov matrices are, in fact, defined for many time-invariant input-output linear systems which are *not* realizable; we shall have more to say about this fact later on.

Evidently (c.f. Example 21.11), not *every* pair of constant realizations for a given realizable input-output linear system are related as in Corollary 21.15. It turns out, however, that if (A, B, C, D) and $(\hat{A}, \hat{B}, \hat{C}, D)$ are any two constant *minimal* realizations for a given system whose McMillan degree is n, then there exists a *unique* invertible $(n \times n)$ matrix P which relates the two realizations in that fashion.

21.16 Theorem: *Let (A, B, C, D) and $(\hat{A}, \hat{B}, \hat{C}, D)$ be two constant minimal realizations for a given m-input, p-output time-invariant continuous- or discrete-time input-output linear system whose McMillan degree is n. Then there exists a unique invertible $(n \times n)$ matrix P such that $\hat{A} = P^{-1}AP$, $\hat{B} = P^{-1}B$, and $\hat{C} = CP$.*

Proof: To economize on notation, we set $Q_r = Q_r(A, B)$, $Q_o = Q_o(A, C)$, $\hat{Q}_r = Q_r(\hat{A}, \hat{B})$, and $\hat{Q}_o = \hat{Q}_o(\hat{A}, \hat{C})$, with notation as in §§13-17. Each of these matrices has (full) rank n by Theorem 21.13 and the results of §§13-17. Define the following four $(n \times n)$ matrices: $R_r = Q_r Q_r^T$; $R_o = Q_o^T Q_o$; $\hat{R}_r = \hat{Q}_r \hat{Q}_r^T$; and $\hat{R}_o = \hat{Q}_o^T \hat{Q}_o$. It is easy to show using the techniques of §12 (c.f. in particular Exercise 12.8) that all four of these matrices are *invertible*.

Now set $P = Q_r \hat{Q}_r^T \hat{R}_r^{-1}$. We claim that P is invertible, and that $P^{-1} = \hat{R}_o^{-1} \hat{Q}_o^T Q_o$. To see this, first note that Fact 21.14 implies that $Q_o Q_r = \hat{Q}_o \hat{Q}_r$; the idea is that since

$$Q_o Q_r = \begin{bmatrix} CB & CAB & CA^2B & . & CA^{n-1}B \\ CAB & CA^2B & . & . & . \\ . & . & . & . & . \\ . & . & . & . & . \\ CA^{n-1}B & CA^nB & CA^{n+1}B & . & CA^{2n-2}B \end{bmatrix},$$

and since the right-hand side does not change its value when A, B, and C are "hatted," we must have $\hat{Q}_o \hat{Q}_r = Q_o Q_r$. Now, multiply P on the left by the proposed P^{-1}, and obtain

$$\begin{aligned} \hat{R}_o^{-1}\hat{Q}_o^T Q_o Q_r \hat{Q}_r^T \hat{R}_r^{-1} &= \hat{R}_o^{-1}\hat{Q}_o^T \hat{Q}_o \hat{Q}_r \hat{Q}_r^T \hat{R}_r^{-1} \\ &= \hat{R}_o^{-1}\hat{R}_o \hat{R}_r \hat{R}_r^{-1} \\ &= I_n . \end{aligned}$$

We show next that $\hat{A} = P^{-1}AP$, $\hat{B} = P^{-1}B$, and $\hat{C} = CP$. This result follows directly from the repeated application of the identity $Q_o Q_r = \hat{Q}_o \hat{Q}_r$. Multiplying both sides of this identity on the left by $\hat{R}_o^{-1}\hat{Q}_o^T$ yields

$$\hat{Q}_r = P^{-1}Q_r ,$$

from which it follows immediately by definition of Q_r that $\hat{B} = P^{-1}B$. Similarly, multiplying the basic identity on the *right* by $\hat{Q}_r^T \hat{R}_r^{-1}$ gives $\hat{Q}_o = Q_o P$, and we conclude that $\hat{C} = CP$.

Finally, to show that $\hat{A} = P^{-1}AP$, note that

$$\hat{Q}_o \hat{A} \hat{Q}_r = Q_o A Q_r ,$$

also by Fact 21.14; each side of this equation, once again, may be written as a "blockwise $(n \times n)$" matrix each of whose $(p \times m)$ blocks is a Markov matrix. Now multiply both sides on the left by $\hat{R}_o^{-1}\hat{Q}_o^T$ and on the right by $\hat{Q}_r^T \hat{R}_r^{-1}$, and the equation $\hat{A} = P^{-1}AP$ results.

Observe that the matrix P is *uniquely defined* by the above argument; for example, $\hat{Q}_o = Q_o P$ implies immediately that $P = R_o^{-1}Q_o^T \hat{Q}_o$. □

Theorem 21.16 is an extremely important result in the realization theory of time-invariant input-output linear systems. It sets up a one-to-one correspondence between the set of all invertible $(n \times n)$ matrices and the set of all constant minimal realizations of a given system whose McMillan degree is n. It should be emphasized, however, that this correspondence is *not uniquely determined*. Loosely speaking, the ambiguity in this correspondence is "isomorphic" to the ambiguity inherent in the choice of a specific *state space linear system*, the dimension of whose state space X is *as small as possible*, and whose input-output behavior models that of the original input-output system. The ambiguity in the choice of *minimal realization* for the input-output system is "isomorphic" to the ambiguity inherent in picking a *specific realization* for the chosen state space linear system, which in turn reflects the ambiguity in the choice of a *specific basis* for the state space X.

We close this section with a result which, at least in principle, enables one to determine whether a given time-invariant input-output linear system is realizable, and, if this be the case, what the system's McMillan degree is. We require a general definition of the *Markov matrices* to which we alluded after Corollary 21.15.

21.17 Definition: (a) For each integer $k \geqslant 0$, the *kth Markov matrix* of a real m-input, p-output time-invariant discrete-time input-output linear system with impulse response $k \to H(k)$, $k \in \mathbb{Z}$, is the real $(p \times m)$ matrix R_k defined by $R_k = H(k+1)$.

(b) Suppose that an m-input, p-output time-invariant continuous-time input-output linear system has impulse response $t \to H(t)$, $t \in \mathbb{R}$, and has transfer function $s \to G(s)$, $s \in ROC(H)$, which extends to a $(p \times m)$ matrix-valued complex function $s \to G_e(s)$ which is analytic on a neighborhood of $s = \infty$ in the complex plane. For each integer $k \geqslant 0$, the *kth Markov matrix* of the system is the (necessarily real) $(p \times m)$ matrix R_k defined by

$$R_k = \lim_{q \to 0} \frac{1}{(k+1)!} \frac{d^{k+1}}{dq^{k+1}} G_e\left(\frac{1}{q}\right). \qquad \square$$

Thus, the Markov matrices are defined for *every* time-invariant discrete-time input-output linear system, and for *some* continuous-time systems; in the continuous-time case, it may be checked that the R_k are coefficients in the *Laurent series expansion* of $s \to G_e(s)$ about $s = \infty$. For reasons which will soon become evident, it is useful to use a system's Markov matrices as "blocks" in much larger matrices associated with the system;

these so-called *Hankel matrices* are defined as follows.

21.18 Definition: Let R_k, $k \geqslant 0$, be the Markov matrices of a given m-input, p-output time-invariant continuous- or discrete-time input-output linear system. For each integer $N \geqslant 0$, the *N th order Hankel matrix* of the system is the real $(Np \times Nm)$ matrix H_N defined as follows:

$$\mathrm{H}_N = \begin{bmatrix} R_0 & R_1 & R_2 & R_3 & \cdot & R_N \\ R_1 & R_2 & R_3 & R_4 & \cdot & R_{N+1} \\ R_2 & R_3 & \cdot & \cdot & \cdot & \cdot \\ \cdot & \cdot & \cdot & \cdot & \cdot & \cdot \\ \cdot & \cdot & \cdot & \cdot & \cdot & \cdot \\ R_{N-1} & R_N & R_{N+1} & R_{N+2} & \cdot & R_{2N-2} \end{bmatrix} \qquad \square$$

The Hankel matrices resemble some of the large matrices which arose in the proof of Theorem 21.16; the following result provides the essential link between the Hankel matrix of an input-output linear system and its realizability.

21.19 Theorem: *Let* H_N, *for* $N \geqslant 0$, *be the N th order Hankel matrix of a given m-input, p-output continuous- or discrete-time input-output linear system. The rank* r_N *of* H_N *is a non-decreasing function of N. Moreover, if the system is realizable and has McMillan degree n, then* $r_N = n$ *for every* $N \geqslant n$.

Proof: The statement that r_N increases with N is obviously true. As for the other assertion about realizability, suppose that the system *is* realizable and that (A, B, C, D) is a constant minimal realization. Observe that since the Markov matrices are given by $R_k = CA^kB$, we have

$$\mathrm{H}_n = Q_o(A, C)Q_r(A, B).$$

(Note that the last equation holds even if the realization is *not* minimal!) Since the realization is assumed minimal, Theorem 21.13 and the results of §§13-17 imply that both $Q_r(A, B)$ and $Q_o(A, C)$ have (full) rank n; their product H_n must therefore also have rank n. Its rank cannot *exceed* n, by Theorem 1.27; moreover, since H_n contains as an $(n \times n)$ submatrix the product of an invertible $(n \times n)$ submatrix of $Q_o(A, C)$ and an invertible $(n \times n)$ submatrix of $Q_r(A, B)$, it contains an invertible $(n \times n)$ submatrix and therefore has rank *equal* to n.

If $N > n$, then we may factor H_N as follows:

$$\mathrm{H}_N = \begin{bmatrix} C \\ CA \\ CA^2 \\ \cdot \\ \cdot \\ \cdot \\ CA^{N-1} \end{bmatrix} \begin{bmatrix} B & AB & A^2B & . & . & . & A^{N-1}B \end{bmatrix};$$

since each of the factors has rank n, their product H_N has rank at most n, hence equal to n by our previous work. □

An alternative statement of Theorem 21.19 is that the rank r_N of a realizable system's Nth order Hankel matrix H_N *stabilizes* at n, the McMillan degree of the system, as N increases. We shall encounter Hankel matrices once again in §25 when we discuss observability indices, controllability indices, and canonical forms.

22. *Polynomial Matrices and Matrix Fraction Descriptions*

In §21, we discovered (c.f. Theorem 21.9) that a real m-input, p-output time-invariant continuous- or discrete-time input-output linear system with impulse response $t \to H(t)$ or $k \to H(k)$ was *realizable* if and only if its transfer function $s \to G(s)$ or $z \to G(z)$ agreed on its region of convergence $ROC(H)$ with a real proper $(p \times m)$ matrix rational function of s or of z. The *constant realizations* for such a system was shown in Facts 21.10 to lie in one-to-one correspondence with matrix quadruples (A, B, C, D) which satisfied

$$G(s) = C(sI_n - A)^{-1}B + D, \qquad s \in ROC(H),$$

or

$$G(z) = C(zI_n - A)^{-1}B + D, \qquad z \in ROC(H).$$

In exploring the concept of a *minimal realization*, we found (Theorem 21.19) that the *McMillan degree* of a realizable system coincided with the *rank of its Hankel matrices* $\mathbf{H}_N$ for N sufficiently large. In the present section, we obtain another characterization of the McMillan degree of a realizable system; this new characterization rests on the notion of a *matrix fraction description* for the transfer function of the system, and enables us, at least in principle, to *compute* the McMillan degree of a realizable system by manipulating its transfer function matrix *algebraically*. As a bonus, the procedure which we describe leads to an *algorithm* for finding a particularly simple minimal realization of the system.

It is fair to say that the sole purpose of the present section is to develop a version of Fact 22.1 below which applies to systems with more than one input and/or more than one output. The statement of the appropriate "multi-input, multi-output" generalization is simple and appealing; nonetheless, it turns out that we shall need to follow an exceedingly circuitous path in order to derive it. Many of the results which we'll encounter along the way are of independent mathematical interest; most of them are *purely algebraic* facts about *polynomial matrices*.

Suppose first that we are given a real single-input, single-output time-invariant continuous-time input-output linear system which is realizable; let $t \to H(t)$ be its impulse response. By Theorem 21.9, we know that the system's transfer function $s \to g(s)$, $s \in ROC(H)$, exists and is given by

$$\begin{aligned} g(s) &= g_{SP}(s) + d \\ &= \frac{p(s)}{q(s)} + d \ , \qquad s \in ROC(H) \ , \end{aligned}$$

where d is a real number, and $p(s)$ and $q(s)$ are *polynomials* in s having real coefficients; in addition, the degree of $q(s)$ exceeds the degree of $p(s)$, so that $g_{SP}(s)$ is a *strictly proper* rational function. We assume, as we may, that the polynomials $p(s)$ and $q(s)$ are *relatively prime;* that is, we assume that they have *no common factors*. We also may assume that the leading coefficient in the polynomial $q(s)$ is 1; in this case, $q(s)$ is said to be a *monic polynomial*.

Let us now construct a *realization* for the system by the method used in the proof of Theorem 21.9. Let

$$q(s) = s^n + q_1 s^{n-1} + \ldots + q_{n-1}s + q_n \ ,$$

and

$$p(s) = p_1 s^{n-1} + p_2 s^{n-2} + \ldots + p_{n-1}s + p_n .$$

Let the matrices A, B, and C be given by

$$A = \begin{bmatrix} 0 & 1 & . & 0 & 0 \\ . & 0 & . & . & . \\ . & . & . & 1 & 0 \\ 0 & 0 & . & 0 & 1 \\ -q_n & -q_{n-1} & . & -q_2 & -q_1 \end{bmatrix}; \quad B = \begin{bmatrix} 0 \\ 0 \\ . \\ 0 \\ 1 \end{bmatrix};$$

$$C = \begin{bmatrix} p_n & p_{n-1} & . & . & p_1 \end{bmatrix}.$$

Then (A, B, C, d) is a *realization* for the system.

We claim, moreover, that the given realization is *minimal*. To see this, suppose that $(\hat{A}, \hat{B}, \hat{C}, d)$ is another realization for the system, and that $\hat{A}$ is $(\hat{n} \times \hat{n})$. Facts 21.10 imply that $g_{SP}(s) = \hat{C}(sI_{\hat{n}} - \hat{A})^{-1}\hat{B}$; by the "determinental" formula for $(sI_n - \hat{A})^{-1}$ (c.f. the discussion following Exercise 1.32), the right-hand side of the last equation is the ratio of two polynomials in s which, when expressed in *lowest terms*, has denominator degree at most $\hat{n}$. We conclude that $n \leqslant \hat{n}$, which implies that (A, B, C, d) is a minimal realization; the McMillan degree of the system is therefore n. The discrete-time version of the foregoing argument is exactly the same, except that z plays the role of s. We have proved the following assertion.

22.1 Fact: *Let there be given a real single-input, single-output continuous- or discrete-time input-output linear system which is realizable. Its transfer function $s \to g(s)$, $s \in ROC(H)$, or $z \to g(z)$, $z \in ROC(H)$, may be expressed on $ROC(H)$ as*

$$g(s) = g_{SP}(s) + d$$

or

$$g(z) = g_{SP}(z) + d ,$$

where g_{SP} is a uniquely determined strictly proper rational function. If the denominator polynomial of g_{SP} has degree n when all common factors between numerator and denominator are canceled, then the McMillan degree of the system is n, and a minimal realization of the system is given by (A, B, C, d) above. □

It should be observed that the role played by the *transfer functions* of the systems in Fact 22.1 is a *purely algebraic one*. It is *irrelevant* that the transfer functions of the systems are complex-valued functions whose "official" domains are $ROC(H) \subset \mathbf{C}$; the rational functions g and g_{SP} might just as well be regarded as *formal expressions* in s or z whose coefficients are to be manipulated. We shall be occupying ourselves throughout the remainder of this section with such *algebraic manipulations* on rational matrix-valued functions. It is, in some sense, easier, at least in the present context, to treat these objects as if they were mere formal expressions rather than to regard them as complex matrix-valued functions having specified domains.

In that spirit, for the time being we shall restrict our attention to real rational matrix functions which are written as *functions of* s rather than of z. Since all of the ensuing results are *purely algebraic*, their proofs are *identical* regardless of whether s or z is the "complex indeterminate."

Before listing our main objectives, a few definitions are in order.

22.2 Definition(s): Let s be a complex indeterminate. A *real* $(k \times l)$ *polynomial matrix* $F(s)$ is a $(k \times l)$ matrix each of whose entries is a polynomial in s having real coefficients. A real $(k \times k)$ (i.e. *square*) polynomial matrix is said to be *nonsingular* if and only if its determinant is a nonzero polynomial is s. A real $(k \times k)$ polynomial matrix is said to be *unimodular* if and only if its determinant is a nonzero real number. □

By the *determinant* of a polynomial matrix $F(s)$, we mean the polynomial which one obtains by computing the determinant of the matrix $F(s)$ using the standard procedure (c.f. Definition 1.30) and the usual rules of polynomial algebra. The concepts of *nonsingularity* and *unimodularity* are significant because a $(k \times k)$ polynomial matrix $F(s)$ which has either of these properties is, at least to some degree, *invertible*. Suppose that $F(s)$ is a real nonsingular $(k \times k)$ polynomial matrix; by using the formula from §1, we may define a matrix $F^{-1}(s)$ by means of

$$F^{-1}(s) = \frac{1}{\det F(s)} \operatorname{adj} F(s),$$

where $\operatorname{adj} F(s)$ is computed as in §1 using the rules of polynomial algebra. Evidently, $F^{-1}(s)$ is a *real* $(k \times k)$ *rational matrix function*, in the sense of the discussion which follows Definition 21.6; it is easily checked that $F^{-1}(s)F(s)$ and $F(s)F^{-1}(s)$ are both equal to I_k. If $F(s)$ is *unimodular*, it is certainly also nonsingular; moreover, the above formula exhibits $F^{-1}(s)$ as a real $(k \times k)$ *polynomial* matrix. Unimodular polynomial matrices are, in this fashion, "polynomially invertible." As an example, consider

$$F_1(s) = \begin{bmatrix} s+1 & s+3 \\ s & s+2 \end{bmatrix}; \quad F_2(s) = \begin{bmatrix} s+1 & s \\ s & s+2 \end{bmatrix}.$$

We have $\det F_1(s) = 2$ and $\det F_2(s) = 3s+2$. Hence, $F_1(s)$ is unimodular and $F_2(s)$ is nonsingular but *not* unimodular. Moreover,

$$F_1^{-1}(s) = \frac{1}{2}\begin{bmatrix} (s+2) & -(s+3) \\ -s & (s+1) \end{bmatrix},$$

and

$$F_2^{-1}(s) = \frac{1}{3s+2}\begin{bmatrix} (s+2) & -s \\ -s & (s+1) \end{bmatrix}.$$

Definition 22.2 gives us some of the verbal machinery which we require in order to state the appropriate multi-input, multi-output version of Fact 22.1. Specifically, suppose that $G(s)$ is a real proper $(p \times m)$ rational matrix function of s; if we define the real $(p \times m)$ matrix D by way of

$$D = \lim_{|s| \to \infty} G(s),$$

then we may write

$$G(s) = Z(s) + D,$$

where $Z(s)$ is *strictly* proper. It turns out that one can find a real $(p \times m)$ polynomial matrix $P(s)$ and a real *nonsingular* $(m \times m)$ polynomial matrix $Q(s)$ for which

$$Z(s) = P(s)Q^{-1}(s).$$

Equally, there exist a real $(p \times m)$ polynomial matrix $N(s)$ and a real nonsingular $(p \times p)$ polynomial matrix $R(s)$ such that

$$Z(s) = R^{-1}(s)N(s).$$

As it happens, many such "factorizations" of $Z(s)$ exist. For example, let $q(s)$ be the lowest common denominator of the (rational) matrix entries in $Z(s)$; then $q(s)Z(s)$ is a *polynomial matrix*. If we set $N(s)$ and $P(s)$ equal to $q(s)Z(s)$, then set $Q(s) = q(s)I_m$ and $R(s) = q(s)I_p$, then we obtain factorizations of the above form.

These factorizations, which exhibit $Z(s)$ as a "ratio" of polynomial matrices, are reminiscent of the representation of $g_{SP}(s)$ in Fact 22.1. The one missing item is a suitable notion of *relative primeness* for polynomial matrices; when, one might ask, is a such a representation "in lowest terms?"

22.3 Definition: (a) Two real polynomial matrices $P(s)$ and $Q(s)$ which have the same number of columns are said to be *right relatively prime* or *right coprime* if and only if the following condition holds: if the real polynomial matrices $\hat{P}(s)$, $\hat{Q}(s)$, and $U(s)$, having respective sizes $(p \times m)$, $(m \times m)$, and $(m \times m)$, are such that

$$P(s) = \hat{P}(s)U(s); \quad Q(s) = \hat{Q}(s)U(s),$$

then $U(s)$ is unimodular.

(b) Two real polynomial matrices $N(s)$ and $R(s)$ which have the same number of rows are said to be *left relatively prime* or *left coprime* if and only if the following condition holds: if the real polynomial matrices $\hat{N}(s)$, $\hat{R}(s)$, and $U(s)$, having respective sizes $(p \times p)$, $(p \times m)$, and $(p \times p)$, are such that

$$N(s) = U(s)\hat{N}(s); \quad R(s) = U(s)\hat{R}(s),$$

then $U(s)$ is unimodular. □

Thus, two polynomial matrices are right relatively prime if and only if they have "no (right) common (polynomial matrix) factors" other than unimodular ones. Similarly, two polynomial matrices are *left* relatively prime if and only if all of their "(left) common (polynomial matrix) factors" are unimodular. It is easy to see that if $P(s)$ and $Q(s)$ are right relatively prime, then so are $P(s)U(s)$ and $Q(s)U(s)$, where $U(s)$ is any unimodular polynomial matrix of the appropriate size. Likewise, if $N(s)$ and $R(s)$ are *left* relatively prime, and $U(s)$ is an appropriately sized unimodular polynomial matrix, then $U(s)N(s)$ and $U(s)R(s)$ are also left relatively prime.

Armed with suitable notions of relative primeness for polynomial matrices, we are ready to state two important definitions.

22.4 Definition: Let $Z(s)$ be a real $(p \times m)$ matrix rational function of s. A *right matrix fraction description*, or *RMFD* of $Z(s)$ is a factorization of the form

$$Z(s) = P(s)Q^{-1}(s), \qquad \text{(RMFD)}$$

where $P(s)$ and $Q(s)$ are real polynomial matrices having respective sizes $(p \times m)$ and $(m \times m)$, with $Q(s)$ nonsingular.

An RMFD of $Z(s)$ is said to be *irreducible* if and only if $P(s)$ and $Q(s)$ are right relatively prime. An RMFD of $Z(s) = P(s)Q^{-1}(s)$ is said to be *minimal* if and only if for every other RMFD $Z(s) = \hat{P}(s)\hat{Q}^{-1}(s)$, the degree of the polynomial $\det \hat{Q}(s)$ is at least as high as the degree of $\det Q(s)$. □

22.5 Definition: Let $Z(s)$ be a real $(p \times m)$ matrix rational function of s. A *left matrix fraction description*, or *LMFD* of $Z(s)$ is a factorization of the form

$$Z(s) = R^{-1}(s)N(s), \qquad \text{(LMFD)}$$

where $N(s)$ and $R(s)$ are real polynomial matrices having respective sizes $(p \times m)$ and $(p \times p)$, with $R(s)$ nonsingular.

An LMFD of $Z(s)$ is said to be *irreducible* if and only if $N(s)$ and $R(s)$ are left relatively prime. An LMFD of $Z(s) = R^{-1}(s)N(s)$ is said to be *minimal* if and only if for every other LMFD $Z(s) = \hat{R}^{-1}(s)\hat{N}(s)$, the degree of the polynomial $\det \hat{R}(s)$ is at least as high as the degree of $\det R(s)$□

Intuitively, then, an RMFD (or LMFD) of $Z(s)$ is irreducible if all the common right (or left) polynomial matrix factors have been *canceled* from the "numerator" and "denominator" polynomial matrices appearing in the RMFD (or LMFD). The property of irreducibility of a matrix fraction description works nicely in the case $m = p = 1$; that is, an expression of the form

$$z(s) = \frac{p(s)}{q(s)}$$

for a *scalar* rational function $z(s)$ cannot be *reduced* if $p(s)$

and $q(s)$ are relatively prime, in which case it is *also* true that the degree of $q(s)$ is less than or equal to the degree of $\hat{q}(s)$ when

$$z(s) = \frac{\hat{p}(s)}{\hat{q}(s)}$$

is any other such representation of $z(s)$. In fact, a matrix fraction description of a (1×1) "matrix" rational function $z(s)$ is irreducible *if and only if* it is minimal. The equivalence of irreducibility and minimality holds in the general $(p\times m)$ case, as we'll see, but is much more difficult to prove.

Furthermore, Fact 22.1 shows that the degree of $q(s)$ in an irreducible representation of the above form *coincides* with the McMillan degree of the single-input, single-output time-invariant linear system whose transfer function is given on its region of convergence by $z(s)$. This equality turns out to hold for multi-input, multi-output systems, as well. We are now ready to state the results which we shall spend the remainder of the section proving.

22.6 Assertions: *Let $G(s)$ be a real proper $(p\times m)$ matrix rational function of s. Set $G(s)=Z(s)+D$, where D is a real $(p\times m)$ matrix and $Z(s)$ is strictly proper. Then*

(a) An RMFD $Z(s)=P(s)Q^{-1}(s)$ or an LMFD $Z(s)=R^{-1}(s)N(s)$ is irreducible if and only if it is minimal.

(b) The degree of $\det Q(s)$ *in a minimal RMFD of $Z(s)$ equals the degree of* $\det R(s)$ *in a minimal LMFD of $Z(s)$.*

(c) The number associated with $Z(s)$ defined by (b) is equal to the McMillan degree of the real m-input, p-output time-invariant (continuous-time) input-output linear system whose transfer function is given on its region of convergence by $Z(s)$. □

As we stated earlier, the proofs of Assertions 22.6 depend on certain fundamental results from the algebra of polynomial matrices. These results depend in turn on a classic fact about polynomials which is known as the *Euclidean algorithm*. The proof is an easy application of the "long division" procedure which should be familiar to most readers from high school algebra. If $a(s)$ and $b(s)$ are two polynomials in s, then we say that $b(s)$ *divides* $a(s)$ if and only if there exists a polynomial $x(s)$ such that $a(s)=x(s)b(s)$. Observe that $b(s)$ cannot divide $a(s)$ if the degree of $b(s)$ is greater than the degree of $a(s)$; if the degree of $b(s)$ is less than that of $a(s)$, then we can always "divide $b(s)$ into $a(s)$ and come up with a remainder;" this remainder is *zero* if and only if $b(s)$ divides $a(s)$. More precisely, we have

22.7 Theorem: (the Euclidean algorithm) *Let $a(s)$ and $b(s)$ be two polynomials in s which have real coefficients; suppose $b(s) \neq 0$. Then there exist polynomials $x(s)$ and $y(s)$ having real coefficients, with the degree of $y(s)$ strictly less than the degree of $b(s)$, such that*

$$a(s) = x(s)b(s) + y(s) . \qquad \square$$

We are observing the convention that a *degree zero polynomial* is a *nonzero (real) number;* the degree of the *zero polynomial* is taken to be $-\infty$. This convention is standard; see, for example, [Jacobson]. We are now ready to present a classic result about polynomial matrices. For the proof of a much more general result, see [Jacobson]. Recall that a *greatest common divisor* of a given family of polynomials is a polynomial which *divides* every polynomial in the family and which is a *multiple* of every other polynomial which divides all the given ones.

22.8 Theorem: *Let $F(s)$ be a real $(p \times m)$ polynomial matrix.*

(a) If $p \geqslant m$, there exist real unimodular polynomial matrices $U(s)$ and $V(s)$, having respective sizes $(p \times p)$ and $(m \times m)$, such that $U(s)F(s)V(s) = \Lambda(s)$, where $\Lambda(s)$ is a $(p \times m)$ matrix having the form

$$\Lambda(s) = \begin{bmatrix} d_1(s) & 0 & . & 0 \\ 0 & d_2(s) & . & . \\ 0 & 0 & . & . \\ . & . & . & . \\ . & . & . & d_m(s) \\ . & . & . & . \\ 0 & 0 & . & 0 \end{bmatrix},$$

where the $d_k(s)$, for $1 \leqslant k \leqslant m$, are polynomials, and $d_k(s)$ divides $d_{k+1}(s)$, $1 \leqslant k < m$. (N.b.: some of the $d_k(s)$ may be zero.

(b) If $p \leqslant m$, there exist real unimodular polynomial matrices $U(s)$ and $V(s)$, having respective sizes $(p \times p)$ and $(m \times m)$, such that $U(s)F(s)V(s) = \Lambda(s)$, where $\Lambda(s)$ is a $(p \times m)$ matrix having the form

$$\Lambda(s) = \begin{bmatrix} d_1(s) & 0 & . & . & 0 & . & 0 \\ 0 & d_2(s) & . & . & . & . & . \\ . & 0 & . & . & . & . & . \\ . & . & . & . & . & . & . \\ 0 & 0 & . & d_p(s) & 0 & . & 0 \end{bmatrix},$$

where the $d_k(s)$, for $1 \leqslant k \leqslant p$, are polynomials, and $d_k(s)$ divides $d_{k+1}(s)$, $1 \leqslant k < p$.

(c) In either case, each product $d_1(s)d_2(s) \ldots d_k(s)$ is a greatest common divisor of the $(k \times k)$ minor determinants of $F(s)$, $1 \leqslant k \leqslant \min(m, p)$.

Proof: The proof is essentially a polynomial version of the Gauss Elimination Algorithm 1.8. We make use of *polynomial versions* of the *elementary matrices* described in Definition 1.6: $P_n(k, l)$ denotes the $(n \times n)$ matrix obtained by exchanging the kth and lth rows of the $(n \times n)$ identity matrix I_n, and $E_n(k, l, f(s))$, for $k \neq l$, is the $(n \times n)$ matrix obtained by substituting the polynomial $f(s)$ for the zero which occupies the (k, l) position in the $(n \times n)$ identity matrix. Observe (c.f. Fact 1.31 and Exercise 1.32) that these matrices are *unimodular;* the determinant of $P_n(k, l)$, when $k \neq l$, is -1, and the determinant of $E_n(k, l, f(s))$ is 1.

We describe the procedure for finding $U(s)$, $V(s)$, and the $d_k(s)$ in step-by-step fashion:

Step 1: Begin by finding in the matrix $F(s)$ the nonzero (polynomial) element $[F(s)]_{kl}$ which has the smallest degree; note that this element could be a nonzero real number. Multiply $F(s)$ on the left by $P_p(1, k)$ and on the right by $P_m(1, l)$; the matrix so obtained has the low degree element in the $(1, 1)$ position; call this new matrix $F_{1A}(s)$.

Now for each i, $2 \leqslant i \leqslant p$, by the Euclidean algorithm 22.6 we may find polynomials $g_i(s)$ and $h_i(s)$ such that

$$[F_{1A}(s)]_{11} g_i(s) + h_i(s) = [F_{1A}(s)]_{i1},$$

where the degree of $h_i(s)$ is less than the degree of $[F_{1A}(s)]_{11}$.

Multiply $F_{1A}(s)$ on the left by the sequence of matrices $E_p(i, 1, -g_i(s))$, $2 \leqslant i \leqslant p$; this maneuver puts $h_i(s)$ in the $(i, 1)$ position of the new matrix $F_{1B}(s)$ which we obtain.

Similarly, for each j, $2 \leqslant j \leqslant m$, we can find $g^j(s)$ and $h^j(s)$ with degree of $h^j(s)$ less than the degree of $[F_{1A}(s)]_{11}$ such that

$$[F_{1A}(s)]_{11} g^j(s) + h^j(s) = [F_{1A}(s)]_{1j} ;$$

multiplying $F_{1B}(s)$ on the *right* by the sequence of $E_m(1, j, -g^j(s))$ puts $h^j(s)$ in the $(1, j)$ position of the new matrix $F_{1C}(s)$.

Now, repeat the procedure outlined in the three preceding paragraphs with $F_{1C}(s)$ in place of $F(s)$; obtain $F_{2A}(s)$, $F_{2B}(s)$, $F_{2C}(s)$, and then $F_{3A}(s)$; continue; and eventually the procedure terminates with a polynomial matrix which has zeroes in every position in the first row and first column save the $(1, 1)$ position.

Step 2: Let the matrix obtained via Step 1 be

$$F_{\text{II}}(s) = \begin{bmatrix} d_1(s) & 0 & . & . & 0 \\ 0 & x & x & x & x \\ . & x & x & x & x \\ . & . & . & . & . \\ 0 & x & x & x & x \end{bmatrix} .$$

We may arrange that $d_1(s)$ divide every element of the $(p-1 \times m-1)$ matrix of x's as follows. Suppose that $d_1(s)$ doesn't divide one of the x's; multiply $F_{\text{II}}(s)$ on the left by an $E_p(1, l, 1)$ to *add* the row containing this element to the first row. Now perform Step 1 above on the new matrix; eventually, an $F_{\text{II}}(s)$ is arrived at whose $d_1(s)$ divides every element in its submatrix of x's.

Step 3: Repeat Steps 1 and 2, but on the matrix of x's in $F_{\text{II}}(s)$.

Eventually, a matrix of the form $\Lambda(s)$ in (a) or (b) is obtained at the end of Step 2, at which point the procedure terminates.

Observe that the matrix $\Lambda(s)$ may be written as $U(s)F(s)V(s)$, where $U(s)$ and $V(s)$ are, respectively, products of $(p \times p)$ and $(m \times m)$ elementary matrices. Since each elementary matrix is unimodular, so are $U(s)$ and $V(s)$. It remains only to prove part (c) of the theorem.

It is clear from the divisibility condition on the $d_k(s)$ that $d_1(s)d_2(s)\ldots d_k(s)$ is the greatest common divisor of the $(k \times k)$ minor subdeterminants of $\Lambda(s)$. Since the rows of $U(s)F(s)$ are linear combinations (with *polynomial* coefficients) of the rows of $F(s)$, the $(k \times k)$ minor determinants of $U(s)F(s)$ are polynomial linear combinations of those of $F(s)$. A similar assertion about the columns of $U(s)F(s)V(s)$ shows that the $(k \times k)$ minors of $\Lambda(s)$ are also linear combinations of those of $F(s)$.

Symmetrically, since $F(s) = U^{-1}(s)\Lambda(s)V^{-1}(s)$, with $U(s)$ and $V(s)$ unimodular, the $(k \times k)$ minors of $F(s)$ are polynomial linear combinations of those of $\Lambda(s)$. Hence the set of common divisors of all the $(k \times k)$ minors of $F(s)$ must be the *same* as the set of common divisors of all the $(k \times k)$ minors of $\Lambda(s)$, and (c) follows. □

The matrix obtained in Theorem 22.8 is called the *Smith form* of the polynomial matrix $F(s)$. The polynomials $d_1(s), d_2(s), \ldots, d_{\min\{m,p\}}(s)$ are called *a set of invariant factors* of $F(s)$. Although the algorithm presented in the proof specifies them completely, their description in part (c), which is usually given as their definition, determines them only up to a nonzero real constant multiple. Their *degrees*, however, are uniquely determined. Observe that if $F(s)$ is a *square* polynomial matrix, then $\det F(s)$ is a nonzero real constant multiple of the product of all the $d_k(s)$. As a consequence, the invariant factors of a real *nonsingular* polynomial matrix must be nonzero polynomials; moreover, the invariant factors of a real *unimodular* polynomial matrix must be nonzero *real numbers*.

As an example, we show how to find the Smith form $\Lambda(s)$ of a very simple polynomial matrix. The numbers above the arrows indicate the step in the algorithm which is being made.

$$F(s) = \begin{bmatrix} s+1 & s \\ s & s+2 \end{bmatrix} \xrightarrow{1} \begin{bmatrix} s+1 & s \\ -1 & 2 \end{bmatrix}$$

$$\xrightarrow{1} \begin{bmatrix} -1 & 2 \\ s+1 & s \end{bmatrix} \xrightarrow{1} \begin{bmatrix} -1 & 0 \\ 0 & 3s+2 \end{bmatrix}.$$

Here, we have

$$U(s) = \begin{bmatrix} -1 & 1 \\ -s & s+1 \end{bmatrix}; \quad V(s) = \begin{bmatrix} 1 & 2 \\ 0 & 1 \end{bmatrix}.$$

The first consequence of Theorem 22.8 is a central result which enables us to take a giant step toward proving Assertions 22.6.

22.9 Theorem: *Let $P(s)$ be a real $(p \times m)$ polynomial matrix, and let $Q(s)$ be a real $(m \times m)$ polynomial matrix. The following three conditions are equivalent:*

(a) $P(s)$ and $Q(s)$ are right relatively prime.

(b) There exist real polynomial matrices $A(s)$ and $B(s)$, which have respective sizes $(m \times m)$ and $(m \times p)$, such that $A(s)Q(s) + B(s)P(s) = I_m$.

(c) The $((p+m) \times m)$ complex matrix

$$F(s_o) = \begin{bmatrix} Q(s_o) \\ P(s_o) \end{bmatrix}$$

has full rank m for every $s_o \in \mathbf{C}$.

Proof: We prove the equivalence by showing that (a) implies (b), then that (b) implies (c), and finally that (c) implies (a). Suppose, then, that (a) holds. We claim , in that case, that the invariant factors $d_1(s), \ldots, d_m(s)$ of the matrix $F(s)$ in (c) are all nonzero real numbers. Suppose not; then, in particular, $d_m(s)$ is either zero or a polynomial with real coefficients having nonzero degree. We factor the Smith form $\Lambda(s) = U(s)F(s)V(s)$ as follows:

$$\Lambda(s) = \hat{\Lambda}(s) \begin{bmatrix} 1 & 0 & . & 0 & 0 \\ 0 & 1 & . & . & . \\ . & 0 & . & . & . \\ . & . & . & 1 & 0 \\ 0 & 0 & . & 0 & d_m(s) \end{bmatrix};$$

call the second matrix in the product $\Delta(s)$; note that $\Delta(s)$ is $(m \times m)$. Observe finally that $\Delta(s)V^{-1}(s)$ is a right common $(m \times m)$ polynomial matrix factor of $P(s)$ and $Q(s)$, since $F(s) = U^{-1}(s)\hat{\Lambda}(s)\Delta(s)V^{-1}(s)$; furthermore, $\det \Delta(s)V^{-1}(s)$ is $d_m(s)$, which is *not* a nonzero real number, and the right relative primeness of $P(s)$ and $Q(s)$ is therefore contradicted.

Thus, condition (a) implies that $\Lambda(s)$ is a real, constant matrix which has an obvious real $(m \times (m+p))$ left inverse Γ (c.f. Theorem 1.28). We conclude that

$$[V(s)\Gamma U(s)]F(s) = I_m ;$$

partitioning the matrix in brackets as $[A(s) \;\; B(s)]$, with $A(s)$ $(m \times m)$ and $B(s)$ $(m \times p)$ yields condition (b).

It is clear that condition (c) follows from (b); to see that (c) implies (a), suppose that the real $(m \times m)$ polynomial matrix $X(s)$ is such that $P(s) = \hat{P}(s)X(s)$ and $Q(s) = \hat{Q}(s)X(s)$, with the hatted matrices polynomial. Since we also have $F(s) = \hat{F}(s)X(s)$, with $\hat{F}(s)$ polynomial, the determinant of $X(s)$ must be a nonzero real number; if $\det X(s)$ were a polynomial of positive degree, then there would exist some $s_o \in \mathbf{C}$ for which $F(s_o)$ would have deficient rank. Thus (c) implies that $X(s)$ is unimodular, and (a) follows. □

There is a "left-handed" version of Theorem 22.9 whose proof is essentially the transpose of the proof of Theorem 22.9.

22.10 Theorem: *Let $N(s)$ be a real $(p \times m)$ polynomial matrix, and let $R(s)$ be a real $(p \times p)$ polynomial matrix. The following three conditions are equivalent:*

(a) $N(s)$ and $R(s)$ are left relatively prime.

(b) There exist real polynomial matrices $A(s)$ and $B(s)$, which have respective sizes $(p \times p)$ and $(m \times p)$, such that $R(s)A(s) + N(s)B(s) = I_p$.

(c) The $(p \times (p+m))$ complex matrix

$$F(s) = \begin{bmatrix} R(s_o) & N(s_o) \end{bmatrix}$$

has full rank p for every $s_o \in \mathbf{C}$. □

We are now equipped to prove Assertion 22.6(a).

22.11 Theorem: *Let $Z(s)$ be a real strictly proper $(p \times m)$ rational matrix. Then the following statements hold.*

(a) An RMFD $Z(s) = P(s)Q^{-1}(s)$ is irreducible if and only if it is minimal. Moreover, if $Z(s) = P(s)Q^{-1}(s)$ and $Z(s) = \hat{P}(s)\hat{Q}^{-1}(s)$ are two minimal RMFD's, then there exists a real $(m \times m)$ unimodular polynomial matrix $U(s)$ such

that $\hat{P}(s) = P(s)U(s)$ *and* $\hat{Q}(s) = Q(s)U(s)$.

(*b*) *An LMFD* $Z(s) = R^{-1}(s)N(s)$ *is irreducible if and only if it is minimal. Moreover, if* $Z(s) = R^{-1}(s)N(s)$ *and* $Z(s) = \hat{R}^{-1}(s)\hat{N}(s)$ *are two minimal LMFD's, then there exists a real* $(p \times p)$ *unimodular polynomial matrix* $U(s)$ *such that* $\hat{N}(s) = U(s)N(s)$ *and* $\hat{R}(s) = U(s)R(s)$.

Proof: We prove only (a); the proof of (b) is similar. It is clear that a minimal RMFD $P(s)Q^{-1}(s)$ must be irreducible; otherwise, there would exist an $(m \times m)$ polynomial matrix $U(s)$ having a nonconstant polynomial as its determinant such that $P(s) = \hat{P}(s)U(s)$ and $Q(s) = \hat{Q}(s)U(s)$ for polynomial matrices $\hat{P}(s)$ and $\hat{Q}(s)$. In that case, $\hat{P}(s)\hat{Q}^{-1}(s)$ would be another RMFD of $Z(s)$, and $\det Q(s)$ would have degree strictly *larger* than $\det\hat{Q}(s)$, contradicting minimality of $P(s)Q^{-1}(s)$.

To show that an irreducible RMFD is minimal, suppose that $Z(s) = P(s)Q^{-1}(s)$ is irreducible and that $Z(s) = \hat{P}(s)\hat{Q}^{-1}(s)$ is minimal (and hence irreducible). By Theorem 22.9, there exist polynomial matrices $A(s)$, $B(s)$, $\hat{A}(s)$, and $\hat{B}(s)$ such that

$$\begin{aligned} A(s)P(s) + B(s)Q(s) &= I_m \\ \hat{A}(s)\hat{P}(s) + \hat{B}(s)\hat{Q}(s) &= I_m . \end{aligned}$$

Now invoke $P(s)Q^{-1}(s) = \hat{P}(s)\hat{Q}^{-1}(s)$; a straightforward manipulation yields

$$\begin{aligned} A(s)\hat{P}(s) + B(s)\hat{Q}(s) &= Q^{-1}(s)\hat{Q}(s) \\ \hat{A}(s)P(s) + \hat{B}(s)Q(s) &= \hat{Q}^{-1}(s)Q(s) , \end{aligned}$$

from which we conclude that the two matrices on the right-hand sides, which are *inverses* of each other, are polynomial matrices, and are therefore unimodular. Set $U(s) = Q^{-1}(s)\hat{Q}(s)$; then $\hat{P}(s) = P(s)U(s)$ and $\hat{Q}(s) = Q(s)U(s)$, so that the degree of $\det Q(s)$ equals the degree of $\det\hat{Q}(s)$, implying that $P(s)Q^{-1}(s)$ is also minimal. The foregoing argument also proves the assertion that any two minimal RMFD's $P(s)Q^{-1}(s)$ and $\hat{P}(s)\hat{Q}^{-1}(s)$ are related as in the theorem statement. □

It should be noted that the preceding results, when taken together, provide a means by which an arbitrary RMFD $Z(s) = P(s)Q^{-1}(s)$ or LMFD $Z(s) = R^{-1}(s)N(s)$ for a real $(p \times m)$ strictly proper rational matrix $Z(s)$ may be *reduced* to a *minimal* one. Suppose that $P(s)Q^{-1}(s)$ is an RMFD for $Z(s)$; as in the proof of Theorem 22.9, set

$$F(s)=\begin{bmatrix} Q(s) \\ P(s) \end{bmatrix}.$$

Suppose that $\Lambda(s)=U(s)F(s)V(s)$ is the Smith form for $F(s)$; define the $(m\times m)$ matrix $\Delta(s)$ to be the diagonal matrix whose (i,i) element is the ith invariant factor $d_i(s)$; alternatively, $[\Delta(s)]_{ii}=[\Lambda(s)]_{ii}$ for $1\leqslant i\leqslant m$. Then

$$F(s)=\hat{F}(s)\Delta(s)V^{-1}(s),$$

where $\hat{F}(s)$ is $U^{-1}(s)$ times the $((m+p)\times m)$ matrix which has an $(m\times m)$ identity matrix as its first m rows and zeroes elsewhere. Evidently, $\hat{F}(s)$ partitions as

$$\hat{F}(s)=\begin{bmatrix} \hat{Q}(s) \\ \hat{P}(s) \end{bmatrix},$$

and $Z(s)=\hat{P}(s)\hat{Q}^{-1}(s)$ is another RMFD. Since the Smith form of $\hat{F}(s)$ is $U(s)\hat{F}(s)$, which has ones as its invariant factors, we may conclude that $\hat{F}(s_o)$ has (full) rank m for every $s_o\in\mathbf{C}$. Thus, by Theorem 22.9, $\hat{P}(s)$ and $\hat{Q}(s)$ are right relatively prime, and the "hatted" RMFD is therefore minimal (c.f. Theorem 22.11). A similar procedure may be employed so as to reduce a nonminimal LMFD to a minimal one. It should be noted that a minimal MFD of $Z(s)$ may also be constructed *directly* by way of the so-called *Smith-McMillan form* for $Z(s)$; see Theorem 22.19 below.

Observe that Theorem 22.11 does *not* assert that if $P(s)Q^{-1}(s)$ is a minimal RMFD of $Z(s)$ and $R^{-1}(s)N(s)$ is a minimal LMFD of $Z(s)$, then the degrees of $\det Q(s)$ and $\det R(s)$ are the same. As it happens, this statement turns out to be true; in order to prove it, we require an elementary algebraic fact.

22.12 Lemma: *Suppose that the $((m+p)\times(m+p))$ polynomial matrix $V(s)$ is nonsingular, and the $(m\times m)$ submatrix $V_1(s)$ is also nonsingular, where*

$$V(s)=\begin{bmatrix} V_1(s) & V_2(s) \\ V_3(s) & V_4(s) \end{bmatrix}.$$

Then

(a) $H(s) = V_4(s) - V_3(s)V_1^{-1}(s)V_2(s)$ *is also nonsingular;*

(b) $\det V(s) = \det H(s) \det V_1(s)$; *and*

(c) $V^{-1}(s)$ *is given by*

$$\begin{bmatrix} V_1^{-1}(s) + V_1^{-1}(s)V_2(s)H^{-1}(s)V_3(s)V_1^{-1}(s) & -V_1^{-1}(s)V_2(s)H^{-1}(s) \\ -H^{-1}(s)V_3(s)V_1^{-1}(s) & H^{-1}(s) \end{bmatrix}.$$

Proof: Parts (a) and (b) follow immediately from the equation

$$\begin{bmatrix} I_m & 0_{(m\times p)} \\ -V_3(s)V_1^{-1}(s) & I_p \end{bmatrix} V(s) = \begin{bmatrix} V_1(s) & V_2(s) \\ 0 & H(s) \end{bmatrix},$$

since the matrix premultiplying $V(s)$ has determinant 1. To get part (c), multiply the last equation on the left by

$$\begin{bmatrix} V_1^{-1}(s) & 0 \\ 0 & H^{-1}(s) \end{bmatrix} \begin{bmatrix} I_m & -V_2(s)H^{-1}(s) \\ 0_{(p\times m)} & I_p \end{bmatrix}.$$ □

We are now equipped to prove Assertion 22.6(b).

22.13 Theorem: *Let* $Z(s) = P(s)Q^{-1}(s)$ *and* $Z(s) = R^{-1}(s)N(s)$ *be, respectively, a minimal RMFD and a minimal LMFD for the real strictly proper* $(p \times m)$ *rational matrix* $Z(s)$. *Then the degree of* $\det Q(s)$ *equals the degree of* $\det R(s)$.

Proof: A simple modification of the first part of the proof of Theorem 22.9 yields a method for constructing a unimodular matrix $U(s)$ having size $((m+p)\times(m+p))$ such that

$$U(s)\begin{bmatrix} Q(s) \\ P(s) \end{bmatrix} = \begin{bmatrix} I_m \\ 0 \end{bmatrix}. \qquad (*)$$

Let $V(s) = U^{-1}(s)$; partition $U(s)$ and $V(s)$ conformably as follows:

$$U(s) = \begin{bmatrix} U_1(s) & U_2(s) \\ U_3(s) & U_4(s) \end{bmatrix}; \quad V(s) = \begin{bmatrix} V_1(s) & V_2(s) \\ V_3(s) & V_4(s) \end{bmatrix}.$$

From equation (*) above, it follows that $V_1(s) = Q(s)$ and $V_3(s) = P(s)$. In particular, $V_1(s)$ is nonsingular; hence, by Lemma 22.12, $U_4(s)$ is *also* nonsingular, and equals $H^{-1}(s)$, where $H(s) = V_4(s) - V_3(s)V_1^{-1}(s)V_2(s)$.

Moreover, since $U(s)V(s) = I_{m+p}$, we have

$$U_3(s)V_2(s) + U_4(s)V_4(s) = I_p ,$$

so that $U_3(s)$ and $U_4(s)$ are left relatively prime, by Theorem 22.10. Furthermore, since

$$U_3(s)V_1(s) + U_4(s)V_3(s) = 0 ,$$

and since $Z(s) = P(s)Q^{-1}(s) = V_3(s)V_1^{-1}(s)$, we have $Z(s) = -U_4^{-1}(s)U_3(s)$.

The last equation gives an LMFD of $Z(s)$; it is irreducible since $U_3(s)$ and $U_4(s)$ are left relatively prime, and hence is minimal by Theorem 22.11. To see that the degrees of $\det U_4(s)$ and $\det Q(s)$ coincide, observe that, by Lemma 22.12, $\det V(s)$ is given by $\det V_1(s)\det H(s)$; since $V_1(s) = Q(s)$ and $U_4(s) = H^{-1}(s)$, we conclude that

$$\det V(s) = \frac{\det Q(s)}{\det U_4(s)} .$$

The equality of the numerator and denominator degrees is a consequence of the unimodularity of $V(s)$; in fact, the two polynomials in question are *nonzero real multiples* of one another, which is an even stronger result. □

Of Assertions 22.6, we have yet to prove (c); before doing that, we'll have to leap over one more significant algebraic hurdle. First of all, Theorem 22.13 enables us to define a new invariant which is associated with any real $(p \times m)$ proper rational matrix.

22.14 Definition: Let $G(s)$ be a real $(p \times m)$ proper rational matrix; let the real $(p \times m)$ matrix D be such that $G(s) = Z(s) + D$, where $Z(s)$ is strictly proper. The *fractional degree* of $G(s)$, which we write $\delta_F(G(s))$, is defined as the degree of $\det Q(s)$ in any minimal RMFD $Z(s) = P(s)Q^{-1}(s)$. Equivalently, by Theorem 22.13, $\delta_F(G(s))$ equals the degree of $\det R(s)$ in any LMFD $Z(s) = R^{-1}(s)N(s)$. □

We shall show eventually that $\delta_F(G(s))$ is the same as the McMillan degree of the m-input, p-output time-invariant continuous-time input-output linear system whose transfer function is given on its region of convergence by $G(s)$. For now, observe that $\delta_F(G(s))$ is equal to the sum of the degrees of the invariant factors $d_i(s)$ appearing in the Smith form of $Q(s)$ whenever $P(s)Q^{-1}(s)$ is a minimal RMFD of $Z(s)=G(s)-D$. To see this, note that $\det Q(s)$ is a nonzero real constant multiple of the *product* of the $d_k(s)$, since

$$\det Q(s) = \det(U^{-1}(s))d_1(s)\dots d_k(s)\det(V^{-1}(s)),$$

with $U(s)$ and $V(s)$ unimodular. (C.f. Theorem 22.8 and the discussion which follows.)

It is easy to show that the fractional degree of $G(s)$ is a *lower bound* for the McMillan degree of $G(s)$; to show that the two degrees are *equal*, however, requires a rather cumbersome constructive argument. That construction has its good side, though; it provides an *algorithm* for constructing a minimal realization (A, B, C, D) for the system whose transfer function is $G(s)$.

To see why $\delta_F(G(s))$ bounds the McMillan degree of $G(s)$ from *below*, suppose that (A, B, C, D) is a minimal realization for $G(s)$ and that A is $(n\times n)$, so the McMillan degree of $G(s)$ is n. We have $G(s)=C(sI_n-A)^{-1}B+D$ (c.f. Facts 21.10); consider the $(n\times m)$ rational matrix $Y(s)=(sI_n-A)^{-1}B$. Observe that $Y(s)$ is strictly proper; furthermore, the expression defining $Y(s)$ gives a *left matrix fraction description* of $Y(s)$, which turns out to be irreducible. To see this, consider the $(n\times(n+m))$ matrix

$$F(s) = \left|(sI_n-A)\;\; B\right|.$$

By Theorem 21.13, (A, B) is a reachable pair; this implies that $F(s_o)$ has full rank n for every $s_o\in\mathbf{C}$. Otherwise, there would exist some $s_o\in\mathbf{C}$ and some $z_o\in\mathbf{C}^n$ such that

$$z_o^TA = s_o z_o^T$$

and $z_o^TB=0$. These two equations taken together tell us that $z_o^TA^kB=0$ for every $k\geqslant 0$, which in turn implies that $z_o^TQ_r(A,B)=0$, contradicting reachability of the pair (A,B).

Thus (c.f. Theorem 22.10), $Y(s)=(sI_n-A)^{-1}B$ is an *irreducible*, hence *minimal*, LMFD of $Y(s)$. Accordingly, $\delta_F(Y(s))$ is n. We claim now that $\delta_F(CY(s))\leqslant\delta_F(Y(s))$; to see this, suppose that $Y(s)=P(s)Q^{-1}(s)$ were a minimal *right* matrix fraction description of $Y(s)$; then $[CP(s)]Q^{-1}(s)$

is an RMFD of $CY(s)$ which could conceivably be non-minimal, implying that $\delta_F(CY(s)) \leqslant n$. Accordingly, $\delta_F(G(s))$, which is the same as $\delta_F(CY(s))$, is at most n.

To show that $\delta_F(G(s)) = n$, we'll construct a realization (A, B, C, D) for the system whose transfer function is $G(s)$; the "A-matrix" in this realization will have size $(\delta_F(G(s)) \times \delta_F(G(s)))$. Constructing such a realization enables us to conclude immediately that the McMillan degree n of $G(s)$ is *less than or equal to* $\delta_F(G(s))$; the idea is that the realization could, conceivably, be non-minimal.

In view of our earlier remarks, however, the realization *must* be minimal, since $\delta_F(G(s)) \leqslant n$. The reachability of the pair (A, B) will be evident upon inspection; for this reason, the special realization (A, B, C, D) is known as the *standard controller realization* for the system whose transfer function is $G(s)$. The pair (A, C) turns out to be observable, as well; we won't check this fact *directly*, since we know it to be true because of the minimality of (A, B, C, D) along with Theorem 21.13. As a consequence of our construction, we are able to conclude that Assertion 22.6(c) is indeed correct.

In order to build the special realization, we require a *special* right matrix fraction description $Z(s) = P(s)Q^{-1}(s)$ for $Z(s)$, the "strictly proper part" of $G(s)$. The RMFD $P(s)Q^{-1}(s)$ is special because the matrix $Q(s)$ has a special form which reflects certain important invariants of the input-output system whose transfer function is $G(s)$. We begin with a definition; for an alternative treatment, see [Forney].

22.15 Definition: Let $Q(s)$ be a real $(m \times m)$ nonsingular polynomial matrix.

(a) For each j, $1 \leqslant j \leqslant m$, the *$j$th unordered column index of* $Q(s)$, written $\hat{l}_j(Q(s))$, is the degree of the highest degree entry in the jth column of $Q(s)$.

(b) The *ordered column indices* of $Q(s)$, denoted by $(l_1(Q(s)), l_2(Q(s)), \ldots, l_m(Q(s)))$, are the $\hat{l}_j(Q(s))$ written in decreasing order.

(c) The *high-order coefficient matrix* of $Q(s)$, written Q_H, is the real $(m \times m)$ matrix whose (i, j)-element is the coefficient of $s^{\hat{l}_j(Q(s))}$ in $[Q(s)]_{ij}$. □

Consider, as an example, the polynomial matrix

$$Q(s) = \begin{bmatrix} 2s^2 & s \\ s & s^3+1 \end{bmatrix}.$$

$Q(s)$ is evidently nonsingular; the unordered column indices are given by $\hat{l}_1(Q(s)) = 2$ and $\hat{l}_2(Q(s)) = 3$, while the ordered column indices are $l_1(s) = 3$ and $l_2(s) = 2$. The high-order coefficient matrix Q_H of $Q(s)$ is given by

$$Q_H = \begin{bmatrix} 2 & 0 \\ 0 & 1 \end{bmatrix}.$$

Suppose now that we are given a minimal RMFD $Z(s) = P(s)Q^{-1}(s)$. We know (c.f. Theorem 22.11) that if $U(s)$ is any unimodular $(m \times m)$ matrix, and we set $\hat{P}(s) = P(s)U(s)$ and $\hat{Q}(s) = Q(s)U(s)$, then $Z(s) = \hat{P}(s)\hat{Q}^{-1}(s)$ is *also* a minimal RMFD. Suppose that the high-order coefficient matrix Q_H is not invertible; we shall construct a unimodular polynomial matrix $U(s)$ such that the high-order coefficient matrix of $Q(s)U(s)$ *is* invertible. The reason for doing so will become apparent shortly.

First choose an $(m \times m)$ matrix P which is a product of *permutation matrices* $P_m(k, l)$ (see Definition 1.6 and Exercise 1.7) so that $Q(s)P$ satisfies $l_j(Q(s)P) = \hat{l}_j(Q(s)P)$ for each j, $1 \leqslant j \leqslant m$. All this means is that, for each j, the highest degree polynomial in the jth column of $Q(s)P$ has degree at least as high as the highest degree polynomial in the $(j+1)$th column.

Since Q_H is not invertible, there exists some nonzero $w \in \mathbf{R}^m$ such that $Q_H w = 0$; we may assume that the first nonzero element in w occurs in the kth position and is a 1. Let $E(s)$ be the $(m \times m)$ polynomial matrix with ones on the diagonal and whose (i, k) element *below* the diagonal (i.e., when $i > k$) is given by the monomial $w_i s^{l_k - l_i}$; set all the remaining elements of $E(s)$ equal to zero. The matrix $E(s)$ is evidently unimodular; moreover, it is not hard to check that $Q(s)PE(s)$ has unordered column indices which satisfy

$$\hat{l}_j(Q(s)PE(s)) = \hat{l}_j(Q(s)P)$$

for all $j \neq k$, and $\hat{l}_k(Q(s)PE(s)) < \hat{l}_k(Q(s)P)$.

If the new matrix $Q(s)PE(s)$ has a singular high-order coefficient matrix, repeat the procedure on the matrix

$Q(s)PE(s)$; otherwise, stop. Since each new step *reduces* one unordered column index and leaves the others the same, the algorithm must terminate after a while; the final product is an $(m \times m)$ nonsingular polynomial matrix $\hat{Q}(s)$ which has an invertible high-order coefficient matrix $\hat{Q}_H$. Note also that $\hat{Q}(s) = Q(s)U(s)$, where the $(m \times m)$ matrix $U(s)$ is unimodular.

We conclude that there exists a unimodular $U(s)$ such that the high-order coefficient matrix of $Q(s)U(s)$ is invertible. It is worth making a few additional comments about the *reduction procedure* outlined above. The algorithm may be viewed as providing a means by which the ordered column indices of a nonsingular polynomial matrix $Q(s)$ may be "reduced" provided that the high-order coefficient matrix Q_H is not invertible.

Specifically, we may impose an *ordering* on the set of m-tuples of ordered column indices as follows: define any m-tuple $l_1, l_2, \ldots, l_m$ of nonnegative integers, arranged in decreasing order, to be *smaller* than another such m-tuple $\hat{l}_1, \hat{l}_2, \ldots, \hat{l}_m$ if and only if there is some k such that $l_k < \hat{l}_k$, and $l_j = \hat{l}_j$ for $j < k$. (Observe that nothing is said about the relationship between l_j and $\hat{l}_j$ if $j > k$.) It is not hard to see that if the "input" to our algorithm is a given $(m \times m)$ nonsingular polynomial matrix $Q(s)$ whose Q_H is singular, then the "output" is another $(m \times m)$ nonsingular polynomial matrix $\hat{Q}(s) = Q(s)\hat{U}(s)$ whose m-tuple of *ordered* column indices is *smaller*, with respect to the aforementioned ordering on m-tuples, than the corresponding m-tuple for $Q(s)$.

If $U(s)$ is allowed to range over the set of *all* $(m \times m)$ unimodular matrices, there will be found an m-tuple of ordered column indices which is *smallest*, with respect to the above ordering, among all such m-tuples for matrices of the form $Q(s)U(s)$. This particular set of column indices has a special importance in linear system theory.

22.16 Definition: Let $Q(s)$ be a real $(m \times m)$ nonsingular polynomial matrix. Let $\hat{Q}(s) = Q(s)\hat{U}(s)$, with $\hat{U}(s)$ real $(m \times m)$ and unimodular, have the *smallest* m-tuple of ordered column indices of any matrix of the form $Q(s)U(s)$, with $U(s)$ real $(m \times m)$ and unimodular. The *jth column Kronecker index* of $Q(s)$, written $n_j(Q(s))$, is defined to be the jth ordered column index of $\hat{Q}(s)$; i.e.,

$n_j(Q(s)) = l_j(\hat{Q}(s))$. □

Thus, the column Kronecker indices of an $(m \times m)$ nonsingular polynomial matrix constitute a *minimal m-tuple* of ordered column indices for matrices obtainable from $Q(s)$ through right multiplication by a unimodular matrix. Recall that if $Z(s) = P(s)Q^{-1}(s)$ is a minimal RMFD for the real $(p \times m)$ strictly proper rational matrix $Z(s)$, then every other minimal RMFD $Z(s) = \hat{P}(s)\hat{Q}^{-1}(s)$ is obtainable as $\hat{P}(s) = P(s)U(s)$, $\hat{Q}(s) = Q(s)U(s)$ for some unimodular $U(s)$ (c.f. Theorem 22.11); thus, the column Kronecker indices may actually be regarded as a property of $Z(s)$.

We saw above that if the high-order coefficient matrix Q_H of $Q(s)$ is *singular*, then the ordered column indices of $Q(s)$ are *not* the column Kronecker indices. As it happens, nonsingularity of the high-order coefficient matrix $\hat{Q}_H$ is *equivalent* to the property that the ordered column indices of $\hat{Q}(s)$ are the column Kronecker indices. In fact, somewhat more may be said.

22.17 Theorem: *Let $Q(s)$ be a real $(m \times m)$ nonsingular polynomial matrix. Let $n_1, n_2, \ldots, n_m$ be the column Kronecker indices for $Q(s)$. Then*

(a) if $U(s)$ is $(m \times m)$ and unimodular, then $\hat{Q}(s) = Q(s)U(s)$ has ordered column indices $l_j(\hat{Q}(s)) = n_j(Q(s))$ if and only if its high-order coefficient matrix $\hat{Q}_H$ is invertible, and

(b) the degree of $\det Q(s)$ *is $n_1 + n_2 + \ldots + n_m$.*

Proof: (a) We have seen that if $\hat{Q}_H$ is not invertible, then there exists a unimodular $U(s)$ such that $\hat{Q}(s)U(s)$ has a "smaller" set of ordered column indices; hence $\hat{Q}_H$ must be invertible if $n_j = l_j(\hat{Q}(s))$ for all j. Conversely, suppose that $\hat{Q}_H$ is invertible. We shall now demonstrate that if $U(s)$ is $(m \times m)$, and if the m-tuple of ordered column indices of $R(s) = \hat{Q}(s)U(s)$ is "smaller" than the corresponding m-tuple for $\hat{Q}(s)$, then $U(s)$ can't be unimodular (in fact, $U(s)$ must be *singular*).

Assume temporarily that the ordered and unordered column indices of $\hat{Q}(s)$ are the same; assume the same for $R(s)$. Suppose that $l_1(\hat{Q}(s)) > l_1(R(s))$. We show, under this assumption, that the first row of $U(s)$ is zero, implying that

$U(s)$ can't be unimodular. Set

$$d = \max_{1 \leqslant k \leqslant m} \{\text{degree}([U(s)]_{k1}) + l_k(\hat{Q}(s))\} .$$

We claim that $l_1(R(s)) = d$; to see this, let $x \in \mathbf{R}^m$ be the vector whose k th component is the coefficient of $s^{d-l_k(Q(s))}$ in $[U(s)]_{k1}$.

By definition, $x \neq 0$; furthermore, since $\hat{Q}_H$ is invertible, $y = \hat{Q}_H x \neq 0$. But y is the vector whose k th component is the coefficient of s^d in $[R(s)]_{k1}$; by definition of d, which implies that $l_1(R(s)) \leqslant d$, we have $l_1(R(s)) = d$.

Now , if $l_1(R(s))$ is to be less than $l_1(\hat{Q}(s))$, then we need, again by definition of d, $[U(s)]_{11} = 0$. A similar argument shows that $l_1(R(s)) < l_1(\hat{Q}(s))$ implies that *every* element of the first row of $U(s)$ must be zero; just perform a similar analysis of each column of $R(s)$, which is simply $\hat{Q}(s)$ times the corresponding column of $U(s)$.

Thus, $l_1(R(s)) = l_1(\hat{Q}(s))$; assume next that $j_o > 1$ is the first value of j such that $l_j(R(s)) < l_j(\hat{Q}(s))$, and that $l_j(R(s)) = l_j(\hat{Q}(s))$ for all $j < j_o$. We can conclude from the above argument that the first $j_o - 1$ rows of $U(s)$ contain a $((j_o-1) \times (j_o-1))$ submatrix of *real numbers* on the left and zeroes in all the remaining columns. Running through the argument again shows that the j_o th row of $U(s)$, in positions to the right of and including the j_o th column, *also* consists of zeroes, implying once again that $\det U(s) = 0$ and that $U(s)$ isn't unimodular.

Finally, let's do away with the assumption that the ordered and unordered column indices for $\hat{Q}(s)$ and $R(s)$ are equal. If the assumption isn't true, find permutation matrices P_1 and P_2 so that the assumption holds for $\hat{Q}(s)P_1$ and $R(s)P_2$. Conclude that if the ordered column indices for $R(s) = \hat{Q}(s)U(s)$ are "smaller" than those for $\hat{Q}(s)$, then $U(s)$ can't be unimodular, since then

$$\hat{Q}(s)P_1[P_1^{-1}U(s)P_2] = R(s)P_2 ,$$

and unimodularity of the bracketed matrix is equivalent to unimodularity of $U(s)$.

(b) Let $U(s)$ be a unimodular matrix such that $\hat{Q}(s) = Q(s)U(s)$ has ordered column indices equal to the column Kronecker indices $n_1(Q(s)), \ldots, n_m(Q(s))$. Since $U(s)$ is unimodular, $\det Q(s)$ has the same degree as $\det\hat{Q}(s)$.

By (a), the high-order coefficient matrix $\hat{Q}_H$ of $\hat{Q}(s)$ is invertible, and hence has nonzero determinant; moreover, the highest degree term in $\det\hat{Q}(s)$ is evidently

$$(\det\hat{Q}_H)s^{n_1+n_2+\ \cdots\ +n_m} \ ,$$

from which the result follows. □

The intricacy of the foregoing manipulations has probably drawn at least a *part* of the reader's mind into an algebraic vortex of sorts; it is worthwhile at this point to pause for a moment and attempt to re-establish the system-theoretic relevance of the major results. We began with a real proper $(p \times m)$ rational matrix $G(s)$, which can be thought of as an expression for the transfer function of some m-input, p-output time-invariant continuous-time input-output linear system. We wrote $G(s) = Z(s) + D$, where $Z(s)$ is *strictly* proper; our objective was to forge a connection between matrix fraction descriptions for $Z(s)$ and realizations for the input-output system.

After defining RMFD's and LMFD's, along with the notions of minimality and irreducibility of such factorizations, we showed, by means of Theorems 22.8 through 22.11, that a matrix fraction description is irreducible if and only if it is minimal, and that the degree of $\det Q(s)$ in any minimal RMFD $P(s)Q^{-1}(s)$ (or of $\det R(s)$ in any minimal LMFD $R^{-1}(s)N(s)$) is equal to the sum of the degrees of any set of invariant factors of $Q(s)$ (or of $R(s)$). Later (c.f. Theorem 22.13), we proved the equality of these two degrees, and denoted their common value by $\delta_F(G(s))$, the *fractional degree* of $G(s)$.

The ensuing portion of the section was designed to pave the way toward a proof of Assertion 22.6(c) that $\delta_F(G(s))$ and the McMillan degree of $G(s)$ are the same. We have shown already (c.f. the discussion which follows Definition 22.14) that $\delta_F(G(s))$ may not *exceed* the McMillan degree of $G(s)$; presently, we shall prove the *equality* of the two degrees by constructing a realization (A, B, C, D) for the associated input-output system whose A-matrix has size $(\delta_F(G(s)) \times \delta_F(G(s)))$.

Accordingly, suppose that $G(s)$ is a real $(p \times m)$ proper rational matrix; let $Z(s) = G(s) - D$, the strictly proper part of $G(s)$, be defined as usual. Let $Z(s) = P(s)Q^{-1}(s)$ be a minimal RMFD of $Z(s)$; assume, as we may by Theorem 22.17, that the ordered column indices $l_1(Q(s)), \ldots, l_m(Q(s))$ are the Kronecker column indices of $Q(s)$; to economize on notation, set $n_j = l_j(Q(s))$ for $1 \leqslant j \leqslant m$.

We may also assume (by multiplying $P(s)$ and $Q(s)$ on the right by appropriate permutation matrices) that the *unordered* column indices of $Q(s)$ coincide with the n_j, as well; thus, the highest-degree element in column j of $Q(s)$ has degree n_j for each j. We also know (Theorem 22.17(b) and Definition 22.14) that

$$\begin{aligned}\delta_F(G(s)) &= \text{degree}(\det(Q(s))) \\ &= n_1+n_2+\ldots+n_m \equiv n .\end{aligned}$$

We'll now construct a triple (A, B, C) of real matrices, having respective sizes $(n \times n)$, $(n \times m)$, and $(p \times n)$, such that $Z(s) = C(sI_n - A)^{-1}B$.

First, let ρ be the largest value of j for which $n_j > 0$; thus $n_j = 0$ if and only if $j > \rho$. Since $P(s) = Z(s)Q(s)$, we have

$$[P(s)]_{ij} = \sum_{k=1}^{m} [Z(s)]_{ik} [Q(s)]_{kj} ;$$

since the degree of $[Q(s)]_{kj}$ is at most n_j for each j, and since $Z(s)$ is *strictly proper*, the degree of $[P(s)]_{ij}$ must be *strictly less* than n_j for each j, $1 \leqslant j \leqslant m$. In particular, $[P(s)]_{ij} = 0$ if $j > \rho$. (As usual, we are observing the convention that the zero polynomial has degree $-\infty$.)

For each integer $k > 0$, define the polynomial k-vector $\mathbf{s}_k$ as follows:

$$\mathbf{s}_k = \begin{bmatrix} 1 \\ s \\ s^2 \\ \cdot \\ \cdot \\ s^{k-1} \end{bmatrix}.$$

Let $\mathbf{S}$ be the $(n \times m)$ polynomial matrix whose jth column $\mathbf{S}^j$ is given as follows: $\mathbf{S}^j = 0$ if $n_j = 0$ (i.e., if $j > \rho$). Otherwise, $\mathbf{S}^j$ has $n_1+n_2+\ldots+n_{j-1}$ zeroes at the top; then the vector $\mathbf{s}_{n_j}$ occupying the next n_j positions; then zeroes in the remaining positions. For example, if $m = 3$, $n_1 = 3$, $n_2 = 2$, and $n_3 = 1$, then $n = 6$, and

$$\mathbf{S} = \begin{bmatrix} 1 & 0 & 0 \\ s & 0 & 0 \\ s^2 & 0 & 0 \\ 0 & 1 & 0 \\ 0 & s & 0 \\ 0 & 0 & 1 \end{bmatrix} .$$

Our earlier observation about the degree constraints on the entries of $P(s)$ guarantees that there exists a unique real $(p \times n)$ matrix C such that $P(s) = C\mathbf{S}$. Thus, $Z(s) = C\mathbf{S}Q^{-1}(s)$. We'll now construct a real $(n \times n)$ matrix A and a real $(n \times m)$ matrix B such that

$$\mathbf{S}Q^{-1}(s) = (sI_n - A)^{-1}B \ .$$

Note that the last equation is equivalent to

$$s\mathbf{S} = A\mathbf{S} + BQ(s) \ .$$

Recall that Q_H, the high-order coefficient matrix of $Q(s)$, is invertible; recall also that the high-degree element in the jth column of $Q(s)$ has degree n_j for each j. We may therefore write the matrix $Q_H^{-1}Q(s)$ as

$$Q_H^{-1}Q(s) = \begin{bmatrix} s^{n_1} & 0 & . & 0 & 0 \\ 0 & s^{n_2} & . & . & . \\ . & 0 & . & . & . \\ . & . & . & s^{n_{m-1}} & 0 \\ 0 & 0 & . & 0 & s^{n_m} \end{bmatrix} + Q_1(s) \ .$$

Denote by $\Gamma(s)$ the diagonal matrix on the right-hand side of the last equation; $Q_1(s)$ is an $(m \times m)$ matrix of polynomials whose degrees satisfy the same "column constraints" as the degrees of the the polynomials in $P(s)$; thus $Q_1(s) = G\mathbf{S}$ for a uniquely determined real $(m \times n)$ matrix G.

It remains to find A and B so that

$$s\mathbf{S} = (A + BQ_H G)\mathbf{S} + BQ_H \Gamma(s) \ . \tag{**}$$

Let B be given by $B = \tilde{B}Q_H^{-1}$, where $\tilde{B}$ is the $(n \times m)$ matrix whose jth column is zero if $j > \rho$ and contains a single one at the $n_1 + n_2 + \ldots + n_j$ position (and zeroes otherwise) if $j \leqslant \rho$. Observe that if $i = n_1 + \ldots + n_j$ for some j, then the ith row of $BQ_H G$ is the same as the jth row of G; for all other values of i, the ith row of $BQ_H G$ is zero.

Now define A as follows: if $i = n_1 + \ldots + n_j$ for some j, the ith row of A is the *negative* of the ith row of $BQ_H G$. For all other values of i, the ith row of A contains a 1 in the $(i, i+1)$ position (just above the main diagonal of A) and zeroes elsewhere.

It is left to the reader to complete the tedious but straightforward verification that the matrices A and B satisfy equation (**) above. We may summarize the foregoing development with the following re-statement of Assertion 22.6(c).

22.18 Theorem: *Let $G(s)$ be a real $(p \times m)$ proper rational matrix; let $G(s) = Z(s) + D$, where D is a real $(p \times m)$ matrix and $Z(s)$ is strictly proper. Let $Z(s) = P(s)Q^{-1}(s)$ (or $Z(s) = R^{-1}(s)N(s)$) be a minimal RMFD (or LMFD) of $Z(s)$, and let $n = \delta_F(G(s))$ be the degree of* $\det Q(s)$ *(or of* $\det R(s)$*). Then n is the McMillan degree of the real m-input, p-output time-invariant continuous-time input-output linear system whose transfer function is given on its region of convergence by $G(s)$, and (A, B, C, D) constructed as above is a minimal realization of the system.* □

There is an additional characterization of McMillan degree which is worth mentioning; it is a direct consequence of polynomial matrix manipulations such as those used in proving Theorem 22.8. As it happens, this final description of McMillan degree resembles closely the original definition of *degree of a rational matrix* which first appeared in [McMillan].

We begin by deriving the so-called *Smith-McMillan form* of a real $(p \times m)$ strictly proper rational matrix $Z(s)$. Given such a matrix, let $q(s)$ be the lowest common denominator of all its entries. Observe that $F(s) = q(s)Z(s)$ is then a real $(p \times m)$ polynomial matrix; thus, by Theorem 22.8, we may find unimodular matrices $U(s)$ and $V(s)$ having respective sizes $(p \times p)$ and $(m \times m)$ such that

$$F(s) = U^{-1}(s)\Lambda(s)V^{-1}(s),$$

where $\Lambda(s)$ is the $(p \times m)$ matrix whose (k, k) elements are the invariant factors $d_k(s)$ for $F(s)$; here, k runs from 1 to the smaller of m and p. The other entries in $\Lambda(s)$ are zeroes.

After dividing both sides of the last equation by $q(s)$, we obtain

$$Z(s) = U^{-1}(s)\Delta(s)V^{-1}(s),$$

where $\Delta(s)$ is the $(p \times m)$ matrix whose (k, k) element is $q^{-1}(s)d_k(s)$ and which has zeroes everywhere else. We may express each nonzero element of $\Delta(s)$ in lowest terms, so that, for each k,

$$[\Delta(s)]_{kk} = \frac{\gamma_k(s)}{\phi_k(s)};$$

if $\gamma_k(s) = 0$, set $\phi_k(s) = 1$. The divisibility conditions on the $d_k(s)$ (c.f. Theorem 22.8) guarantee that *none* of the $\gamma_k(s)$ has a nontrivial factor in common with *any* of the $\phi_j(s)$. We are left with the following result.

22.19 Theorem: *Let $Z(s)$ be a real $(p \times m)$ strictly proper rational matrix.*

(a) If $m \leqslant p$, then there exist unimodular matrices $U(s)$ and $V(s)$ having respective sizes $(p \times p)$ and $(m \times m)$ such that

$$U(s)Z(s)V(s) = \Gamma(s)\Phi^{-1}(s),$$

where $\Phi(s)$ is an $(m \times m)$ diagonal polynomial matrix with (k, k) element $\phi_k(s)$ and $\Gamma(s)$ is a $(p \times m)$ polynomial matrix with zeroes at every position except the (k, k) positions, where polynomials $\gamma_k(s)$ appear. Moreover, $\phi_{k+1}(s)$ divides $\phi_k(s)$ and $\gamma_k(s)$ divides $\gamma_{k+1}(s)$ for $1 \leqslant k < m$.

(b) If $p \leqslant m$, then there exist unimodular matrices $U(s)$ and $V(s)$ having respective sizes $(p \times p)$ and $(m \times m)$ such that

$$U(s)Z(s)V(s) = \Phi^{-1}(s)\Gamma(s),$$

where $\Phi(s)$ is a $(p \times p)$ diagonal polynomial matrix with (k, k) element $\phi_k(s)$ and $\Gamma(s)$ is a $(p \times m)$ polynomial matrix with zeroes at every position except the (k, k) positions, where polynomials $\gamma_k(s)$ appear. Moreover, $\phi_{k+1}(s)$ divides $\phi_k(s)$ and $\gamma_k(s)$ divides $\gamma_{k+1}(s)$ for each k, $1 \leqslant k < p$. □

The matrix product appearing on the right-hand side of the relevant equation in Theorem 22.19 is called the *Smith-McMillan form* for $Z(s)$. Observe that the divisibility conditions on the various polynomials imply that the lowest common denominator of the *nonzero* $(k \times k)$ minors of the Smith-McMillan form for $Z(s)$ is precisely $\phi_1(s) \ldots \phi_k(s)$ for each k, $1 \leqslant k \leqslant \min\{m, p\}$. A similar argument to the one used in the proof of Theorem 22.8 shows that this product is *also* the lowest common denominator of all the nonzero $(k \times k)$ minors of $Z(s)$. The idea is that the $(k \times k)$ minors of $Z(s)$ and the $(k \times k)$ minors of the Smith-McMillan form for $Z(s)$ are *polynomial linear combinations of each other*; hence, any common denominator for one set is a common denominator for the other.

Supposing for the moment that $m \leqslant p$, if we set $P(s) = U^{-1}(s)\Gamma(s)$ and $Q(s) = V(s)\Phi(s)$, then $Z(s) = P(s)Q^{-1}(s)$ is an RMFD of $Z(s)$, which turns out to be minimal. To see why $P(s)Q^{-1}(s)$ is minimal, put

$$F(s) = \begin{bmatrix} Q(s) \\ P(s) \end{bmatrix} = \begin{bmatrix} V(s) & 0 \\ 0 & U^{-1}(s) \end{bmatrix} \begin{bmatrix} \Phi(s) \\ \Gamma(s) \end{bmatrix};$$

the relative primeness conditions on the $\phi_k(s)$ and the $\gamma_k(s)$ tell us that $F(s_o)$ has full rank m for all $s_o \in \mathbf{C}$, implying (c.f. Theorem 22.9) that $P(s)Q^{-1}(s)$ is minimal. A similar argument using an LMFD holds when $p \leqslant m$.

The degree of $\det Q(s)$ (or of $\det R(s)$) in the minimal RMFD (or LMFD) of $Z(s)$ constructed from the Smith-McMillan form is, evidently, the sum of the degrees of the $\phi_k(s)$. Theorem 22.18 implies that for any real $(p \times m)$ matrix D, the McMillan degree of the input-output linear system with transfer function $G(s) = Z(s) + D$ is equal to the sum of the degrees of the $\phi_k(s)$ appearing in the Smith-McMillan form for $Z(s)$. Since this sum is precisely the degree of the lowest common denominator of *all* the nonzero minors of $Z(s)$, we have the following convenient characterization of the McMillan degree.

22.20 Theorem: *Let $G(s)$ be a real proper $(p \times m)$ rational matrix. Let $G(s) = Z(s) + D$, where D is a real $(p \times m)$ matrix and $Z(s)$ is strictly proper. The McMillan degree of the real m-input, p-output time-invariant continuous-time input-output linear system with transfer function $G(s)$ is equal to the degree of the lowest common denominator of all of the minor subdeterminants of $Z(s)$.* □

An example would seem to be in order at this point.

22.21 Example: Let $G(s)$ be given by

$$G(s) = \begin{bmatrix} \dfrac{-s}{s+1} & \dfrac{1}{s+2} \\ \dfrac{1}{s} & \dfrac{1}{s+1} \end{bmatrix}.$$

We may write $G(s) = Z(s) + D$, with $Z(s)$ and D given by

$$Z(s) = \begin{bmatrix} \dfrac{1}{s+1} & \dfrac{1}{s+2} \\ \dfrac{1}{s} & \dfrac{1}{s+1} \end{bmatrix}; \qquad D = \begin{bmatrix} -1 & 0 \\ 0 & 0 \end{bmatrix}.$$

The lowest common denominator of all the *entries* in $Z(s)$ is $q(s) = s(s+1)(s+2)$; the lowest common denominator of all the *minors* of $Z(s)$ is $s(s+1)^2(s+2)$, from which we conclude that the system whose transfer function is $G(s)$ has McMillan degree four.

We'll get a minimal RMFD for $Z(s)$ by putting $Z(s)$ in Smith-McMillan form. Let $F(s) = q(s)Z(s)$; putting $F(s)$ in Smith form gives

$$U(s)F(s)V(s) = \begin{bmatrix} -2 & 0 \\ 0 & \tfrac{1}{2}s(s+2) \end{bmatrix},$$

where $U(s)$ and $V(s)$ are the unimodular matrices

$$U(s) = \begin{bmatrix} -1 & 1 \\ (1+\tfrac{1}{2}s) & -\tfrac{1}{2}s \end{bmatrix}; \quad V(s) = \begin{bmatrix} -1 & (-\tfrac{1}{2}s) \\ 1 & (\tfrac{1}{2}s+1) \end{bmatrix}.$$

Dividing through by $q(s)$ yields

$$U(s)Z(s)V(s) = \begin{bmatrix} \dfrac{-2}{s(s+1)(s+2)} & 0 \\ 0 & \dfrac{1}{2(s+1)} \end{bmatrix}.$$

We obtain a minimal RMFD $Z(s) = P(s)Q^{-1}(s)$ by setting

$$P(s) = U^{-1}(s)\begin{bmatrix} -2 & 0 \\ 0 & \tfrac{1}{2} \end{bmatrix} = \begin{bmatrix} -s & \tfrac{1}{2} \\ -(s+2) & \tfrac{1}{2} \end{bmatrix},$$

and

$$Q(s) = V(s)\begin{bmatrix} s(s+1)(s+2) & 0 \\ 0 & s+1 \end{bmatrix}$$
$$= \begin{bmatrix} -s(s+1)(s+2) & -\tfrac{1}{2}(s+1) \\ s(s+1)(s+2) & \tfrac{1}{2}(s+1)(s+2) \end{bmatrix}.$$

Note that the high-order coefficient matrix Q_H of $Q(s)$ is not full rank; we employ the reduction procedure outlined after Definition 22.15 to get

$$\hat{Q}(s) = Q(s)\begin{bmatrix} 1 & 0 \\ -2s & 1 \end{bmatrix}$$
$$= \begin{bmatrix} -2s(s+1) & -\tfrac{1}{2}s(s+1) \\ 0 & \tfrac{1}{2}(s+1)(s+2) \end{bmatrix}.$$

The inverse of the high-order coefficient matrix $\hat{Q}_H$ of $\hat{Q}(s)$ is given by

$$\hat{Q}_H^{-1} = \begin{bmatrix} -\tfrac{1}{2} & -\tfrac{1}{2} \\ 0 & 2 \end{bmatrix},$$

so that

$$\hat{Q}_H^{-1}\hat{Q}(s) = \begin{bmatrix} s^2 & 0 \\ 0 & s^2 \end{bmatrix} + \begin{bmatrix} 0 & 1 & -\tfrac{1}{2} & -\tfrac{1}{2} \\ 0 & 0 & 2 & 3 \end{bmatrix}\mathbf{S},$$

where

$$\mathbf{S} = \begin{bmatrix} 1 & 0 \\ s & 0 \\ 0 & 1 \\ 0 & s \end{bmatrix};$$

the Kronecker column indices of $\hat{Q}(s)$ are $(2, 2)$.

We may now construct a minimal realization (A, B, C, D) for our given system. Let

$$A = \begin{bmatrix} 0 & 1 & 0 & 0 \\ 0 & -1 & \tfrac{1}{2} & \tfrac{1}{2} \\ 0 & 0 & 0 & 1 \\ 0 & 0 & -2 & -3 \end{bmatrix}$$

and

$$B = \begin{bmatrix} 0 & 0 \\ -1/2 & -1/2 \\ 0 & 0 \\ 0 & 2 \end{bmatrix}.$$

As for C, we need to alter the original $P(s)$ to $\hat{P}(s)$ so $\hat{P}(s)\hat{Q}^{-1}(s) = Z(s)$. We find that

$$\hat{P}(s) = P(s)\begin{bmatrix} 1 & 0 \\ -2s & 1 \end{bmatrix},$$

and hence

$$\hat{P}(s) = \begin{bmatrix} -2s & 1/2 \\ -2(s+1) & 1/2 \end{bmatrix}.$$

Writing $\hat{P}(s) = C\mathbf{S}$ gives

$$C = \begin{bmatrix} 0 & -2 & 1/2 & 0 \\ -2 & -2 & 1/2 & 0 \end{bmatrix}.$$

We leave it to the reader to verify that $G(s) = C(sI_n - A)^{-1}B + D$, as desired. □

We close the section by mentioning once again the following important and rather obvious fact: *every result in §22 holds if s is replaced by z and "continuous-time" is replaced by "discrete-time."*

PART IV
STABILITY AND FEEDBACK

23. *Stability of State Space Linear Systems*

In this section, we introduce the fundamental definitions and results which lie at the core of *stability theory* for state space linear systems. We shall see that all of the stability results for state space systems are merely system-theoretic restatements of parallel results for *differential equations* and *difference equations*. The stability theory of differential equations is perhaps the most well-developed part of the theory of differential equations as a whole. Over the last century or so, it has occupied the attention of numerous physical scientists and mathematicians; in particular, a great many Soviet mathematicians have made seminal contributions. There is an enormous mathematical literature on stability theory; our attempt will be to provide a fleeting glimpse at the way the theory specializes to *linear* differential and difference equations.

At the heart of stability theory lies the notion of an *equilibrium* of a differential or difference equation. Roughly speaking, $\bar{x} \in \mathbf{R}^n$ is an equilibrium of a differential or difference equation if and only if the unique solution to the equation starting from initial condition $\bar{x}$ at *any* initial time $t_o \in \mathbf{R}$ is identically equal to $\bar{x}$. That is, if the equation is *started* at $\bar{x}$, then the solution *stays* at $\bar{x}$ forever.

23.1 Definition: Let $t \to A(t)$, $t \in \mathbf{R}$, be a continuous real $(n \times n)$ matrix function. $\bar{x} \in \mathbf{R}^n$ is said to be an *equilibrium* of the differential equation

$$\dot{x}(t) = A(t)x(t), \quad t \in \mathbf{R}, \qquad \text{(DE)}$$

if and only if for every $t_o \in \mathbf{R}$, the unique solution $t \to x(t)$ to the differential equation along with the initial condition $x(t_o) = \bar{x}$ is given by $x(t) \equiv \bar{x}$, $t \in \mathbf{R}$. □

23.2 Definition: Let $k \to A(k)$, $k \in \mathbf{Z}$, be a real $(n \times n)$ discrete-time matrix function. $\bar{x} \in \mathbf{R}^n$ is said to be an *equilibrium* of the difference equation

$$x(k+1) = A(k)x(k) , \quad k \in \mathbf{Z} , \qquad \text{(DfcE)}$$

if and only if for every $k_o \in \mathbf{Z}$, the unique solution $k \to x(k)$ to the difference equation along with the initial condition $x(k_o) = \bar{x}$ is given by $x(k) \equiv \bar{x}$, $k \geqslant k_o$. □

The following observations are immediate consequences of Definitions 22.1 and 22.2.

23.3 Observations: *(a) Let $t \to A(t)$, $t \in \mathbf{R}$, be a continuous real $(n \times n)$ matrix function. Then $\bar{x} \in \mathbf{R}^n$ is an equilibrium of equation (DE) above if and only if $A(t)\bar{x} = 0$ for every $t \in \mathbf{R}$.*

(b) Let $k \to A(k)$, $k \in \mathbf{Z}$, be a real $(n \times n)$ discrete-time matrix function. Then $\bar{x} \in \mathbf{R}^n$ is an equilibrium of equation (DfcE) above if and only if $A(k)\bar{x} = \bar{x}$ for every $k \in \mathbf{Z}$.

Proof: $\bar{x}$ is an equilibrium of (DE) if and only if for every $t_o \in \mathbf{R}$, $t \to x(t) \equiv \bar{x}$ solves (DE) subject to the initial condition $x(t_o) = \bar{x}$. Plugging this $t \to x(t)$ into (DE) yields

$$0 = A(t)\bar{x} , \quad t \in \mathbf{R} .$$

A similar argument proves the discrete-time result (b). □

Recall from §§2-3 that the unique solutions to (DE) and (DfcE) may be written in terms of the initial conditions x_o and the *transition matrices* $\Phi(t, t_o)$ and $\Phi(k, k_o)$ corresponding to the A-matrices in (DE) and (DfcE). Specifically, define $\Phi(t, \tau)$, $t, \tau \in \mathbf{R}$, as the unique $(n \times n)$ matrix solution to

$$\frac{d}{dt}\Phi(t, \tau) = A(t)\Phi(t, \tau) , \quad t, \tau \in \mathbf{R} ,$$

subject to

$$\Phi(\tau, \tau) = I_n ,$$

and $\Phi(k, l)$, for integers $k \geqslant l$, as the solution to

$$\Phi(k+1, l) = A(k)\Phi(k, l) , \quad k \geqslant l ,$$

with

$$\Phi(l, l) = I_n .$$

The unique real n-vector function $t \to x(t)$, $t \in \mathbf{R}$, which satisfies (DE) along with $x(t_o) = x_o$ is then given by

$$x(t) = \Phi(t, t_o)x_o , \quad t \in \mathbf{R};$$

and the real discrete-time n-vector function $k \to x(k)$, $k \geqslant k_o$, which satisfies (DfcE) along with $x(k_o) = x_o$ is

$$x(k) = \Phi(k, k_o)x_o , \quad k \geqslant k_o .$$

We conclude that $\bar{x} \in \mathbf{R}^n$ is an equilibrium of (DE) if and only if $\Phi(t, t_o)\bar{x} = \bar{x}$ for all $t, t_o \in \mathbf{R}$, and that $\bar{x}$ is an equilibrium of (DfcE) precisely when $\Phi(k, k_o)\bar{x} = \bar{x}$ for every pair of integers k, k_o satisfying $k \geqslant k_o$.

Evidently, $\bar{x} = 0$ is *always* an equilibrium of (DE) and (DfcE); in what follows, we shall assume for the most part that $\bar{x} = 0$ is the *only* equilibrium. If $\bar{x}$ is a *nonzero* equilibrium of (DE) or (DfcE), observe that there exists an entire (nontrivial) *subspace* of $\mathbf{R}^n$ consisting of equilibria for the equation in question. To see this, note that by Observation 23.2(a), the equilibria of (DE) are precisely those vectors which lie in the nullspace of $A(t)$ for every $t \in \mathbf{R}$; this family of vectors is clearly a subspace of $\mathbf{R}^n$, since it is the *intersection* of all the $N(A(t))$ for $t \in \mathbf{R}$. Similarly, by Observation 23.2(b), $\bar{x}$ is an equilibrium of (DfcE) if and only $\bar{x}$ is an eigenvector of $A(k)$, corresponding with eigenvalue 1, for every $k \in \mathbf{Z}$. The set of such vectors is also a subspace of $\mathbf{R}^n$.

The differential and difference equations which we have been considering are supposed to serve as models for the dynamical behavior of real-world processes. Intuitively, an equilibrium of a differential or difference equation is a *stationary condition* or *state of rest* for the process which the equation models. The idea is that if an unforced process *starts* in a state of rest, then it will *remain* forever in that same state of rest.

Naturally, it is impossible to "set" the initial condition of a real-world process *exactly* in a state of rest. Suppose that such a setting is attempted, and that a small error is made; one might wonder whether the error will have an important *long-term effect* on the behavior of the process. Does the process remain "close" to being in a state of rest, or does the error in the initial setting influence the future behavior of the process somewhat more radically?

In terms of differential and difference equations, we say that an equilibrium $\bar{x}$ as in Definitions 22.1 and 22.2 is *stable* if an initial condition which is close to $\bar{x}$ leads to an x-function which is forever close to $\bar{x}$. If equilibria $\bar{x}$ are thought of as modeling *states of rest* for real-world processes, then the *stable* equilibria are, in some sense, the only *physically meaningful* states of rest; intuitively, it is somewhat unlikely that such a process will be "found" in an equilibrium condition which is not stable.

Consider, for example, a pendulum with a stiff support hanging from a pivot about which it is allowed to move freely in two dimensions. The conditions "straight down" and "straight up" are both evidently *states of rest* for the pendulum; of the two, however, the former is the only *stable* one. The following definitions should serve to make the stability concept more precise. Throughout the remainder of this section, the notation $\| x \|$, where $x \in \mathbf{R}^n$, refers to the *Euclidean norm* of x (c.f. §5); i.e.,

$$\| x \| = \Big(\sum_{i=1}^{n} x_i^{\,2} \Big)^{1/2}, \qquad x \in \mathbf{R}^n$$

23.4 Definition: Let $t \to A(t)$, $t \in \mathbf{R}$, be a continuous real $(n \times n)$ matrix function. An equilibrium $\bar{x} \in \mathbf{R}^n$ for

$$\dot{x}(t) = A(t)x(t), \quad t \in \mathbf{R}, \tag{DE}$$

is said to be *stable* if and only if the following condition holds: for every $\epsilon > 0$ there exists for each $t_0 \in \mathbf{R}$ a $\delta > 0$, possibly depending on t_0, such that if $\| x_0 - \bar{x} \| < \delta$, then

$$\| \Phi(t, t_0)x_0 - \bar{x} \| < \epsilon \tag{*}$$

for every $t \geqslant t_0$.

The equilibrium $\bar{x}$ is said to be *uniformly stable* if and only if it is stable and the δ in the definition of stability may be chosen *independently of t_0*; that is, for all $\epsilon > 0$ there is some $\delta > 0$ such that for every $t_0 \in \mathbf{R}$, inequality (*) holds for all $t \geqslant t_0$. □

23.5 Definition: Let $k \to A(k)$, $k \in \mathbf{Z}$, be a real $(n \times n)$ discrete-time matrix function. An equilibrium $\bar{x} \in \mathbf{R}^n$ for

$$x(k+1) = A(k)x(k), \quad k \in Z, \tag{DfcE}$$

is said to be *stable* if and only if the following condition holds: for every $\epsilon > 0$ there exists for each $k_o \in \mathbf{R}$ a $\delta > 0$, possibly depending on k_o, such that if $\| x_o - \bar{x} \| < \delta$, then

$$\| \Phi(k, k_o)x_o - \bar{x} \| < \epsilon \tag{**}$$

for every $k \geq k_o$.

The equilibrium $\bar{x}$ is said to be *uniformly stable* if and only if it is stable and the δ in the definition of stability may be chosen *independently of* k_o; that is, for all $\epsilon > 0$ there is some $\delta > 0$ such that for every $k_o \in \mathbf{R}$, inequality (**) holds for all $k \geq k_o$. □

Our next two definitions give precise characterizations of a *stronger* kind of stability. Consider once again the pendulum example. If the pendulum's support is mounted on a perfectly frictionless pivot, then the pendulum, once in motion, will swing forever at the same amplitude regardless of the proximity of its initial setting to the straight down (stable) equilibrium position. If, on the other hand, there is friction in the pivot, then the oscillations of the pendulum starting from any initial position will damp out; asymptotically, the pendulum will *approach* a (stable) state of rest, although it will never *get there*.

23.6 Definition: In the notation of Definition 23.4, an equilibrium $\bar{x}$ of (DE) is said to be *asymptotically stable* if and only if it is stable and, in addition, there exists for every $t_o \in \mathbf{R}$ some $R > 0$, possibly depending on t_o, such that if $\| x_o - \bar{x} \| < R$, then

$$\lim_{t \to \infty} \Phi(t, t_o)x_o = \bar{x}$$

The equilibrium $\bar{x}$ is said to be *uniformly asymptotically stable* if and only if it is uniformly stable, asymptotically stable, and, in addition, the R in the definition of asymptotic stability *and* the convergence to $\bar{x}$ are *independent of* t_o in the following sense: for every $\epsilon > 0$, there exist a $T > 0$ and an $R > 0$ such that if $\| x_o - \bar{x} \| < R$ and $\tau > T$, then for every $t_o \in \mathbf{R}$, $\| \Phi(t_o + \tau, t_o)x_o - \bar{x} \| < \epsilon$. □

The definition of asymptotic stability for an equilibrium $\bar{x}$ of (DfcE) is similar.

23.7 Definition: In the notation of Definition 23.5, an equilibrium $\bar{x}$ of (DfcE) is said to be *asymptotically stable* if and only if it is stable and, in addition, there exists for every $k_o \in Z$ some $R > 0$, possibly depending on k_o, such that if $\| x_o - \bar{x} \| < R$, then

$$\Phi(k, k_o)x_o \to \bar{x}$$

as $k \to \infty$.

The equilibrium $\bar{x}$ is said to be *uniformly asymptotically stable* if and only if it is uniformly stable, asymptotically stable, and, in addition, the R in the definition of asymptotic stability *and* the convergence to $\bar{x}$ are *independent of* k_o in the following sense: for every $\epsilon > 0$, there exist a $K > 0$ and an $R > 0$ such that if $\| x_o - \bar{x} \| < R$ and $l > K$, then for every $k_o \in Z$,

$$\| \Phi(k_o + l, k_o)x_o - \bar{x} \| < \epsilon . \qquad \square$$

It is important to note that $\bar{x} = 0$ *is the only possible asymptotically stable equilibrium for (DE) or (DfcE)*. Moreover, if $\bar{x} = 0$ is asymptotically stable, then *there can be no other equilibria*. We noted earlier that there is a whole nontrivial subspace of equilibria for (DE) or (DfcE) if even a single nonzero equilibrium exists; arbitrarily close to each equilibrium in the subspace lie an infinity of other equilibria, and hence *none* of the equilibria in the subspace may be asymptotically stable.

There is a bit of superfluity in Definitions 23.6 and 23.7; nonetheless, we have stated the definitions as they are given because their *nonlinear* counterparts are almost exactly the same. The interested reader is referred to [Hahn] for an exhaustive treatment of stability theory for nonlinear differential and difference equations. As for the excess baggage in the last two definitions, observe that if $\bar{x} = 0$ is the only equilibrium for (DE) or (DfcE), and is asymptotically stable, then *any* $R > 0$ works in the definition of asymptotic stability. To see this, note that if $\Phi(t, t_o)x_o \to 0$ as $t \to \infty$ whenever $\| x_o \| < R$, then for *arbitrary* $x_o \in R^n$, $\Phi(t, t_o)x_o \to 0$ as $t \to \infty$, since

$$\Phi(t, t_o)x_o = \frac{1+\|x_o\|}{R}\{\Phi(t, t_o)\frac{Rx_o}{1+\|x_o\|}\},$$

and the quantity in braces goes to zero as $t \to \infty$. The argument in the discrete-time case is exactly the same.

We are now equipped to define the stability of *abstract* continuous- and discrete-time state space linear systems as defined in §6. Given such a system, suppose we have chosen a basis **x** for the state space X, and have constructed the corresponding realization as in §7. It is assumed that if we are dealing with a continuous-time system, then the system satisfies the realizability Assumptions 7.3.

Consider the differential equation

$$\dot{x}(t) = A(t)x(t) + B(t)u(t), \quad t \geqslant t_o,$$

or the difference equation

$$x(k+1) = A(k)x(k) + B(k)u(k), \quad k \geqslant k_o,$$

which arises in the given realization; this equation governs the time-evolution of the *coordinate vector* with respect to the basis **x** of the *state* of the system under the influence of the input $u \in \mathrm{U}$. If a given initial condition $x_o \in \mathbf{R}^n$ is specified at time t_o or k_o, and the identically zero input is applied for all $t \geqslant t_o$ or $k \geqslant k_o$, then we are left with an *unforced* differential or difference equation such as that appearing in Definition 23.1 or 23.2.

We shall say that the given state space system possesses a certain *stability property* if and only if the equilibrium $\bar{x} = 0$ of this differential or difference equation possesses that *same stability property* in the sense of Definitions 23.4 through 23.7. More precisely, we have

23.8 Definition: (a) Let there be given a real m-input, p-output continuous-time state space linear system which satisfies Assumptions 7.3 and has an n-dimensional state space X. We say that the system is *stable*, *uniformly stable*, *asymptotically stable*, or *uniformly asymptotically stable* if and only if the same property holds for the equilibrium $\bar{x} = 0 \in \mathbf{R}^n$ of the differential equation

$$\dot{x}(t) = A(t)x(t)$$

arising as above in any realization of the system.

(b) Let there be given a real m-input, p-output discrete-time state space linear system with an n-dimensional state space X. We say that the system is *stable*, *uniformly stable*, *asymptotically stable*, or *uniformly asymptotically stable* if and only if the same property holds for the equilibrium $\bar{x} = 0 \in \mathbf{R}^n$ of the difference equation

$$x(k+1) = A(k)x(k)$$

arising as above in any realization of the system. □

Of course, Definition 23.8 is *loaded* in the sense that we have defined the various stability properties as *features of the system itself*, whereas it might appear, *a priori*, that they depend on a choice of realization for the system. The crucial fact which makes the stability of a state space system well-defined is the relationship between the matrix functions appearing in different realizations of the system (c.f. Exercise 7.11).

Suppose for the moment that we are given a continuous-time system. If $\mathbf{x}$ and $\hat{\mathbf{x}}$ are two bases for the system's state space X, then there must exist an invertible $(n \times n)$ matrix P such that the $(n \times n)$ matrix functions $t \to A(t)$ and $t \to \hat{A}(t)$ appearing, respectively, in the realizations of the system with respect to the bases $\mathbf{x}$ and $\hat{\mathbf{x}}$, are related by

$$\hat{A}(t) = P^{-1}A(t)P , \quad t \in \mathbf{R} .$$

Thus, the solution $t \to \hat{x}(t)$, $t \in \mathbf{R}$, to

$$\dot{\hat{x}}(t) = \hat{A}(t)\hat{x}(t) , \quad t \in \mathbf{R} ,$$

$$\hat{x}(t_o) = P^{-1}x_o ,$$

is given for every $x_o \in \mathbf{R}^n$ by $\hat{x}(t) = P^{-1}x(t)$, $t \in \mathbf{R}$, where x solves

$$\dot{x}(t) = A(t)x(t) , \quad t \in \mathbf{R} ,$$

$$x(t_o) = x_o .$$

If $\bar{x} = 0$ is a stable equilibrium of either differential equation, it is also a stable equilibrium of the other. To see this, note that

$$\|\hat{x}(t)\| \leqslant \|P^{-1}\| \, \|x(t)\| \leqslant \|P^{-1}\| \, \|P\| \, \|\hat{x}(t)\|$$

for every $t \in \mathbf{R}$; here, $\|P\|$ and $\|P^{-1}\|$ indicate, respectively, the norms of the linear transformations $x \to Px$ and $x \to P^{-1}x$ induced by the Euclidean norm on $\mathbf{R}^n$ (c.f.

Definition 5.12 and Exercise 5.13(b)). It follows easily from the above chain of inequalities that the stability, asymptotically stability, uniformly stability, or uniformly asymptotically stability of $\bar{x} = 0$ as an equilibrium holds either for both differential equations simultaneously or for neither one.

23.9 Exercise: Prove carefully the assertion at the end of the preceding paragraph. □

The above argument works for discrete-time systems, as well; simply substitute $k \to A(k)$ and $k \to \hat{A}(k)$, respectively, for $t \to A(t)$ and $t \to \hat{A}(t)$, and substitute difference equations and their solutions for the corresponding continuous-time objects. The upshot is that the stability (or asymptotic stability, or uniform (asymptotic) stability) of an "abstract" state space linear system is a well-defined property *of the system itself*.

Throughout the remainder of the section, we shall be deriving *stability criteria* for state space linear systems. As the reader might have guessed, stability criteria for time-invariant systems are much simpler and more easily tested than those for general time-varying systems. It turns out that the stability properties of a state space linear system are determined by the behavior of its state transition mapping $x^\circ \to \phi(t, t_o, x^\circ, 0)$ or $x^\circ \to \phi(k, k_o, x^\circ, 0)$ for large t or k. Once a realization for the system has been selected, the behavior of the above mapping is in turn reflected in the behavior of $x_o \to \Phi(t, t_o)x_o$ or $x_o \to \Phi(k, k_o)x_o$, where Φ is the state transition matrix arising from the chosen realization.

Once again, the notation $\|M\|$, where M is an $(n \times n)$ matrix, refers to the norm of the linear transformation $x \to Mx$, $x \in \mathbf{R}^n$, which is induced by the standard Euclidean norm on $\mathbf{R}^n$ (c.f. Definition 5.12 and the discussion which accompanies it). Equivalently, by Exercise 12.13, $\|M\|$ is equal to the the square root of the largest eigenvalue of the non-negative definite $(n \times n)$ matrix $M^T M$.

23.10 Theorem: *Let there be given a real, m-input, p-output continuous-time state space linear system which satisfies the realizability Assumptions 7.3 and has an n-dimensional state space X. Let (A, B, C, D) be the matrix functions appearing in a realization of the system; given $t_o \in \mathbf{R}$, let*

$t \to \Phi(t, t_o)$ satisfy

$$\frac{d}{dt}\Phi(t, t_o) = A(t)\Phi(t, t_o), \quad t \in \mathbf{R},$$

$$\Phi(t_o, t_o) = I_n .$$

The system is

(a) stable if and only if $t \to \|\Phi(t, t_o)\|$ is bounded on $[t_o, \infty)$ for every $t_o \in \mathbf{R}$;

(b) asymptotically stable if and only if $\|\Phi(t, t_o)\| \to 0$ as $t \to \infty$ for every $t_o \in \mathbf{R}$;

(c) uniformly stable if and only if there exists some $M > 0$ such that $\|\Phi(t, t_o)\| \leqslant M$ for every $t_o \in \mathbf{R}$ and for every $t \geqslant t_o$; and

(d) uniformly asymptotically stable if and only if for every $\epsilon > 0$ there exists some $T > 0$ such that for every $t_o \in \mathbf{R}, t \geqslant t_o + T$ implies that $\|\Phi(t, t_o)\| < \epsilon$.

Proof: We show that each condition on $\|\Phi(t, t_o)\|$ is equivalent to the corresponding stability property for $\bar{x} = 0$ as an equilibrium of the differential equation (DE) in Definition 23.1; the assertion about the stability of the given *system* then follows from Definition 23.8. We begin with (a) and (c).

Suppose that $\bar{x} = 0$ is a stable equilibrium of (DE); let $\epsilon = 1$ in the definition of stability, and find for each $t_o \in \mathbf{R}$ a $\delta(t_o)$ such that whenever $\|x_o\| \leqslant \delta(t_o)$, $\|\Phi(t, t_o)x_o\| < 1$ for every $t \geqslant t_o$. Then when $x_o \leqslant 1$, we have $\|\Phi(t, t_o)x_o\| < [\delta(t_o)]^{-1}$, from which it follows that

$$\|\Phi(t, t_o)\| \leqslant M(t_o) = \frac{1}{\delta(t_o)}$$

for every $t \geqslant t_o$. Thus if $\bar{x} = 0$ is stable, then $t \to \|\Phi(t, t_o)\|$ is *bounded in t* for every t_o; if $\bar{x} = 0$ is *uniformly stable*, then $\delta(t_o)$, and hence $M(t_o)$, may be chosen independently of t_o, and the condition on Φ in (c) holds.

Conversely, if for each $t_o \in \mathbf{R}$ there exists $M(t_o)$ such that $\|\Phi(t, t_o)\| \leqslant M(t_o)$ for every $t \geqslant t_o$, we can conclude (c.f. Exercise 5.13(b)) that $\|\Phi(t, t_o)x_o\| \leqslant \epsilon$ for every $t \geqslant t_o$ provided that $x_o \leqslant \delta(t_o) = \epsilon M^{-1}(t_o)$; the stability of $\bar{x} = 0$ follows immediately. If there exists $M > 0$ such that *for all* t_o we have $\|\Phi(t, t_o)\| \leqslant M$ when $t > t_o$, then $\|x_o\| \leqslant \epsilon M^{-1}$ guarantees that for every t_o

$\|\Phi(t, t_o)x_o\| < \epsilon$ when $t \geqslant t_o$, and $\bar{x} = 0$ is therefore *uniformly* stable. This completes the proofs of (a) and (c).

Now for (b) and (d). Suppose that for all t_o, it is true that $\|\Phi(t, t_o)\| \to 0$ as $t \to \infty$. This means that not only is $t \to \|\Phi(t, t_o)\|$ *bounded* for each $t_o \in \mathbf{R}$ (implying stability of the equilibrium $\bar{x} = 0$), but also that for every $\epsilon > 0$ there exists for each t_o some $T(t_o)$ such that when $t \geqslant t_o + T(t_o)$, we have $\|\Phi(t, t_o)\| < \epsilon$. Take $R = 1$ in Definition 23.6 of asymptotic stability, and conclude that $\|\Phi(t, t_o)x_o\| < \epsilon$ for every x_o satisfying $\|x_o\| \leqslant R$ and every $t \geqslant t_o + T(t_o)$, from which the asymptotic stability of $\bar{x} = 0$ follows. If $T(t_o)$ in the foregoing argument can be chosen independently of t_o, we clearly have *uniform* asymptotic stability of $\bar{x} = 0$.

Conversely, suppose that $\bar{x} = 0$ is asymptotically stable; for each $t_o \in \mathbf{R}$, given $\epsilon > 0$ find $R(t_o) > 0$ and $T(t_o) \geqslant 0$ such that when $\|x_o\| \leqslant R(t_o)$ and $t \geqslant t_o + T(t_o)$, we have $\|\Phi(t, t_o)x_o\| < \epsilon$. The last inequality implies that for $t \geqslant t_o + T(t_o)$, it is true that

$$\|\Phi(t, t_o)\| \leqslant \frac{\epsilon}{R(t_o)},$$

from which it follows that $\|\Phi(t, t_o)\| \to 0$ as $t \to \infty$. If $\bar{x} = 0$ is *uniformly* asymptotically stable, then (given ϵ) $R(t_o)$ and $T(t_o)$ may be chosen independently of t_o, and the condition in (d) is an immediate consequence. □

Observe that in the course of proving Theorem 23.10, we ignored the fact that the $R > 0$ in the definition of asymptotic stability is *irrelevant* for linear systems; the proofs really aren't any harder because of this omission. Moreover, the arguments as presented bear a stronger resemblance to their nonlinear counterparts than they would otherwise. We turn now to the statement and proof of the discrete-time version of Theorem 23.10.

23.11 Theorem: *Let there be given a real, m-input, p-output discrete-time state space linear system which has an n-dimensional state space X. Let (A, B, C, D) be the discrete-time matrix functions appearing in a realization of the system; given $k_o \in \mathbf{Z}$, let $t \to \Phi(k, k_o)$ satisfy*

$$\Phi(k+1, k_o) = A(k)\Phi(k, k_o), \quad k \geqslant k_o,$$

$$\Phi(k_o, k_o) = I_n \ .$$

The system is

(a) stable if and only if $k \to \|\Phi(k, k_o)\|$ *is bounded on* $\{k_o, k_o+1, k_o+2, \ldots\}$ *for every* $k_o \in \mathrm{Z}$;

(b) asymptotically stable if and only if $\|\Phi(k, k_o)\| \to 0$ *as* $k \to \infty$ *for every* $k_o \in \mathrm{Z}$;

(c) uniformly stable if and only if there exists some $M > 0$ *such that* $\|\Phi(k, k_o)\| \leqslant M$ *for every* $k_o \in \mathrm{Z}$ *and for every* $k \geqslant k_o$; *and*

(d) uniformly asymptotically stable if and only if for every $\epsilon > 0$ *there exists some* $K > 0$ *such that for every* $k_o \in \mathrm{Z}, k \geqslant k_o + K$ *implies that* $\|\Phi(k, k_o)\| < \epsilon$.

Proof: We show that each condition on $\|\Phi(k, k_o)\|$ is equivalent to the corresponding stability property for $\bar{x} = 0$ as an equilibrium of the difference equation (DfcE) in Definition 23.2; the assertion about the stability of the given *system* then follows from Definition 23.8. We begin with (a) and (c).

Suppose that $\bar{x} = 0$ is a stable equilibrium of (DfcE); let $\epsilon = 1$ in the definition of stability, and find for each $k_o \in \mathrm{Z}$ a $\delta(k_o)$ such that whenever $\|x_o\| \leqslant \delta(k_o)$, we have $\|\Phi(k, k_o)x_o\| < 1$ for every $k \geqslant k_o$. Then when $x_o \leqslant 1$, we have $\|\Phi(k, k_o)x_o\| < [\delta(k_o)]^{-1}$, from which it follows that

$$\|\Phi(k, k_o)\| \leqslant M(k_o) = \frac{1}{\delta(k_o)}$$

for every $k \geqslant k_o$. It follows that if $\bar{x} = 0$ is stable, then $k \to \|\Phi(k, k_o)\|$ is *bounded in* k for every k_o; if $\bar{x} = 0$ is *uniformly stable*, then $\delta(k_o)$, and hence $M(k_o)$, may be chosen independently of k_o, and the condition on Φ in (c) holds.

Conversely, if for each $k_o \in \mathrm{R}$ there exists $M(k_o)$ such that $\|\Phi(k, k_o)\| \leqslant M(k_o)$ for every $k \geqslant k_o$, we can conclude (c.f. Exercise 5.13(b)) that $\|\Phi(k, k_o)x_o\| \leqslant \epsilon$ for every $k \geqslant k_o$ provided that $x_o \leqslant \delta(k_o) = \epsilon M^{-1}(k_o)$; the stability of $\bar{x} = 0$ follows immediately. If there exists $M > 0$ such that *for all* k_o we have $\|\Phi(k, k_o)\| \leqslant M$ when $k > k_o$, then $\|x_o\| \leqslant \epsilon M^{-1}$ guarantees that $\|\Phi(k, k_o)x_o\| < \epsilon$ when $k \geqslant k_o$, for every k_o, and $\bar{x} = 0$ is therefore *uniformly* stable. This completes the proofs of (a) and (c).

Now for (b) and (d). Suppose that for all k_o, it is true that $\|\Phi(k, k_o)\| \to 0$ as $k \to \infty$. This means that not only is $k \to \|\Phi(k, k_o)\|$ *bounded* for each $k_o \in \mathbf{Z}$ (implying stability of the equilibrium $\bar{x} = 0$), but also that for every $\epsilon > 0$ there exists for each k_o some $K(k_o)$ such that when $k \geqslant k_o + K(k_o)$, we have $\|\Phi(k, k_o)\| < \epsilon$. Take $R = 1$ in Definition 23.7 of asymptotic stability, and conclude that $\|\Phi(k, k_o)x_o\| < \epsilon$ for every x_o whose norm is at most R and every $k \geqslant k_o + K(k_o)$, from which the asymptotic stability of $\bar{x} = 0$ follows. If $K(k_o)$ in the foregoing argument can be chosen independently of k_o, we clearly have *uniform* asymptotic stability of $\bar{x} = 0$.

Conversely, suppose that $\bar{x} = 0$ is asymptotically stable; for each $k_o \in \mathbf{Z}$, given $\epsilon > 0$ find $R(k_o) > 0$ and $K(k_o) \geqslant 0$ such that when $\|x_o\| \leqslant R(k_o)$ and $k \geqslant k_o + K(k_o)$, we have $\|\Phi(k, k_o)x_o\| < \epsilon$. The last inequality implies that for $k \geqslant k_o + K(k_o)$, it is true that

$$\|\Phi(k, k_o)\| \leqslant \frac{\epsilon}{R(k_o)},$$

from which it follows that $\|\Phi(k, k_o)\| \to 0$ as $k \to \infty$. If $\bar{x} = 0$ is *uniformly* asymptotically stable, then (given ϵ) $R(k_o)$ and $K(k_o)$ may be chosen independently of k_o, and the condition in (d) is an immediate consequence. □

It is often the case in applications of linear system theory to real-world problems that knowing the *rate* at which some quantity decays to zero is just as important as knowing merely that the quantity decays. It turns out that if $\bar{x} = 0$ is a *uniformly asymptotically stable* equilibrium of (DE) or (DfcE), then $\|\Phi(t, t_o)x_o\|$ or $\|\Phi(k, k_o)x_o\|$ decays *exponentially* to zero as t or k increases. This important consequence of linearity, which follows from the *semigroup property* (c.f. §2) of Φ, is an exceptionally strong result about convergence rates.

23.12 Theorem: *The equilibrium $\bar{x} = 0$ of (DE) is uniformly asymptotically stable if and only if there exist $M > 0$ and $\alpha > 0$ such that for every $t_o \in \mathbf{R}$ and $t \geqslant t_o$,*

$$\|\Phi(t, t_o)\| \leqslant Me^{-\alpha(t-t_o)}.$$

Proof: It is clear that $\bar{x} = 0$ is a uniformly asymptotically stable equilibrium if M and α may be found; in this case, we have, for each $x_o \in \mathbf{R}^n$ and $t_o \in \mathbf{R}$,

$$\|\Phi(t, t_o)x_o\| \leqslant Me^{-\alpha(t-t_o)}\|x_o\|, \quad t \geqslant t_o .$$

Given $\epsilon > 0$, let $T = \alpha^{-1}\ln(M\epsilon^{-1})$ and $R = 1$ in Definition 23.6, and the uniform asymptotic stability of $\bar{x} = 0$ follows.

Conversely, suppose that $\bar{x} = 0$ is uniformly asymptotically stable. Since $\bar{x}$ is uniformly stable, by Theorem 23.10(c) we can find $\hat{M} > 0$ such that $\|\Phi(t, t_o)\| \leqslant \hat{M}$ for every $t_o \in \mathbf{R}$ and every $t \geqslant t_o$. In addition, by Theorem 23.10(d), we may pick an arbitrary $\beta > 0$ and find $T > 0$ so that $\|\Phi(t_o+T, t_o)\| \leqslant e^{-\beta}$ for every $t_o \in \mathbf{R}$. Set $M = \hat{M}e^{\beta}$ and $\alpha = \beta T^{-1}$; it will follow that $\|\Phi(t, t_o)\| \leqslant Me^{\alpha(t-t_o)}$ for all $t \geqslant t_o$.

Given $t_o \in \mathbf{R}$ and $t \geqslant t_o$, we may write $t = qT + t_o + \tau$, where q is a nonnegative integer and $\tau \in [0, T)$. Then $\|\Phi(t, t_o)\|$ is given by

$$\|\Phi(t, t_o+qT) \ldots \Phi(t_o+2T, t_o+T)\Phi(t_o+T, t_o)\| ,$$

which is bounded above by

$$\|\Phi(t, t_o+qT)\| \ldots \|\Phi(t_o+2T, t_o+T)\| \, \|\Phi(t_o+T, t_o)\| .$$

Each of the last k terms is at most $e^{-\beta}$; the first is bounded above by $\hat{M}$, which is in turn equal to $Me^{-\alpha T}$. Thus

$$\begin{aligned} \|\Phi(t, t_o)\| &\leqslant Me^{-\alpha T}e^{-q\beta} \\ &\leqslant Me^{-\alpha\tau}e^{-\alpha qT} = Me^{-\alpha(t-t_o)} . \end{aligned}$$

□

The discrete-time version of Theorem 23.12 is proved similarly.

23.13 Theorem: *The equilibrium $\bar{x} = 0$ of (DfcE) is uniformly asymptotically stable if and only if there exist $M > 0$ and $\gamma \in (0, 1)$ such that for every $k_o \in \mathrm{Z}$ and $k \geqslant k_o$,*

$$\|\Phi(k, k_o)\| \leqslant M\gamma^{k-k_o} .$$

Proof: We prove only that uniform asymptotic stability *implies* the exponential decay of Φ. Pick $\beta \in (0, 1)$; find $\hat{M} > 0$ and $K > 0$ such that $\|\Phi(k, k_o)\| \leqslant \hat{M}$ and $\|\Phi(k_o+K, k_o)\| \leqslant \beta$ for every $k_o \in \mathrm{Z}$ and $k \geqslant k_o$. Set

$\gamma^K = \beta$ and $M = \hat{M}\gamma^{-K}$; we'll show that $\|\Phi(k, k_o)\| \leqslant M\gamma^{k-k_o}$ for every $k \geqslant k_o$.

First of all, given $k \geqslant k_o$, write $k = qK + l$, where $0 \leqslant l < K$. Then $\|\Phi(k, k_o)\|$ is bounded above by the q-fold product

$$\|\Phi(k, k_o + qK)\| \ldots \|\Phi(k_o + 2K, k_o + K)\| \, \|\Phi(k_o + K, k_o)\| ,$$

from which it follows as in the proof of Theorem 23.12 that

$$\begin{aligned} \|\Phi(k, k_o)\| &\leqslant M\gamma^K \beta^q \\ &\leqslant M\gamma^l \gamma^{qK} = M\gamma^{k-k_o} . \end{aligned}$$

□

Theorems 23.12 and 23.13 attest to the fact that the uniform asymptotic stability of the equilibrium $\bar{x} = 0$ of (DE) or (DfcE) is, indeed, a very strong property. It is worth examining an example where $\bar{x} = 0$ fails to be a *uniformly* asymptotically stable equilibrium even though it is asymptotically stable. Consider the differential equation

$$\dot{x}(t) = \frac{-2t}{1+t^2} x(t), \qquad t \in \mathbf{R};$$

the state transition "matrix" $\Phi(t, t_o)$ satisfies

$$\frac{d}{dt}\Phi(t, t_o) = \frac{-2t}{1+t^2}\Phi(t, t_o), \qquad t \in \mathbf{R}$$

$$\Phi(t_o, t_o) = 1 .$$

It is easily checked that

$$\Phi(t, t_o) = \frac{1+t_o^2}{1+t^2}$$

for every $t_o \in \mathbf{R}$ and for every $t > t_o$. Evidently, $\Phi(t, t_o) \to 0$ as $t \to \infty$ for every $t_o \in \mathbf{R}$, so that the equilibrium $\bar{x} = 0$ is asymptotically stable by Theorem 23.10(b). On the other hand, it is true for every $t_o \in \mathbf{R}$ and every $T \geqslant 0$ that

$$\Phi(t_o + T, t_o) = \frac{1+t_o^2}{1+(t_o+T)^2};$$

hence, given $\epsilon > 0$, there is *no* $T > 0$ such that $\|\Phi(t_o + \tau, t_o)\| < \epsilon$ for every $\tau > T$ and every $t_o \in \mathbf{R}$. The point is that if $t_o < 0$, then $\|\Phi(t_o + T, t_o)\|$ attains the value

$1+t_o^2$ when $T = -t_o$. We conclude, by Theorem 23.10(d), that $\bar{x} = 0$ is not a uniformly asymptotically stable equilibrium.

The differential equation in the preceding example is *time varying;* we shall prove in a moment that if the matrix function $t \to A(t)$ in (DE) (or $k \to A(k)$ in (DfcE)) is *constant*, then the (asymptotic) stability of the equilibrium $\bar{x} = 0$ is equivalent to its uniform (asymptotic) stability. By means of Definition 23.8, along with Exercise 7.11 and the discussion following Definition 7.2, it may be concluded that a given real m-input, p-output *time-invariant* continuous- or discrete-time state space linear system is stable if and only if it is *uniformly* stable, and asymptotically stable if and only if it is *uniformly* asymptotically stable.

23.14 Theorem: *If the real $(n \times n)$ matrix function $t \to A(t)$ in (DE) (or $k \to A(k)$ in (DFcE)) is identically equal to a constant $(n \times n)$ matrix A, then the equilibrium $\bar{x} = 0$ of (DE) (or of (DfcE)) is stable if and only if it is uniformly stable, and is asymptotically stable if and only if it is uniformly asymptotically stable.*

Proof: By definition, uniform (asymptotic) stability of $\bar{x} = 0$ *implies* (asymptotic) stability of $\bar{x} = 0$; consequently, we shall prove only that the reverse implications hold. Consider (DE) first; by the results of §2, the matrix $\Phi(t, t_o)$ is given for every $t, t_o \in \mathbf{R}$ by

$$\Phi(t, t_o) = e^{(t-t_o)A} .$$

If $\bar{x} = 0$ is stable, then by Theorem 23.10(a) we may find some $M > 0$ such that $\| e^{tA} \| \leqslant M$ for every $t \geqslant 0$. It follows that for every $t_o \in \mathbf{R}$, $\| e^{(t-t_o)A} \| \leqslant M$ for every $t \geqslant t_o$, and the uniform stability of $\bar{x} = 0$ is then a consequence of Theorem 23.10(c). For the difference equation (DfcE), a similar argument using the identity

$$\Phi(k, k_o) = A^{k-k_o} , \quad k \geqslant k_o ,$$

shows that stability of the equilibrium $\bar{x} = 0$ implies its uniform stability.

Assume now that $\bar{x} = 0$ is an asymptotically stable equilibrium of (DE). By Theorem 23.10(b), we may conclude that $\| e^{tA} \| \to 0$ as $t \to \infty$. Given $\epsilon > 0$, find $T > 0$ such that $\| e^{tA} \| < \epsilon$ whenever $t > T$. For any $t_o \in \mathbf{R}$, we have

$\Phi(t_o+t, t_o) = e^{tA}$; thus, $\|\Phi(t_o+t, t_o)\| < \epsilon$ whenever $t > T$. Thus, by Theorem 23.10(d), $\bar{x} = 0$ is uniformly asymptotically stable. A similar argument works in discrete time. □

The proof of Theorem 23.14 shows, among other things, that the stability properties of a time-invariant state space linear system (or, equivalently, of the equilibrium $\bar{x} = 0$ of (DE) or (DfcE) in the case of constant A) are determined by the time behavior of e^{tA} for $t \geqslant 0$ or of A^k for $k \geqslant 0$. The analysis in §§8–11 shows that this behavior is determined, in turn, by the *eigenvalues* of the $(n \times n)$ matrix A. It is therefore natural to expect that conditions for time-invariant stability may be stated conveniently in terms of these eigenvalues.

23.15 Theorem: *If the matrix function $t \to A(t)$ in (DE) is identically equal to a real constant $(n \times n)$ matrix A, then*

(a) The equilibrium $\bar{x} = 0$ of (DE) is asymptotically stable (hence uniformly asymptotically stable) if and only if all the eigenvalues of A have negative real parts, and

(b) $\bar{x} = 0$ is stable (hence uniformly stable) if and only if no eigenvalue of A has a positive real part, and, in addition, whenever λ_o is an eigenvalue of A having a zero real part, the algebraic multiplicity of λ_o is equal to its geometric multiplicity.

Proof: (a) Suppose that A has an eigenvalue λ_o with a nonnegative real part; write $\lambda_o = \mu_o + i\omega_o$, with $\mu_o \geqslant 0$. Let $z_o \in \mathbf{C}^n$ be an eigenvector of A corresponding with eigenvalue λ_o. Then $\mathrm{Re}\{e^{tA} z_o\} = \mathrm{Re}\{e^{\lambda_o t} z_o\}$; since e^{tA} is real, we have

$$e^{tA}\mathrm{Re}\{z_o\} = e^{\mu_o t}\cos(\omega_o t)\mathrm{Re}\{z_o\} - \sin(\omega_o t)\mathrm{Im}\{z_o\} .$$

Since $\mathrm{Re}\{z_o\}$ and $\mathrm{Im}\{z_o\}$ are linearly independent real n-vectors (c.f. the discussion following Corollary 8.5), and since $\mu_o \geqslant 0$, $e^{tA}\mathrm{Re}\{z_o\}$ does not go to zero as $t \to \infty$, and $\bar{x} = 0$ is not asymptotically stable.

Conversely, if every eigenvalue of A has a negative real part, Procedure 9.11 enables us to write

$$e^{tA} = P e^{t\Lambda} e^{t\tilde{N}} P^{-1} ,$$

where P is an invertible complex $(n \times n)$ matrix, $e^{t\tilde{N}}$ has entries which are polynomials in t, and $e^{t\Lambda}$ is a diagonal matrix whose diagonal entries are decaying exponentials. It is then immediate that $e^{tA}x_o \to 0$ as $t \to \infty$ for *every* $x_o \in \mathbf{R}^n$, and the asymptotic stability of $\bar{x} = 0$ follows.

(b) If $\bar{x} = 0$ is stable, then an argument similar to that used above in the proof of the first part of (a) shows that no eigenvalue of A can have positive real part; if $z_o \in \mathbf{C}^n$ were an eigenvector corresponding with such an eigenvalue, then $\| e^{tA}\,\mathrm{Re}\{z_o\} \|$ would become unbounded as $t \to \infty$, implying that $\| e^{tA} \|$ would *also* become unbounded as $t \to \infty$, contradicting stability by Theorem 23.10(a). Thus stability implies that all of A's eigenvalues have nonpositive real parts; the analysis in §9 shows that we may write

$$e^{tA} = Pe^{t\Lambda}e^{t\tilde{N}}P^{-1},$$

where P is an invertible complex $(n \times n)$ matrix, and

$$\Lambda = \begin{bmatrix} \Lambda_- & 0 \\ 0 & \Lambda_0 \end{bmatrix}$$

and

$$\tilde{N} = \begin{bmatrix} \tilde{N}_- & 0 \\ 0 & \tilde{N}_0 \end{bmatrix},$$

where Λ_- and Λ_0 are diagonal matrices whose diagonal elements have negative and zero real parts, respectively. The condition $\tilde{N}_0 = 0$ is *equivalent* to the condition that the algebraic multiplicity of each pure imaginary eigenvalue of A (including 0) equals its geometric multiplicity.

We have

$$P^{-1}e^{tA}P = \begin{bmatrix} e^{t\Lambda_-}e^{t\tilde{N}_-} & 0 \\ 0 & e^{t\Lambda_0}e^{t\tilde{N}_0} \end{bmatrix};$$

the terms in the upper left block of the matrix on the right *decay exponentially* as $t \to \infty$, and those in the lower right block are all *bounded* on $[0, \infty)$ if and only if $\tilde{N}_0 = 0$. If $\tilde{N}_0 \neq 0$, we conclude that some element of e^{tA} becomes unbounded (and hence so does $\| e^{tA} \|$) as $t \to \infty$, contradicting stability.

The foregoing argument is easily reversed to show that the eigenvalue conditions *imply* stability. If the conditions hold, then all the entries in matrix on the right-hand side of the last equation are *bounded* functions on $[0, \infty)$; thus there exists some $\hat{M} > 0$ such that the magnitude of every entry in e^{tA} is bounded by $\hat{M}$ for every $t \in [0, \infty)$. It is then a simple matter to show that $\|e^{tA}\|$ is bounded on $[0, \infty)$, which implies stability of the equilibrium $\bar{x} = 0$ of (DE). □

The following is the discrete-time analogue of Theorem 23.15.

23.16 Theorem: *If the discrete-time matrix function $k \to A(k)$ in (DfcE) is identically equal to a real constant $(n \times n)$ matrix A, then*

(a) The equilibrium $\bar{x} = 0$ of (DfcE) is asymptotically stable (hence uniformly asymptotically stable) if and only if all the eigenvalues of A have magnitudes less than one, and

(b) $\bar{x} = 0$ is stable (hence uniformly stable) if and only if no eigenvalue of A has a magnitude greater than one, and, in addition, whenever λ_o is an eigenvalue of A having a magnitude one, the algebraic multiplicity of λ_o is equal to its geometric multiplicity.

Proof: The proof is similar to that of Theorem 23.15; we present an abbreviated version. Suppose that $\bar{x} = 0$ is a stable equilibrium of (DfcE). If some eigenvalue λ_o of A has magnitude greater than one, and $z_o \in \mathbf{C}^n$ is a corresponding eigenvector, then $\|A^k z_o\| = \|\lambda_o\|^k \|z_o\|$, which *grows* as $k \to \infty$, contradicting stability. Hence stability implies that all the eigenvalues of A have magnitudes at most one. Moreover, *asymptotic* stability implies that each eigenvalue λ_o of A must have magnitude *less than* one, since if $\|\lambda_o\| = 1$, then $\|e^{tA} z_o\|$, with z_o a corresponding eigenvector, does not decay as $k \to \infty$.

Again by the formula which follows Procedure 9.11, we may write

$$P^{-1}A^k P = \sum_{l=0}^{k} \binom{k}{l} \Lambda^l \tilde{N}^{k-l} ,$$

where Λ is a diagonal matrix whose diagonal entries are the eigenvalues of A, each appearing as many times as its algebraic multiplicity, and $\tilde{N}^n = 0$. If all eigenvalues have magnitudes less than one, then it is clear from the last equation that

$A^k \to 0$ as $k \to \infty$, implying asymptotic stability (and completing the proof of (a)). To see this, note that for each k, the sum contains at most $n+1$ nonzero terms; furthermore, the entries in the $\tilde{N}^j$ are polynomials in k, and the smallest power of Λ appearing in the sum is Λ^{k-n+1}.

If some of A's eigenvalues have magnitudes *equal* to one, then we may assume that Λ and $\tilde{N}$ are partitioned conformably as

$$\Lambda = \begin{bmatrix} \Lambda_- & 0 \\ 0 & \Lambda_1 \end{bmatrix}$$

and

$$\tilde{N} = \begin{bmatrix} \tilde{N}_- & 0 \\ 0 & \tilde{N}_1 \end{bmatrix},$$

where the diagonal elements of Λ_- are less than one in magnitude, and those of Λ_1 have magnitudes *equal* to one. The multiplicity condition on the eigenvalues given in (b) is equivalent to the statement that $\tilde{N}_1 = 0$.

If $\tilde{N}_1 \neq 0$, then there is some $z_o \in \mathbf{C}^n$ such that $(A - \lambda_o I_n)^2 z_o = 0$ and $(A - \lambda_o I_n) z_o \neq 0$, where $\| \lambda_o \| = 1$. It's not hard to demonstrate in this case that

$$A^k z_o = \lambda_o^k z_o + k \lambda_o^{k-1} (A - \lambda_o I_n) z_o \ ;$$

the second term grows in magnitude as $k \to \infty$ and is linearly independent of z_o, contradicting stability.

Conversely, if $\tilde{N}_1 = 0$, then the formula for A^k shows that every entry of $k \to A^k$ is bounded on $\{0, 1, 2, \ldots\}$; the same must therefore hold for $k \to \| A^k \|$, and stability of $\bar{x} = 0$ follows from Theorem 23.10. □

Theorems 23.15 and 23.16 tell essentially the whole story regarding stability of time-invariant state space linear systems. The reader might wonder whether these results may be applied in any way to the analysis of the stability of *time-varying* state space linear systems. Were it not for the example preceding Theorem 23.14, one might be tempted to conjecture that if the eigenvalues of $A(t)$ in (DE) have, for every $t \in \mathbf{R}$, strictly negative real parts, then $\bar{x} = 0$ is a stable equilibrium of (DE). Observe that $A(t)$ in the aforementioned example goes to zero as

$t \to \infty$; thus its single eigenvalue is not *bounded away from the imaginary axis*. Perhaps if this *stronger* condition on the eigenvalues of $A(t)$ held, then a stability assertion would follow. The following example serves to dash any such hopes.

23.17 Example: Consider the differential equation

$$\dot{x}(t) = \begin{bmatrix} -1 & e^{2t} \\ 0 & -1 \end{bmatrix} x(t), \qquad t \in \mathbf{R}.$$

The transition matrix $\Phi(t, t_o)$ is easily found to be

$$\Phi(t, t_o) = \begin{bmatrix} e^{-(t-t_o)} & \frac{1}{2} e^{-(t-t_o)}(e^{2t} - e^{2t_o}) \\ 0 & e^{-(t-t_o)} \end{bmatrix}.$$

The sole eigenvalue of the matrix $A(t)$ in this example is -1 for *all* $t \in \mathbf{R}$; still, $\Phi(t, t_o)$ is not even *bounded* on $[t_o, \infty)$. □

We close this section by discussing another way in which the stability of state space linear systems may be analyzed. This technique, which is known as the *second method of Lyapunov*, is especially useful in the analysis of the stability of equilibria of *nonlinear* differential equations. Although the method and its attendant auxiliary results are almost trivialities in the context of linear systems, it is worth describing them; an understanding of their intuitive content, in particular, is quite valuable.

Think of (DE), for a moment, as modeling some physical process; each $x \in \mathbf{R}^n$ corresponds with some particular *state* of the physical process. Suppose that with each state x there is associated a specific nonnegative value of the *energy* of the process; assume that the unique state of *zero energy* is the equilibrium state $\bar{x} = 0$. Suppose it is observed that when the system is initialized at time zero in a state $x_o \in \mathbf{R}^n$ with very small energy $V(x_o)$, then there results a time-evolution $t \to x(t)$, $t \geq 0$, of the state for which $V(x(t))$ *does not increase beyond* $V(x_o)$. In that case, it might be expected that the state $x(t)$ stays close to $\bar{x}$ for all $t \geq 0$, implying that $\bar{x}$ is, at least roughly speaking, a *stable* equilibrium state for the process. If, in addition, the process is in some sense *dissipative*, then $V(x(t))$ ought to converge to zero as $t \to \infty$. Since the unique state of zero energy is $\bar{x} = 0$, we could conclude that $x(t) \to 0$ as $t \to \infty$, implying a sort of *asymptotic* stability for the equilibrium state $\bar{x}$.

In 1892, the Soviet mathematician A. M. Lyapunov published the famous paper [Lyapunov], in which he made rigorous the connection between *energy considerations*, such as those just described, and the stability of equilibria of differential equations.

23.18 Definition: A function $V : \mathbf{R}^n \to \mathbf{R}$ is called a *Lyapunov function for the equilibrium* $\bar{x} = 0$ of (DE) if and only if

(a) V is continuous;

(b) $V(0) = 0$, and $V(x) > 0$ for all nonzero $x \in \mathbf{R}^n$; and

(c) for all $t_o \in \mathbf{R}$, and for all $x_o \in \mathbf{R}^n$, if $t \to x(t)$, $t \geqslant t_o$, satisfies (DE) along with the condition $x(t_o) = x_o$, then the function $t \to V(x(t))$ is monotonically decreasing on $[t_o, \infty)$.

If, in (c), $t \to V(x(t))$ is *strictly* decreasing on $[t_o, \infty)$, then V is said to be a *strict* Lyapunov function for $\bar{x} = 0$. □

23.19 Definition: A function $V : \mathbf{R}^n \to \mathbf{R}$ is called a *Lyapunov function for the equilibrium* $\bar{x} = 0$ of (DfcE) if and only if

(a) V is continuous;

(b) $V(0) = 0$, and $V(x) > 0$ for all nonzero $x \in \mathbf{R}^n$; and

(c) for all $k_o \in \mathrm{Z}$ and all $x_o \in \mathbf{R}^n$, if $k \to x(k)$, $k \geqslant k_o$, satisfies (DfcE) along with the condition $x(k_o) = x_o$, then the discrete-time function $k \to V(x(k))$ is monotonically decreasing on $\{k_o, k_o+1, k_o+2, \ldots\}$.

If, in (c), $k \to V(x(k))$ is *strictly* decreasing on $\{k_o, k_o+1, k_o+2, \ldots\}$, then V is said to be a *strict* Lyapunov function for $\bar{x} = 0$. □

It turns out that $\bar{x} = 0$ is a uniformly stable equilibrium of (DE) or (DfcE) *if* there exists a Lyapunov function for $\bar{x}$. Moreover, if the A-matrix function in (DE) or (DfcE) is *constant*, then $\bar{x} = 0$ is asymptotically stable *if and only if* there exists a strict Lyapunov function for $\bar{x}$. The preceding assertions are easy to prove using the definitions of stability and Lyapunov functions. Neither the second assertion nor its converse generalizes to time-varying differential or difference equations unless further assumptions are made; see Theorem 23.21 below. It should be mentioned that there do, in fact, exist ways of defining *time-varying Lyapunov functions* for the equilibrium $\bar{x} = 0$ in (DE) or (DfcE) when $t \to A(t)$ or $k \to A(k)$ is not constant; the reader is referred to [Hahn] or [Lefschetz] for a complete account.

We shall formalize the preceding remarks in a moment; first, let us consider examples of time-varying differential equations for which the second assertion fails. For instance, no strict Lyapunov exists for the (asymptotically stable) equilibrium $\bar{x} = 0$ in the example which was described just before Theorem 23.14. If the differential equation is started from initial condition $x_o = 1$ at time $t_o \in \mathbf{R}$, the solution $t \to x(t)$ is given by

$$x(t) = \frac{1+t_o^2}{1+t^2}, \qquad t \geqslant t_o .$$

If a strict Lyapunov function $V : \mathbf{R} \to \mathbf{R}$ existed, then taking taking $t_o = -1$ and $t = 1$ would require that $V(1) < V(x(t_o)) = V(1)$ which is, of course, impossible.

Recall that $\bar{x} = 0$ in this example is not *uniformly* asymptotically stable; as it happens, uniform asymptotic stability of $\bar{x} = 0$ *still* does not imply the existence of strict Lyapunov function for $\bar{x}$. Consider the differential equation

$$\dot{x}(t) = a(t)x(t), \qquad t \in \mathbf{R},$$

where $a(t) = -1$ except over the interval $[-\pi, \pi]$, over which $a(t) = \cos t$. It happens that $\bar{x} = 0$ is a uniformly asymptotically stable equilibrium; nonetheless, if the equation is started from any $x_o \in \mathbf{R}$ at some time $t_o \in [-(1/2)\pi, (1/2)\pi]$, then $x(t) = x_o$ for some time $t > t_o$, and thus no function $V : \mathbf{R} \to \mathbf{R}$ can satisfy $V(x(t)) < V(x_o)$ for all $x_o \in \mathbf{R}$, $t_o \in \mathbf{R}$, and $t > t_o$.

The next example is one for which $\bar{x} = 0$ is *not* an asymptotically stable equilibrium despite the fact that there exists a strict Lyapunov function for $\bar{x}$, in the sense of Definition 23.18. The differential equation is

$$\dot{x}(t) = -e^{-t}x(t), \qquad t \in \mathbf{R};$$

the transition "matrix" $\Phi(t, t_o)$ is given by

$$\Phi(t, t_o) = e^{[e^{-t} - e^{-t_o}]}, \qquad t, t_o \in \mathbf{R}.$$

Observe that for every $t_o \in \mathbf{R}$,

$$\Phi(t, t_o) \to e^{-e^{-t_o}}$$

as $t \to \infty$; thus, by Theorem 23.10, $\bar{x} = 0$ is *not* an asymptotically stable equilibrium. Nonetheless, the function $V : \mathbf{R} \to \mathbf{R}$ defined by $V(x) = x^2$, $x \in \mathbf{R}$, is a strict Lyapunov function for $\bar{x} = 0$ since $\|\Phi(t, t_o)x_o\|$ decreases monotonically on $[t_o, \infty)$ for every $t_o \in \mathbf{R}$.

The following result, which formalizes two of our three previous assertions about Lyapunov functions, is known as *Lyapunov's Theorem*.

23.20 Theorem (Lyapunov): *If there exists a Lyapunov function $V : \mathbf{R}^n \to \mathbf{R}$ for the equilibrium $\bar{x} = 0$ of (DE) or (DfcE), then $\bar{x} = 0$ is uniformly stable. If there exists a strict Lyapunov function for $\bar{x} = 0$, and $t \to A(t)$ or $k \to A(k)$ is constant, then $\bar{x} = 0$ is (uniformly) asymptotically stable.*

Proof: We consider the continuous-time case only; the discrete-time argument is similar. Let $\epsilon > 0$ be given; let M be the minimum of $\{V(x) : \|x\| = \epsilon\}$. M exists since V is continuous, and is positive by definition of a Lyapunov function. Consider

$$W = \{x \in \mathbf{R}^n : V(x) < M \text{ and } \|x\| < \epsilon\};$$

the set W contains $\bar{x} = 0$, is contained in the sphere of radius ϵ about $\bar{x} = 0$, and, since V is continuous, contains some sphere of radius δ about 0.

Observe that if $\|x_o\| < \delta$, then for every $t_o \in \mathbf{R}$ and $t \geqslant t_o$, $\|\Phi(t, t_o)x_o\| < \epsilon$; otherwise, $\|\Phi(t_1, t_o)x_o\|$ would have to equal ϵ for some $t_1 \geqslant t_o$. This is impossible since $x_o \in W$ implies that $V(x_o) < M$, and $\|\Phi(t, t_o)x_o\| = \epsilon$ implies that

$$V(\Phi(t, t_o)x_o) \geqslant M;$$

however, since V is a Lyapunov function, we need $V(\Phi(t_1, t_o)x_o) \leqslant V(x_o)$. Thus, if $\|x_o\| < \delta$, then $\|\Phi(t, t_o)x_o\| < \epsilon$ for all $t \geqslant t_o$, and $\bar{x} = 0$ is a stable equilibrium since $\epsilon > 0$ was arbitrary. Note that $\bar{x} = 0$ is *uniformly* stable because the δ did not depend on t_o.

Suppose, in addition, that $t \to A(t)$ in (DE) is identically equal to a real $(n \times n)$ matrix A, and that V is a *strict* Lyapunov function for $\bar{x} = 0$. Pick ϵ and find δ as in the preceding paragraph; since $\| e^{tA} x_o \| < \epsilon$ for all $t \geqslant 0$, there exists a sequence of times $\{t_k\}$ going to infinity such that $e^{t_k A} x_o$ converges to some $z \in \mathbf{R}^n$. Note that $V(e^{t_k A} x_o) > V(z)$ for all k, since V is a strict Lyapunov function. We claim that $z = 0$; otherwise, $V(e^{\tau A} z) < V(z)$ for every $\tau > 0$; since $e^{(t_k + \tau)A} x_o$ converges to $e^{\tau A} z$, and since V is continuous, it follows that $V(e^{(t_k + \tau)A} x_o)$ converges to some number *less than* $V(z)$. This, however, is impossible; otherwise, there would be some k_o such that $V(e^{t_{k_o} A} x_o) < V(z)$.

Hence, $z = 0$; we have therefore shown that every convergent sequence $\{e^{t_k A} x_o\}$ converges to $\bar{x} = 0$; it may be concluded that $e^{tA} x_o \to 0$ as $t \to \infty$ for every x_o satisfying $\| x_o \| < \delta$ (and hence for *every* $x_o \in \mathbf{R}^n$); consequently, $e^{(t - t_o)A} x_o \to 0$ for every $t_o \in \mathbf{R}$ and every $x_o \in \mathbf{R}^n$, and $\bar{x} = 0$ is an asymptotically stable equilibrium of (DE). $\bar{x}$ is also uniformly asymptotically stable by Theorem 23.14. □

Even in the time-varying case, if the rate of decay of a strict Lyapunov function along solutions of (DE) or (DfcE) is sufficiently fast, then it may be concluded that $\bar{x} = 0$ is an asymptotically stable equilibrium. Specifically, suppose that a given strict Lyapunov function $V : \mathbf{R}^n \to \mathbf{R}$ for the equilibrium $\bar{x} = 0$ of (DE) is *differentiable;* for each $z_o \in \mathbf{R}^n$, denote by $DV(z)$ the $(1 \times n)$ matrix whose $(1, j)$-element is the partial derivative of V with respect to x_j, evaluated at the point z. Suppose it is known that for every $t \in \mathbf{R}$, $DV(z)A(t)z \leqslant -W(z)$ for every $z \in \mathbf{R}^n$, where $W : \mathbf{R}^n \to \mathbf{R}$ satisfies $W(z) > 0$ for $z \neq 0$ and $W(0) = 0$. For every $x_o \in \mathbf{R}^n$ and $t_o \in \mathbf{R}$, we then have

$$\frac{d}{dt} V(\Phi(t, t_o)x_o) \leqslant -W(\Phi(t, t_o)x_o), \quad t \geqslant t_o ,$$

from which it may be deduced that

$$\begin{aligned} V(\Phi(t, t_o)x_o) &\leqslant V(x_o) - \int_{t_o}^{t} W(\Phi(\tau, t_o)x_o) d\tau \\ &\leqslant V(x_o) - (t - t_o)\min\{W(\Phi(\tau, t_o)x_o) : t_o \leqslant \tau \leqslant t\} . \end{aligned}$$

Since $V(x) \geqslant 0$ for all $x \in \mathbf{R}^n$, it is necessary that

$$\lim_{t \to \infty} \min\{W(\Phi(\tau, t_o)x_o) : t_o \leqslant \tau \leqslant t\} = 0;$$

thus, there exists a sequence $\{t_k\}$ of times increasing to infinity for which $\Phi(t_k, t_o)x_o \to 0$. In particular, $V(\Phi(t_k, t_o)x_o)$ converges to zero.

We conclude that $\Phi(t, t_o)x_o \to 0$ as $t \to \infty$. To see this, recall that $V(x) = 0$ if and only if $x = 0$; moreover, $t \to V(\Phi(t, t_o)x_o)$ must decrease, as $t \to \infty$, to its infimum over all $t \geqslant t_o$, and we have just shown this infimum to be zero.

The foregoing argument proves the following variant of Theorem 23.20.

23.21 Theorem: *Suppose that $V : \mathbf{R}^n \to \mathbf{R}$ is a strict Lyapunov function for the equilibrium $\bar{x} = 0$ in (DE). Suppose, in addition, that V is differentiable, and that for every $z \in \mathbf{R}^n$ and $t \in \mathbf{R}$, we have $DV(z)A(t)z \leqslant -W(z)$, where $W : \mathbf{R}^n \to \mathbf{R}$ is nonnegative and satisfies $W(z) = 0$ if and only if $z = 0$. Then $\bar{x} = 0$ is an asymptotically stable equilibrium of (DE).* □

The following discrete-time analogue of Theorem 23.21 is proved similarly.

23.22 Theorem: *Suppose that $V : \mathbf{R}^n \to \mathbf{R}$ is a strict Lyapunov function for the equilibrium $\bar{x} = 0$ in (DfcE). Suppose, in addition, that for every $z \in \mathbf{R}^n$ and $k \in \mathbf{Z}$, we have $V(A(k)z) - V(z) \leqslant -W(z)$, where $W : \mathbf{R}^n \to \mathbf{R}$ is nonnegative and satisfies $W(z) = 0$ if and only if $z = 0$. Then $\bar{x} = 0$ is an asymptotically stable equilibrium of (DfcE).* □

We turn now to the problem of *constructing* Lyapunov functions for the equilibrium $\bar{x} = 0$ of (DE) or (DfcE) when the matrix function A is constant and $\bar{x}$ is stable. Suppose for the moment that $\bar{x} = 0$ is *asymptotically* stable; this assumption is equivalent, by Theorems 23.15 and 23.16, to the assertion that all the eigenvalues of the A-matrix in (DE) have strictly negative real parts or that all the eigenvalues of A in (DfcE) have magnitudes less than one. Our objective is to find a strict Lyapunov function $V : \mathbf{R}^n \to \mathbf{R}$ for $\bar{x}$; it is only natural to attempt to find a V which is as simple as possible.

Accordingly, let us try to construct a strict Lyapunov function which is *quadratic;* that is, our Lyapunov function $V:\mathbf{R}^n \to \mathbf{R}$ should satisfy $V(x) = x^T Rx$, $x \in \mathbf{R}^n$, for some $(n \times n)$ positive definite matrix R. Let $x_o \in \mathbf{R}^n$ be given; let $t \to x(t)$, $t \geqslant t_o$, and $k \to x(k)$, $k \geqslant k_o$, denote the solutions to (DE) and (DfcE) which satisfy the initial conditions $x(t_o) = x_o$ and $x(k_o) = x_o$, respectively. In the case of (DE), we have

$$\frac{d}{dt}V(x(t)) = x^T(t)[A^T R + RA]x(t), \qquad t \geqslant t_o;$$

for (DfcE), the corresponding equation is

$$V(x(k+1)) - V(x(k)) = x^T(k)[A^T RA - R]x(k),$$

which holds for $k \geqslant k_o$. In either case V will be a strict Lyapunov function if the matrix Q in the brackets is *negative definite*. In that case, the sizes of the eigenvalues of Q reflect the *rate of decay* of the Lyapunov function V along solutions to the differential or difference equation.

The following result, which is sometimes referred to as the *Lyapunov Lemma*, makes it possible to construct strict quadratic Lyapunov functions by the method just described.

23.23 Lemma: (*a*) *If the eigenvalues of the real* $(n \times n)$ *matrix* A *have negative real parts, then for every real* $(n \times n)$ *matrix* Q *there exists a unique real* $(n \times n)$ *matrix* R *which satisfies*

$$A^T R + RA = -Q .$$

Moreover, if Q *is positive definite, then so is* R.

(*b*) *If the eigenvalues of the real* $(n \times n)$ *matrix* A *have magnitudes less than one, then for every real* $(n \times n)$ *matrix* Q *there exists a unique real* $(n \times n)$ *matrix* R *which satisfies*

$$A^T RA - R = -Q .$$

Moreover, if Q *is positive definite, then so is* R.

Proof: (a) Under the assumption on the eigenvalues of A, the entries of $t \to e^{tA}$ decay exponentially as $t \to \infty$; c.f. the proof of Theorem 23.15. Hence the integral

$$R = \int_0^\infty e^{tA^T} Q e^{tA}\, dt$$

exists; moreover,

$$\begin{aligned} A^T R + RA &= \int_0^\infty \frac{d}{dt}[e^{tA^T} Q e^{tA}]dt \\ &= -Q \ , \end{aligned}$$

where the last equality follows from the fact that $e^{tA} \to 0$ as $t \to \infty$.

As for (b), the proof of Theorem 23.16 shows that since the eigenvalues of A have magnitudes less than one, $A^k \to 0$, exponentially in k, as $k \to \infty$. Thus, the following sum converges:

$$R = \sum_{k=0}^{\infty} (A^T)^k Q A^k \ ;$$

a term-by-term subtraction yields $A^T RA - R = -Q$.

In either case, given Q, the solution R is unique; first, observe that we may consider the mappings $R \to A^T R + RA$ and $R \to A^T RA - R$ as linear transformations on the vector space of $(n \times n)$ matrices. We have just shown that, under the eigenvalue assumptions, these mappings are *surjective;* we may therefore conclude that they are *injective*, as well, from Theorem 4.24. Their injectivity, however, is equivalent to the uniqueness of R given Q.

The reader is invited to complete the proof by showing, in both cases, that R is positive definite if Q is. □

The equations in the statement of Lemma 23.23 are often called *Lyapunov equations*. In view of our earlier remarks, Lemma 23.23 enables us to find a strict Lyapunov function for the equilibrium $\bar{x} = 0$ of (DE) or (DfcE) (provided, of course, that it is asymptotically stable) by solving a *linear matrix equation*. All we have to do is pick *any* positive definite matrix Q and solve the corresponding Lyapunov equation for R; the function $x \to x^T Rx$, $x \in \mathbf{R}^n$, is then a strict Lyapunov function for $\bar{x} = 0$.

A generalization of Lemma 23.23 makes it possible to construct Lyapunov functions for $\bar{x} = 0$ in (DE) or (DfcE) when $\bar{x}$ is merely a *stable* equilibrium and the matrix function A is constant. Consider the continuous-time case first. Begin by finding an invertible $(n \times n)$ matrix P such that $P^{-1}AP$ takes the form

$$\begin{bmatrix} A_- & 0 \\ 0 & A_0 \end{bmatrix},$$

where the eigenvalues of A_- have negative real parts and those of A_0 have zero real parts. Since $\bar{x} = 0$ is stable, Theorem 23.15 guarantees that $z^T[A_0 + A_0^T]z = 0$ for every $z \in \mathbf{R}^n$; the idea is that A_o has pure imaginary eigenvalues and is semi-simple by the given multiplicity conditions.

Suppose that A_0 is $(n_0 \times n_0)$; let the $(n \times n)$ matrix $\hat{R}$ be given by

$$\hat{R} = \begin{bmatrix} R_1 & 0 \\ 0 & I_{n_0} \end{bmatrix},$$

where $A_-^T R_1 + R_1 A_- = -Q$ for some positive definite $([n - n_0] \times [n - n_0])$ matrix Q. It is then not hard to verify that if $R = P^T \hat{R} P$, then $x^T Rx > 0$ for every nonzero $x \in \mathbf{R}^n$, so that $x \to V(x) = x^T Rx$ is a Lyapunov function for the stable equilibrium $\bar{x} = 0$ of (DE).

23.24 Exercise: Do the discrete-time version of the preceding calculation. That is, assuming that $\bar{x} = 0$ is a stable equilibrium of (DfcE), and that A is constant, find an $(n \times n)$ matrix R such that $x \to x^T Rx$ is a Lyapunov function for $\bar{x}$. [*Suggestion*: First find P so that $P^{-1}AP$ takes the form

$$\begin{bmatrix} A_- & 0 \\ 0 & A_1 \end{bmatrix},$$

where the eigenvalues of A_1 have magnitude one. Use stability to show that $z^T A_1 z = \|z\|^2$ for every $z \in \mathbf{R}^2$.] □

24. *Stability of Input-Output Linear Systems*

The stability theory of input-output linear systems has developed largely over the last forty years. Recently, the theory has begun to take on a sophisticated functional analytic flavor. It has been recognized that many well-known stability definitions and results may be expressed conveniently in terms of *linear mappings* between *normed vector spaces* of input and output functions. In what follows, we shall endeavor to follow a rudimentary version of this functional analytic approach to input-output stability theory.

We begin by recalling Definitions 18.1 and 18.2 of continuous- and discrete-time input-output linear systems. An m-input, p-output continuous-time input-output linear system consists of a vector space $\mathbf{U}$ of $\mathbf{R}^m$-valued input functions, a family of vector spaces $\{\mathbf{Y}_{t_o} : t_o \in \mathbf{R}\}$ of vector spaces of $\mathbf{R}^p$-valued output functions, and a family $\{\sigma_{t_o} : \mathbf{U} \to \mathbf{Y}_{t_o}\}$ of linear mappings between the input function space and the output function spaces. In discrete time, we have a vector space $\mathbf{U}$ of $\mathbf{R}^m$-valued *discrete-time* input functions, vector spaces $\{\mathbf{Y}_{k_o} : k_o \in \mathbf{Z}\}$ of $\mathbf{R}^p$-valued discrete-time output functions, and a linear input-output mapping $\sigma_{k_o} : \mathbf{U} \to \mathbf{Y}_{k_o}$ for each $k_o \in \mathbf{Z}$. The mappings σ_{t_o} and σ_{k_o} are assumed to satisfy natural consistency and causality conditions.

Our first input-output stability definitions are prompted by an elementary and intuitive view of what it should mean for an input-output system to be stable. In §23, we promoted the the stability of a *state space* linear system as a sort of insurance against the possibility that a small nonzero initial condition could have a major effect on the system's long-term behavior. One might expect that a stable input-output system would have the property that an input function $u \in \mathbf{U}$ which is always small should give rise to output functions $\sigma_{t_o}(u)$ or $\sigma_{k_o}(u)$ which are also small on their domains of definition. In particular, we should expect that a *bounded* input function $u \in \mathbf{U}$ should give rise, for each $t_o \in \mathbf{R}$ or $k_o \in \mathbf{Z}$, to a *bounded* output function on $[t_o, \infty)$ or $\{k_o, k_o + 1, \ldots\}$. In Definitions 24.1 and 24.2, and throughout the remainder of the section, we shall use the word *bounded* (c.f. §5) to mean *bounded with respect to the (usual) Euclidean norm on* $\mathbf{R}^n$; we shall also use $\| \ \|_2$, or simply $\| \ \|$ when context permits, to denote the Euclidean norm.

24.1 Definition: A real m-input, p-output continuous-time input-output linear system as in Definition 18.1 is said to be *bounded-input bounded-output stable* (or *BIBO stable*) if and only if for every $t_o \in \mathbf{R}$, $\sigma_{t_o}(u) \in \mathbf{Y}_{t_o}$ is a bounded $\mathbf{R}^p$-valued function on $[t_o, \infty)$ whenever $u \in \mathbf{U}$ is a bounded $\mathbf{R}^m$-valued function on $\mathbf{R}$.

The system is said to be *uniformly BIBO stable* if and only if for every bounded $u \in \mathbf{U}$ there exists a single number $M(u) > 0$ such that $\| \sigma_{t_o}(u)(t) \| \leqslant M(u)$ for every $t_o \in \mathbf{R}$ and $t \geqslant t_o$. □

24.2 Definition: A real m-input, p-output discrete-time input-output linear system as in Definition 18.2 is said to be *bounded-input bounded-output stable* (or *BIBO stable*) if and only if for every $k_o \in \mathbf{R}$, $\sigma_{k_o}(u) \in \mathbf{Y}_{k_o}$ is a bounded $\mathbf{R}^p$-valued function on $\{k_o, k_o+1, k_o+2, \ldots\}$ whenever $u \in \mathbf{U}$ is a bounded $\mathbf{R}^m$-valued function on $\mathbf{Z}$.

The system is said to be *uniformly BIBO stable* if and only if for every bounded $u \in \mathbf{U}$ there exists a single number $M(u) > 0$ such that $\| \sigma_{k_o}(u)(k) \| \leqslant M(u)$ for every $k_o \in \mathbf{R}$ and $k \geqslant k_o$. □

As might be expected, the stability conditions in Definitions 24.1 and 24.2 may be stated conveniently in terms of the *weighting pattern* of the system in question; in the case of time-invariant systems, there are equivalent statements in terms of the system's *impulse response*. Recall from §18 (c.f. Definition 18.4) that a given m-input, p-output discrete-time input-output linear system possesses a real $(p \times m)$ matrix-valued weighting pattern $(k, l) \to W(k, l)$, defined on the set of all integer pairs (k, l) satisfying $k \geqslant l$; W reflects the input-output behavior of the system in the sense that

$$\sigma_{k_o}(u)(k) = \sum_{l = k_o}^{k} W(k, l) u(l), \quad k \geqslant k_o,$$

for every $u \in \mathbf{U}$. The system is BIBO stable if and only if for every $k_o \in \mathbf{Z}$ and for every bounded $u \in \mathbf{U}$ there exists a real number $M(u, k_o)$ such that

$$\left\| \sum_{l = k_o}^{k} W(k, l) u(l) \right\| \leqslant M(u, k_o), \quad k \geqslant k_o;$$

the system is, in addition, *uniformly* BIBO stable if $M(u, k_o)$ may be chosen independently of k_o for every bounded $u \in \mathbf{U}$.

Similarly, in the light of Lemma 18.6 and Definition 18.8, a given m-input, p-output continuous-time input-output linear system satisfying Assumptions 18.5 also possesses a weighting pattern. More precisely, given such a system one may find for each $t \in \mathbf{R}$ sequences of times $\{t_k(t)\}$ and real $(p \times m)$ matrices $\{D_k(t)\}$, along with a $(p \times m)$ matrix-valued function $(t, \tau) \to W(t, \tau)$ defined on $\{(t, \tau) \in \mathbf{R}^2 : t \geqslant \tau\}$, such that for every $u \in \mathrm{U}$,

$$\sigma_{t_o}(u)(t) = \int_{t_o}^{t} W(t, \tau)u(\tau)d\tau + \sum_{[t_o, t]} D_k(t)u(t_k(t)), \quad t \geqslant t_o ;$$

the summation is taken over $\{k : t_k(t) \in [t_o, t]\}$. The weighting pattern of the system is defined to be the real $(p \times m)$ matrix "function" $(t, \tau) \to \tilde{W}(t, \tau)$ given by

$$\tilde{W}(t, \tau) = W(t, \tau) + \sum_{(-\infty, t]} D_k(t)\delta(\tau - t_k(t)), \quad t \geqslant \tau .$$

According to Definition 24.1, the system is BIBO stable precisely when one may find, for every bounded $u \in \mathrm{U}$ and every $t_o \in \mathbf{R}$, a real number $M(u, t_o)$ such that

$$\left\| \int_{t_o}^{t} W(t, \tau)u(\tau)d\tau + \sum_{[t_o, t]} D_k(t)u(t_k(t)) \right\| \leqslant M(u, t_o)$$

for every $t \geqslant t_o$. The system is *uniformly* BIBO stable if and only if $M(u, t_o)$ may be chosen independently of t_o for every $u \in \mathrm{U}$.

As we have defined them, the concepts of BIBO stability and uniform BIBO stability for input-output linear systems do not have attached to them any notion of "uniformity in the input u." It turns out that such a notion may be made precise in terms of some fairly simple functional analytic constructions. A glance at §5, particularly the material between Definition 5.8 and Lemma 5.14, might serve the reader well at this point.

Recall (c.f. Examples 4.2(e) and (f)) the definitions of the L^q and l^q spaces of functions; the vector space L^q, where $1 \leqslant q < \infty$, was defined as the set of all $f : \mathbf{R} \to \mathbf{R}$ satisfying

$$\int_{-\infty}^{\infty} |f(t)|^q dt < \infty ,$$

while l^q was the set of all $f : \mathrm{Z} \to \mathbf{R}$ satisfying

$$\sum_{k=-\infty}^{\infty} |f(k)|^q < \infty .$$

By analogy, we may define L^∞ as the set of $f : \mathbf{R} \to \mathbf{R}$ which are *bounded*, and, similarly, l^∞ as the set of bounded $f : \mathbf{Z} \to \mathbf{R}$. Norms on these last two vector spaces are given, respectively, by

$$\| f \|_\infty = \sup\{ |f(t)| : -\infty < t < \infty\} , \qquad f \in L^\infty ,$$

and

$$\| f \|_\infty = \sup\{ |f(k)| : -\infty < k < \infty\} , \qquad f \in l^\infty .$$

By virtue of Lemma 5.9 on the equivalence of norms on $\mathbf{R}^n$, we may also conclude that $f : \mathbf{R} \to \mathbf{R}^n$ is bounded (with respect to the usual Euclidean norm on $\mathbf{R}^n$) if and only for every i, $1 \leqslant i \leqslant n$, the ith component function of f is in L^∞ for every i, $1 \leqslant i \leqslant n$. Likewise, $f : \mathbf{Z} \to \mathbf{R}^n$ is bounded if and only if each $k \to f_i(k)$ is in l^∞. A suitable norm on the vector space L_n^∞ of all *bounded* functions $f : \mathbf{R} \to \mathbf{R}^n$ is

$$\| f \|_\infty = \sup\{ \| f(t) \|_2 : -\infty < t < \infty\} ;$$

similarly, we may set

$$\| f \|_\infty = \sup\{ \| f(k) \|_2 : -\infty < k < \infty\}$$

for bounded $f : \mathbf{Z} \to \mathbf{R}^n$, which together form a normed vector space l_n^∞.

To say that an m-input, p-output continuous-time input-output linear system is BIBO stable is the same as saying that for every $t_o \in \mathbf{R}$, the mapping $\sigma_{t_o} : \mathbf{U} \to \mathbf{Y}_{t_o}$ maps the vector space $\mathbf{U} \cap L_m^\infty$ into the vector space of all *bounded* $y \in \mathbf{Y}_{t_o}$, which is in general a *proper* subspace of $\mathbf{Y}_{t_o}$. Similarly, each input-output mapping σ_{k_o} of an m-input, p-output discrete-time input-output linear system which is BIBO stable must map $\mathbf{U} \cap l_m^\infty$ into the vector space of all bounded $y \in \mathbf{Y}_{k_o}$.

Now, the families $\mathbf{U} \cap L_m^\infty$ and $\mathbf{U} \cap l_m^\infty$ are *normed* vector spaces, where the norms are specified as above; similarly, the subspaces of bounded y's in $\mathbf{Y}_{t_o}$ and $\mathbf{Y}_{k_o}$ are also normed vector spaces, with norms given by appropriate modifications of the L_p^∞ and l_p^∞ norms, respectively. It is *not* immediately clear that the input-output mappings σ_{t_o} and σ_{k_o} are, in the case of BIBO stable systems, *bounded linear mappings* with respect to the given norms (c.f. the discussion preceding Lemma 5.11); this assertion, however, turns out to be true. The boundedness of σ_{t_0} and σ_{k_0} as linear mappings between normed vector spaces of functions gives us a notion of input-output stability which is *uniform in the input function u*.

Recall (c.f. [Royden] or [Rudin]) that a *complete* normed vector space is a normed vector space V in which every *Cauchy sequence* possesses a limit; that is, V is complete if and only if whenever a sequence $\{v^k\}$ of vectors in V satisfies

$$\lim_{m,n \to \infty} \| v^m - v^n \| = 0 ,$$

there exists some vector $\hat{v} \in V$ such that

$$\lim_{n \to \infty} \| v^n - \hat{v} \| = 0 .$$

For example, $\mathbf{R}^n$ and $\mathbf{C}^n$, for any n, are complete normed vector spaces with respect to any norm; moreover, the vector spaces l^∞ and l_n^∞ are complete. Some special "functions" which take on infinite values on discrete sets of real numbers must be added to L^∞ in order to make it complete, but, for our purposes, L^∞ and L_n^∞ may be regarded as "essentially complete" with respect to the norms introduced above. We need the following important result from functional analysis, known as the Uniform Boundedness Theorem; once again, see [Royden] or [Rudin].

24.3 Fact: *Let X be a family of bounded linear mappings $T : V \to W$, where V and W are complete normed vector spaces with respective norms $\| \ \|_V$ and $\| \ \|_W$. Suppose that for each $v \in V$, there exists some $K_v > 0$ such that every $T \in X$ satisfies*

$$\| T(v) \|_W \leqslant K_v .$$

Then there exists a single $K > 0$ such that $\| T \|_{V,W} \leqslant K$ for every $T \in X$. □

The Uniform Boundedness Theorem says that a subset of $\mathrm{Hom}_{\mathbf{B}}(V, W)$ (see §5 for notation) which is *pointwise bounded* (as a set of mappings) is also a *uniformly bounded* subset of $\mathrm{Hom}_{\mathbf{B}}(V, W)$. In the present context, suppose that we are given an m-input, p-output discrete-time input-output linear system. For a fixed $k_o \in \mathbf{Z}$, consider the family of linear mappings $\gamma_{k_o}^k : U \cap l_m^\infty \to \mathbf{R}^p$ defined by

$$\gamma_{k_o}^{k}(u) = \sigma_{k_o}(u)(k), \qquad u \in \mathbf{U} \cap l_m^{\infty}.$$

Note that each $\gamma_{k_o}^{k}$ is a *bounded* linear mapping from $\mathbf{U} \cap l_m^{\infty}$ to $\mathbf{R}^p$; indeed,

$$\|\gamma_{k_o}^{k}(u)\| \leqslant \|u\|_{\infty} \max\{\|W(k,l)\| : k_o \leqslant l \leqslant k\}$$

for every $u \in \mathbf{U} \cap l_m^{\infty}$, where $W(k,l)$ is the weighting pattern of the system and $\|u\|_{\infty}$ is given above. If, in addition, the system is BIBO stable, then there exists for each u some $M(u,k_o) > 0$ such that $\|\gamma_{k_o}^{k}(u)\| \leqslant M(u,k_o)$ for *every* $k \in \mathbf{Z}$.

In that case, the Uniform Boundedness Theorem implies that there is some single $M(k_o) > 0$ such that $\|\gamma_{k_o}^{k}(u)\| \leqslant M(k_o)\|u\|_{\infty}$ for *every* $k \geqslant k_o$ and $u \in \mathbf{U} \cap l_m^{\infty}$. If the BIBO stability is uniform, in which case the $M(u,k_o)$'s may be chosen independently of k_o, then we may assert in addition that there is an $M > 0$ such that

$$\sup\{\|\sigma_{k_o}(u)(k)\|_2 : k \geqslant k_o\} \leqslant M \sup\{\|u(l)\|_2 : l \in \mathbf{Z}\}$$

for every bounded $u : \mathbf{Z} \to \mathbf{R}^m$.

The argument for a continuous-time BIBO stable system is similar; first, define the mappings $\gamma_{t_o}^{t}$ for $t \geqslant t_o$ by

$$\gamma_{t_o}^{t}(u) = \sigma_{t_o}(u)(t), \qquad u \in \mathbf{U} \cap L_m^{\infty};$$

next, show that each $\gamma_{t_o}^{t}$ is a bounded linear mapping from $\mathbf{U} \cap L_m^{\infty}$ to $\mathbf{R}^p$, then use BIBO stability to put a "u-wise" bound on the $\gamma_{t_o}^{t}$, from which a global bound follows from the Uniform Boundedness Theorem. More precisely, BIBO stability implies that for every $t_o \in \mathbf{R}$ there exists some $M(t_o) > 0$ such that

$$\sup\{\|\sigma_{t_o}(u)(t)\| : t \geqslant t_o\} \leqslant M(t_o) \sup\{\|u(\tau)\|_2 : -\infty < \tau < \infty\},$$

and *uniform* BIBO stability implies that a single $M > 0$ works as $M(t_o)$ in the last inequality for every $t_o \in \mathbf{R}$ in the sense that for every $t_o \in \mathbf{R}$,

$$\sup\{\|\sigma_{t_o}(u)(t)\| : t \geqslant t_o\} \leqslant M \sup\{\|u(\tau)\|_2 : -\infty < \tau < \infty\},$$

for some $M > 0$.

The foregoing discussion leads to a concise necessary and sufficient condition for an input-output linear system to be uniformly BIBO stable; the condition takes an especially nice form for time-invariant systems.

24.4 Theorem: *Let the weighting pattern of a real m-input, p-output continuous-time input-output linear system satisfying Assumptions 18.5 be given by*

$$\tilde{W}(t,\tau) = W(t,\tau) + \sum_k D_k(t)\delta(t - t_k(t)), \quad t \geqslant \tau .$$

The system is BIBO stable if and only if there exists for each $t_o \in \mathbf{R}$ an $M(t_o) > 0$ such that

$$\int_{t_o}^{t} \| W(t,\tau) \| \, d\tau + \sum_k \| D_k(t) \| \leqslant M(t_o)$$

for every $t \geqslant t_o$. Moreover, if the system is uniformly BIBO stable, then $M(t_o)$ may be chosen independently of t_o.

Proof: Suppose first that the system is BIBO stable. It is not hard to show that if the conclusion of the theorem fails, then there exists a $t_o \in \mathbf{R}$ and a pair (i, j), where $1 \leqslant i \leqslant p$ and $1 \leqslant j \leqslant m$, such that

$$\int_{t_o}^{t} |[W(t,\tau)]_{ij}| \, d\tau + \sum_k |[D_k(t)]_{ij}| \to \infty$$

as $t \to \infty$. The sum in the last expression is taken over $\{k : t_k(t) \geqslant t_o\}$. If the integral blows up as $t \to \infty$, then for each $t \geqslant t_o$ define $u_t : \mathbf{R} \to \mathbf{R}^m$ as follows:

$$u_t(\tau) = \begin{cases} e^j \operatorname{sgn}([W(t,\tau)]_{ij}) & t_o \leqslant \tau \leqslant t \\ 0 & \text{otherwise} , \end{cases}$$

where $e^j \in \mathbf{R}^m$ is the jth standard basis vector for $\mathbf{R}^m$ and "sgn" means "sign." Since $\| u_t(\tau) \|_2 = 1$, by BIBO stability there must be some $M > 0$ such that $\sigma_{t_o}(u_t)(t) \leqslant M$ for every $t \geqslant t_o$, but this is clearly not the case, and we have a contradiction. A similar argument works if the sum of the $|[D_k(t)]_{ij}|$ blows up. Note that we have allowed for the existence of *discontinuous* input functions u; if U is assumed to contain only continuous input functions, then the definition of u_t may be adjusted so that, for each $t \geqslant 0$, u_t is a continuous approximation to the sgn function.

Conversely, if the conclusion of the theorem holds, then the system must be BIBO stable; the idea is that for every $t_o \in \mathbf{R}$ and for every $t \geqslant t_o$, we have

$$\| \sigma_{t_0}(u)(t) \|_2 = \| \int_{t_0}^{t} W(t,\tau)u(\tau)d\tau + \sum_k D_k(t)u(t_k(t)) \|_2$$
$$\leqslant M(t_0) \sup_{t_0 \leqslant \tau \leqslant t} \{ \| u(\tau) \|_2 \} ,$$

which is equivalent (in view of the foregoing discussion) to BIBO stability of the given system. We leave it to the reader to show that uniform BIBO stability is equivalent to the possibility of choosing $M(t_0)$ independently of t_0. □

The proof of the discrete-time result corresponding to Theorem 24.4 is left as another exercise for the reader.

24.5 Theorem: *Let the weighting pattern of a real m-input, p-output discrete-time input-output linear system be given by $W(k,l)$, $k \geqslant l$. The system is BIBO stable if and only if for every $k_0 \in \mathbb{Z}$ there exists an $M(k_o) > 0$ such that*

$$\sum_{l=k_0}^{k} \| W(k,l) \| \leqslant M(k_0)$$

for every $k \geqslant k_0$. The system is uniformly BIBO stable if and only if $M(k_0)$ may be chosen independently of k_0. □

If a BIBO stable input-output linear system is time-invariant, then the criteria given in Theorems 24.4 and 24.5 take on an especially simple form. Moreover, Theorems 24.4 and 24.5 enable us to show easily that BIBO stability and *uniform* BIBO stability are equivalent properties of a time-invariant system. Suppose that a given discrete-time input-output linear system is BIBO stable. By Theorem 24.5, we may find for each $k_0 \in \mathbb{Z}$ a positive $M(k_0)$ such that the system's weighting pattern satisfies

$$\sum_{l=k_0}^{k} \| W(k,l) \| \leqslant M(k_0)$$

for all $k \geqslant k_0$. Let $k \to H(k)$ denote the impulse response of the system (c.f. Definition 18.16(b)); thus, $H(k) = W(k,0)$ for every $k \geqslant 0$, and $W(k,l) = H(k-l)$ whenever $k \geqslant l$. Taking $k_0 = 0$ in the BIBO stability condition implies that

$$\sum_{l=0}^{k} \| H(l) \| \leqslant M(0)$$

for every $k \geqslant 0$; thus

$$\sum_{k=0}^{\infty} \|H(k)\| \leqslant M(0). \tag{*}$$

The last inequality enables us to conclude that for all $k_o \in \mathbf{Z}$, for every bounded $u \in \mathbf{U}$, and for all $k \geqslant k_o$,

$$\begin{aligned} \|\sigma_{k_o}(u)(k)\|_2 &= \Big\| \sum_{l=k_o}^{k} H(k-l)u(l) \Big\| \\ &\leqslant M(0)\max\{\|u(l)\|_2 : k_o \leqslant l \leqslant k\}, \end{aligned}$$

from which uniform BIBO stability follows immediately upon taking limits as $k_o \to -\infty$ and $k \to \infty$. The argument in the continuous-time case is similar; Theorem 24.4 tells us that the impulse response $t \to H(t)$ of a BIBO stable time-invariant continuous-time input-output linear system must satisfy

$$\int_0^{\infty} \|H(t)\| dt \leqslant M \tag{**}$$

for some $M > 0$, from which *uniform* BIBO stability of the system follows.

Since the impulse responses of time-invariant input-output systems are zero for negative valuse of "time," inequalities (*) and (**) above tell us that the impulse response of a BIBO stable time-invariant continuous-time system is "in L^1," in the sense that its *norm* is an L^1 function. Similarly, the impulse response of a BIBO stable time-invariant discrete-time system is "in l^1." A glance at the argument above should convince the reader that BIBO stability of a time-invariant continuous- (respectively, discrete-) time input-output linear system is *equivalent* to its impulse response's being L^1 (respectively, l^1).

We may also conclude from the foregoing discussion that the steady-state response (c.f. Definitions 18.13 and 18.14) of a time-invariant BIBO stable system to any *bounded* input function u is well-defined. To see this, consider the case of a discrete-time system; we need to show that for all k, $\sigma_{k_o}(u)(k)$ approaches limit as $k_o \to -\infty$ if u is bounded. We know that if $k_o \leqslant k_1$, then

$$\|\sigma_{k_o}(u)(k) - \sigma_{k_1}(u)(k)\| \leqslant \|u\|_\infty \sum_{k_o}^{k_1} \|H(k-l)\|,$$

and that the sum must go to zero as k_o and k_1 go to $-\infty$ by the

l^1 nature of H; thus $\{\sigma_{k_o}(u)(k)\}$ forms a Cauchy sequence (as $k_o \to -\infty$) for every $k \in \mathbf{Z}$, and therefore converges. The continuous-time argument is similar.

It should by now be apparent why BIBO stability of time-invariant continuous-time input-output linear systems is often referred to as L^∞-*stability*, while BIBO stability of time-invariant discrete-time systems is sometimes called l^∞ *stability*. The steady-state response $\sigma(u)$ of a BIBO stable time-invariant continuous-time input-output linear system to any $u \in \mathrm{U} \cap L_m^\infty$ is well-defined; moreover, $\sigma(u)$ is in L_p^∞, and

$$\| \sigma(u) \|_\infty \leqslant M \| u \|_\infty$$

for some $M > 0$. Likewise, if u is a bounded input to a BIBO stable time-invariant discrete-time system, then $\sigma(u)$, the steady state response to u, is in l_p^∞ and respects a similar bound on its l_p^∞-norm. The following generalizations suggest themselves immediately.

24.6 Definition: A real m-input, p-output time-invariant continuous-time input-output linear system with input function space U is said to be L^q-*stable* (for some $q \geqslant 1$) if and only if the steady-state response $\sigma(u)$ to any $u \in \mathrm{U} \cap L^q$ is defined and in L_p^q, and there exists an $M > 0$ such that $\| \sigma(u) \|_q \leqslant M \| u \|_q$ for all $u \in \mathrm{U} \cap L_m^q$. □

24.7 Definition: A real m-input, p-output time-invariant discrete-time input-output linear system with input function space U is said to be l^q-*stable* (for some $q \geqslant 1$) if and only if the steady-state response $\sigma(u)$ to any $u \in \mathrm{U} \cap l_m^q$ is defined and in l_p^q, and there exists an $M > 0$ such that $\| \sigma(u) \|_q \leqslant M \| u \|_q$ for all $u \in \mathrm{U} \cap l_m^q$. □

Here, we are employing the standard notation

$$\| f \|_q = \left(\int_{-\infty}^{\infty} (\| f(t) \|_2)^q \, dt \right)^{1/q}$$

for $f \in L_n^q$, and

$$\| f \|_q = \left(\sum_{k=-\infty}^{\infty} (\| f(k) \|_2)^q \right)^{1/q}$$

for $f \in l_n^q$. With Definitions 24.6 and 24.7, we have touched on the notation and terminology of a vast body of fairly recent research on the stability theory of input-output linear systems. For an extensive discussion of more general functional analytic approaches to input-output stability theory, see [Desoer and Vidyasagar].

The integrability conditions (*) and (**) on the impulse responses of BIBO stable time-invariant input-output linear systems make it possible to state BIBO stability criteria conveniently in terms of the *transfer functions* of such systems (c.f. Definitions 19.6 and 20.6). It follows from (*) above that every BIBO stable time-invariant discrete-time input-output linear system possesses a transfer function; simply note that since the system's impulse response satisfies (*), $R^{-k}H(k) \to 0$ as $k \to \infty$ for any $R \geqslant 1$. Similarly, $e^{-\sigma t}H(t)$ must decay as $t \to \infty$ for every $\sigma \geqslant 0$ if H is the impulse response of a BIBO stable time-invariant continuous-time system.

It is also true that every BIBO stable time-invariant input-output linear system possesses a frequency response. This follows for continuous-time systems (see the discussion preceding Definition 19.1) from the fact that $t \to \cos(\omega_o t)$ and $t \to \sin(\omega_o t)$ are bounded input functions, and therefore give rise to well-defined steady-state responses. Similarly, the steady-state response of a BIBO stable discrete-time system to the bounded input functions $k \to \cos \Omega_o k$ and $k \to \sin \Omega_o k$ is well-defined, and such a system therefore possesses a frequency response. We could have drawn the same conclusions from the fact that $t \to H(t)$ (respectively, $k \to H(k)$), the impulse response of a continuous- (respectively, discrete-) time system, is in L^1 (respectively, l^1), and is therefore Fourier transformable; c.f. Definitions 19.1 and 20.1.

The fact that the impulse response of a BIBO stable time-invariant system decays as t (or k) becomes large guarantees that the region of convergence for the system's transfer function includes the entire half-plane $\text{Re}\{s\} > 0$ for a continuous-time system and the entire annulus $|z| > 1$ for a discrete-time system. It is *not* in general true that the imaginary axis (or the unit circle for a discrete-time system) is included in the region of convergence $ROC(H)$ for the transfer function of a BIBO stable system. Consider, for example, the single-input single-output time-invariant continuous-time system whose impulse response is $H(t) = t^{-2}1(t-1)$, $t \in \mathbf{R}$. Evidently, by Theorem 24.4 and the discussion which follows, the system is (uniformly) BIBO stable; nonetheless, $ROC(H)$ is given by

$\mathrm{Re}\{s\} > 0$, since $\sigma_a(h) = 0$ (see Definition 19.7 for the definition of $\sigma_a(H)$).

It turns out, however, that any *realizable* BIBO stable time-invariant continuous-time system has a transfer function whose region of convergence includes the imaginary axis; correspondingly, the transfer function of a realizable BIBO stable discrete-time system includes the unit circle. To see this, recall from Theorem 21.9 that the transfer function $s \to G(s)$, $s \in ROC(H)$, of a realizable continuous-time system agrees on $ROC(H)$ with a $(p \times m)$ proper rational matrix function of s. Likewise, the transfer function $z \to G(z)$ of a discrete-time system is given on its region of convergence by a $(p \times m)$ proper rational matrix function of z. By a *pole* of such a transfer function, we mean a pole of one of the rational matrix elements in G. In order to prove the above assertions, we'll need to construct realizations of the given input-output systems. The following result is proved implicitly in §21, but we give it here for completeness.

24.8 Lemma: *If the A-matrix in a realization (A, B, C, D) of a real realizable m-input, p-output time-invariant input-output linear system has an eigenvalue which is not a pole of the system's transfer function, then (A, B, C, D) is not a minimal realization.*

Proof: Let λ_o be such an eigenvalue; by the results of §9, we may find an invertible matrix P such that $P^{-1}AP$ takes the form

$$P^{-1}AP = \begin{bmatrix} A_1 & 0 \\ 0 & A_o \end{bmatrix},$$

where λ_o is the sole eigenvalue of A_o. Partition $\hat{C} = CP$ and $\hat{B} = P^{-1}B$ conformably as

$$CP = \begin{bmatrix} C_1 & C_o \end{bmatrix}; \quad P^{-1}B = \begin{bmatrix} B_1 \\ B_o \end{bmatrix}.$$

The transfer function of the system (if it a continuous-time system) is then given on its region of convergence by

$$C_1(sI_n - A_1)^{-1}B_1 + C_o(sI_n - A_o)^{-1}B_o + D\ ;$$

for a discrete-time system, the same decomposition holds with z in place of s. The second term is a strictly proper $(p \times m)$ rational matrix whose only possible pole is at λ_o; since λ_o is not a pole of the first term nor of the transfer function, the second term in the decomposition must be a *constant* $(p \times m)$ matrix, which must be zero since it is strictly proper. Thus (A_1, B_1, C_1, D) is another realization for the system, with a smaller A-matrix that (A, B, C, D), and the conclusion of the Lemma follows. □

Suppose now that a given real m-input, p-output time-invariant continuous-time input-output linear system is BIBO stable and realizable. Let $t \to H(t)$ be the system's impulse response and $s \to G(s)$, $s \in ROC(H)$, the system's transfer function; we've seen already that $ROC(H)$ includes the *open* right half-plane $\mathrm{Re}\{s\} > 0$. Let (A, B, C, D) be a minimal realization for the system. $G(s)$ agrees on its region of convergence with real proper rational $(p \times m)$ matrix function $C(sI_n - A)^{-1}B + D$. Moreover, since

$$H(t) = Ce^{tA}B\,1(t) + D\,\delta(t),$$

every element of H is a linear combination of impulses and terms of the form $t^k e^{\lambda_o t}\,1(t)$, where λ_o is an eigenvalue of A.

If some element of $G(s)$ has a pair of complex conjugate poles $\pm i\omega_o$ on the imaginary axis, it is easily checked that the corresponding element of H must contain a term which is a constant multiple of $t \to t^k \cos(\omega_o t)1(t)$ or $t \to t^k \sin(\omega_o t)1(t)$ for some nonnegative integer k. This contradicts the fact that $H(t) \to 0$ as $t \to \infty$, which we have already seen to be a consequence of BIBO stability.

We conclude that $G(s)$ has no poles on the imaginary axis. Since (A, B, C, D) is a minimal realization for the system, the eigenvalues of A must lie among the poles of $G(s)$ by Lemma 24.8; thus, all of A's eigenvalues lie in the open left half-plane $\mathrm{Re}\{s\} < 0$. By the formula given above for H, all the exponentials appearing in the matrix elements of H have exponents whose real parts are negative. $ROC(H)$, therefore, *extends* into the open left half-plane, and therefore *includes the imaginary axis*.

The argument is almost identical for discrete time systems, modulo the appropriate discrete-time substitutions. We summarize the foregoing discussion as follows.

24.9 Theorem: *(a) A real m-input, p-output time-invariant continuous-time realizable input-output linear system is*

BIBO stable if and only if the region of convergence of its transfer function includes the closed right half-plane $\mathrm{Re}\{s\} \geqslant 0$.

(b) A real m-input, p-output time-invariant discrete-time realizable input-output linear system is BIBO stable if and only if the region of convergence of its transfer function includes the closed annulus $|z| \geqslant 1$. □

It should be emphasized that for a time-invariant continuous-time system which is not realizable, the statement that the system's transfer function has no poles in the half-plane $\mathrm{Re}\{s\} \geqslant 0$ is in general *weaker* than the statement that the region of convergence of the transfer function includes $\mathrm{Re}\{s\} \geqslant 0$; the point is that a non-rational function may possess singularities other than poles. For realizable systems, the two assertions are equivalent, and coincide with BIBO stability. A similar remark holds true for discrete-time systems.

We close the section with a brief discussion of some relationships between the stability properties of input-output linear systems and the corresponding properties of state space systems with which the input-output systems are associated. Lemma 24.8 is one result along those lines; it asserts that the eigenvalues in the A-matrix of a realization (A, B, C, D) for a given realizable time-invariant input-output linear system must lie among the poles of the system's transfer function. One may regard such a realization as defining a state space system with state space $\mathbf{R}^n$, where A is assumed to be $(n \times n)$. Theorems 23.15 and 23.16, along with Definition 23.8, imply that if the input-output system is BIBO stable, then the state space system so constructed is uniformly asymptotically stable. In other words, BIBO stability of the input-output system implies uniform asymptotic stability of the equilibrium $\bar{x} = 0$ of

$$\dot{x}(t) = Ax(t), \quad t \in \mathbf{R},$$

or

$$x(k+1) = Ax(k), \quad k \in \mathbf{Z}.$$

Conversely, suppose that an m-input, p-output time-invariant state space linear system is given; consider its associated input-output linear system, specified as in Definition 18.18 or Definition 18.19. If (A, B, C, D) is a (constant) realization for the state space system, then asserting that the state space system is uniformly asymptotically stable is the same as saying that all the eigenvalues of A have negative real parts (in continuous time) or magnitudes less than one (in discrete time). The associated input-output linear system is time-invariant, realizable, and has transfer function given on its region of convergence by $C(sI_n - A)^{-1}B + D$ or $C(zI_n - A)^{-1}B + D$; it follows that the input-output system is uniformly BIBO stable, since the poles of its transfer function, by the last two formulas, must lie among the eigenvalues of A.

25. *Feedback, Observers, and Canonical Forms*

It is said that the idea of using *feedback* to improve the performance of an electrical signal amplifier first occurred to Harold S. Black one Saturday morning in 1927 while he was riding a ferry across the Hudson River to his office in New York City. There is no question that Black's fundamental insight, coupled with the concurrent and subsequent pioneering work of Hendrick Bode, Harry Nyquist, and others, changed the course of modern engineering. Perhaps more than any other theoretical development, the concept of using feedback to modify the behavior of input-output processes has served to establish the identity of system theory as a distinguished subfield of engineering modeling.

Feedback systems are everywhere; Norbert Wiener has stated (see [Wiener]) that the human being, in the course of performing everyday tasks, is perhaps the quintessential "feedback system." Imagine a person attempting to insert a plug into an electrical outlet while blindfolded; the "trajectory" which the plug follows en route to the socket is certainly *not* precisely determined before the task is accomplished. Rather, after having figured out an approximately correct path, the person in question completes the maneuver by using his knowledge of the physical environment along with "measurements" which he obtains through his sense of touch to *compensate* for errors in his *a priori* strategy. This simple example is, in fact, quite illustrative of some of the key ideas which underlie the mathematicization of feedback in the context of linear system theory. There are often "desired" paths which a designer would like certain variables in a system to follow; effective strategies for making the system follow these desired paths often involve using feedback of the *deviation* of the "actual" paths from the "nominal" ones.

We begin our treatment of feedback in linear systems with a short discussion of some of the reasons for its effectiveness in improving the behavior of input-output processes. For the time being, we'll focus on continuous-time systems; most of the results have obvious analogues in the discrete-time context. We shall often use *block diagram* notation in an effort to streamline the development.

A diagram such as

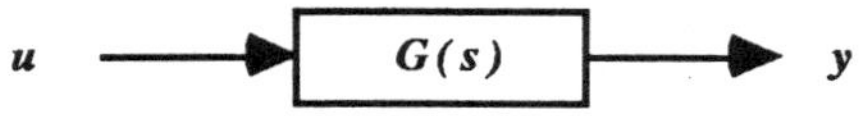

Figure 25.1 Block diagram notation

is meant to denote an m-input, p-output time-invariant continuous-time linear system whose $(p \times m)$ transfer function matrix is given on its region of convergence by $G(s)$; the u denotes an input function for which the steady-state response of the system is defined, and the y represents the steady-state response. Of course, questions about Laplace transformability and regions of convergence are glossed over by this notation, but in most of the situations considered here, $G(s)$ will be a strictly proper $(p \times m)$ rational matrix function of s, and will therefore represent the transfer function of an input-output linear system whose steady-state response to a large variety of input functions is well-defined.

The following block diagram should be kept in mind during the discussion below.

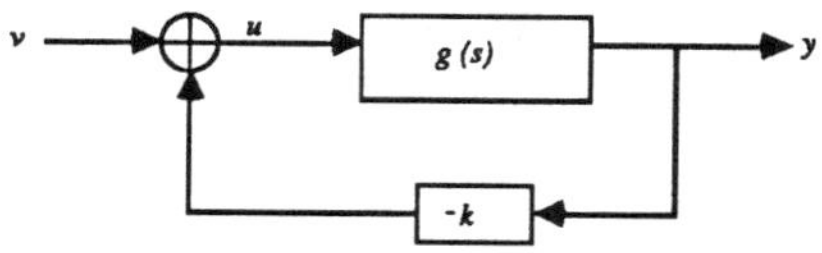

Figure 25.2 Feedback compensation

Consider a single-input, single-output system with strictly proper rational transfer function $g(s)$. We might want to take the given system and connect it in some fashion with other time-invariant linear systems so as to *compensate* for certain properties of the system which we find undesirable. Once having done so, we've generated a new time-invariant system whose transfer function is a *compensated* version of the original system's transfer function $g(s)$.

In Figure 25.2, k is a real number; the diagram means that under appropriate assumptions on the existence of Laplace transforms, the Laplace transform $\tilde{y}(s)$ of the steady-state response of the system to an input function $t \to v(t)$ is determined by the two formulas

$$\begin{aligned} \tilde{y}(s) &= g(s)\tilde{u}(s) \\ \tilde{u}(s) &= \tilde{v}(s) - k\tilde{y}(s) . \end{aligned}$$

The minus sign in front of the k is in deference to the historical development of the theory; the last two formulas determine the "transfer function from v to y" to be

$$g_d(s) = g(s)(1 + kg(s))^{-1} .$$

We can think of $u \to g(s) \to y$ as being the original system, while $v \to g_d(s) \to y$ is the compensated system, which has *desired* transfer function $g_d(s)$. The formula for $g_d(s)$ reveals $g_d(s)$ as the *product* of $g(s)$ and $(1+kg(s))^{-1}$; as a result, the same transfer function from v to y could have been achieved by a *cascade* compensation scheme of the form

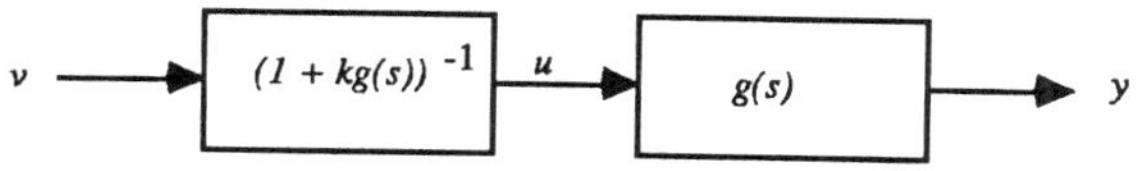

Figure 25.3 Cascade compensation

instead of the *feedback compensation scheme* in Figure 25.2.

A consideration of certain practical issues provides a strong rationale for choosing the feedback compensation scheme over the one in Figure 25.3. In practice, it is safe to assume that one's knowledge of the parameters of a given system is approximate, at best; thus, even if one *thinks* that $g(s)$ is the transfer function of the "true system" to be compensated, it is probably the case that the actual system in question has transfer function $\hat{g}(s) \neq g(s)$. Moreover, in any real engineering situation, one must cope with *disturbances* of various kinds; a common way to model inaccuracies in real-world measuring devices, for example, is to incorporate into the above diagrams an *additive disturbance* $t \to e(t)$ at the output end of the system whose transfer function is $g(s)$.

Let us compare the performance of the two proposed schemes when these important practical issues are taken into account. In the cascade compensation scheme, our design of the cascade compensator $(1+kg(s))^{-1}$ is based on the supposition that the system to be compensated has transfer function $g(s)$; if, in fact, the real system's transfer function is $\hat{g}(s)$, then the transfer function of the compensated system is actually $\hat{g}(s)(1+kg(s))^{-1}$. If we denote by $g_c(s)$ the transfer function of this cascade compensated system, then

$$g_d(s)-g_c(s) = [g(s)-\hat{g}(s)](1+kg(s))^{-1} .$$

Now, let $g_f(s)$ be the transfer function of the feedback system in Figure 25.2; since $\hat{g}(s)$ is really what's there instead of $g(s)$, we have $g_f(s) = \hat{g}(s)(1+k\hat{g}(s))^{-1}$, and a few lines of algebra show that

$$g_d(s)-g_f(s) = [g(s)-\hat{g}(s)](1+kg(s))^{-1}(1+k\hat{g}(s))^{-1} .$$

It may be concluded that the difference between the *desired* transfer function $g_d(s)$ and the *actual* compensated transfer function is different in the two cases; moreover, if k is large, then the difference between the desired and actual compensated transfer functions might be significantly less over a wide range of s-values if the feedback strategy is used instead of the cascade strategy of Figure 25.3.

Furthermore, let us consider the effect of an additive disturbance $t \to e(t)$ at the output. In the cascade scheme, we see that the Laplace transform of the output is given in terms of the transforms of the input $t \to v(t)$ and disturbance e by

$$\tilde{y}(s) = g_c(s)\tilde{v}(s)+\tilde{e}(s) ,$$

while in the feedback scheme of Figure 25.2, we have

$$\begin{aligned} \tilde{y}(s) &= \hat{g}(s)\tilde{u}(s)+\tilde{e}(s) \\ \tilde{u}(s) &= \tilde{v}(s)-k\tilde{y}(s) , \end{aligned}$$

so that

$$\tilde{y}(s) = g_f(s)\tilde{v}(s)+(1+k\hat{g}(s))^{-1}\tilde{e}(s) .$$

Again, the effect on the output of the disturbance e is *attenuated* by a factor of $(1+k\hat{g}(s))^{-1}$ if the feedback compensation strategy is used in an effort to achieve $g_d(s)$.

The quantity $(1+k\hat{g}(s))$ is often referred to as the *return difference* of the feedback system; we have glossed over various issues in promoting the return difference as an *attenuation factor*, most notably the fact that $1+k\hat{g}(s)$ might be zero for some values of s. Suffice it to say that a large number of "practical" systems have input-output properties which are well-reflected in transfer functions $\hat{g}(s)$ whose real parts are *strictly positive* throughout the right half-plane $\text{Re}\{s\} \geqslant 0$; such transfer functions are said to be *strictly positive real*. Roughly speaking, such transfer functions describe the input-output behavior of *passive* systems; see, for example, [MacMillan]. If $\hat{g}(s)$ is strictly positive real and $k > 0$, the return difference $1+k\hat{g}(s)$ can never be zero.

The foregoing comments should have convinced the reader that feedback can be a good thing; moreover, it might appear that *high-gain feedback*, that is, feedback of the form in Figure 25.2 with a large positive value of k, would perform even better with respect to reducing the effects on overall system performance of parameter uncertainty in the uncompensated transfer function $\hat{g}(s)$ and of disturbances e at the output. A little thought, however, dispels this hope, at least in part. Consider the system whose transfer function is

$$g(s) = \frac{s+1}{(s+2)(s+3)} .$$

This system is BIBO stable, by the results of §24, since all of the poles of its transfer function lie in the half-plane $\text{Re}\{s\} < 0$. Moreover, it is easily checked that $g(s)$ is strictly positive real. Using this $g(s)$ in Figure 25.2 gives rise to the "closed-loop" compensated transfer function

$$g_f(s) = \frac{(s+1)}{s^2+(5+k)s+(6+k)} .$$

Consider the behavior of the poles of $g_f(s)$ as $k \geqslant 0$ increases. At $k = 0$, the poles lie at -2 and -3, just as one might expect. At $k = 1$, the poles are at $3 \pm \sqrt{2}$; both are still in the "stable" region $\text{Re}\{s\} < 0$. At $k = 2$, however, it is easily checked that one of the poles of $g_f(s)$ is in the right half-plane $\text{Re}\{s\} > 0$, so that the closed-loop system is unstable by the results of §24.

We conclude that feedback may have the effect of *destabilizing* an otherwise stable system; the design of feedback systems must therefore be done carefully in order to circumvent any such difficulties. We have all seen (or heard) instances of feedback destabilization; for example, if the microphone or pickup of an amplified musical instrument is brought too close to the amplifier output, then there results the familiar earsplitting hum which many people call "feedback."

How might feedback destabilization be prevented or, at least, predicted? For a realizable single-input, single-output time-invariant continuous-time input-output linear system whose transfer function $g(s)$ has a denominator polynomial with very high degree, the computation of the poles of the closed-loop system (using feedback as in Figure 25.2) might be prohibitively difficult. There is, however, a fundamental result due to Nyquist which answers the stability question for feedback systems without recourse to an actual computation of poles.

Recall from §19 that one of the main selling points of frequency domain techniques for input-output linear systems is that they appeal to *empirical intuition*. More precisely, it is possible, at least in principle, to obtain an approximation to the frequency response of a given time-invariant continuous-time input-output linear system by "driving" the system with many sinusoidal inputs of the form $t \to e^{i\omega_0 t}$ and seeing what comes out. *Nyquist's Criterion* for the stability of the system in Figure 25.2 is phrased in terms of the *frequency response* of the system whose transfer function is $g(s)$; in fact, generalizations of Nyquist's Criterion exist for systems which don't have frequency responses at all, at least in the sense in which we have defined them. Nyquist's Criterion is based on a fundamental result from complex variable theory known as the *Argument Principle*; we state here without proof a restricted version which suits our needs.

25.1 Theorem (The Argument Principle): *Suppose that $\omega \to \gamma(\omega)$, $\omega \in [a, b]$, is a parametrized closed curve in the complex plane, directed in the clockwise sense. If the complex function $s \to f(s)$ is analytic along the curve γ and throughout the region which γ encloses, and if f does not take on the value zero on γ, then the number of times that f takes on the value zero in the region enclosed by γ is equal to the number of clockwise encirclements of the point zero by the parametrized closed curve $\omega \to f(\gamma(\omega))$, $\omega \in [a, b]$.* □

The intuition behind the Argument Principle is fairly simple. Imagine the function f as mapping the complex plane to itself; in particular, f maps the curve traced out by γ to another parametrized closed curve $f(\gamma)$, while mapping the region enclosed by γ to the region enclosed by $f(\gamma)$. Suppose that $f(\gamma)$ encircles the point zero n times, clockwise; this means, in particular, that f assumes the value zero somewhere in the region enclosed by γ. Moreover, the number of "trips" which $f(s)$ makes around the curve $f(\gamma)$ as s goes once around γ is the same as the net change in the *argument* of $f(s)$ as γ is traversed once, clockwise, by s.

Inversely, as γ is traversed one n th of the way around by s, $f(s)$ makes a full circuit of $f(\gamma)$. One can therefore think of f as mapping each of n wedge-shaped subregions of the region enclosed by γ onto the *entire* region enclosed by $f(\gamma)$; in this way, f may be viewed as an "n-to-one" mapping of the region enclosed by γ onto that enclosed by $f(\gamma)$. It is important to note that we are *counting multiplicities* in "the number of times f assumes the value zero" to which the statement of Theorem 25.1 alludes; we shall not, however, need to consider this issue any more deeply in what follows.

Back, now, to feedback systems. Suppose that the transfer function of a given real single-input, single-output time-invariant continuous-time input-output linear system is the *strictly proper* rational function $s \to g(s)$; suppose that all of the poles of $g(s)$ lie in the open left half-plane $\mathrm{Re}\{s\} < 0$, so that the system is BIBO stable. Consider a semicircular, clockwise-directed closed curve $C(\omega_o)$ in the complex plane which has as its straight side the segment $[-\omega_o, \omega_o]$ of the imaginary axis, and for its curved side the semicircle $|s| = \omega_o$ in the right half-plane. We can choose ω_o large enough so that $|g(s)|$ remains as small as we want while s runs over the curved part of $C(\omega_o)$; the point is that because $g(s)$ is strictly proper, $g(s) \to 0$ as $|s| \to \infty$. Thus, as $\omega_o \to \infty$, the curve traced out by $g(s)$ as s traverses $C(\omega_o)$ approaches the curve traced out by the *frequency response* $\omega \to g(i\omega)$, $-\infty < \omega < \infty$, of the system; here, ω runs from $-\infty$ to ∞, and the frequency response's existence is guaranteed by BIBO stability.

The curve $\omega \to g(i\omega)$, $-\infty < \omega < \infty$, is often called the *Nyquist locus* of the system with transfer function $g(s)$. The Nyquist Criterion for stability of the feedback system of Figure 25.2 is phrased in terms of the Nyquist locus.

25.2 Theorem (The Nyquist Criterion): *Suppose that $s \to g(s)$ is a strictly proper rational function of s and has no poles in the half-plane* $\mathrm{Re}\{s\} \geqslant 0$. *Suppose also that $kg(s)$ does not take on the value -1 on the imaginary axis. The number of poles in* $\mathrm{Re}\{s\} \geqslant 0$ *of the rational function $s \to g(s)(1+kg(s))^{-1}$, where $k > 0$, is equal to the net number of clockwise encirclements of the point $-k^{-1}$ by the curve $\omega \to g(i\omega)$ as ω runs from $-\infty$ to ∞. Accordingly, the closed loop system in Figure 25.2 is BIBO stable if and only if $\omega \to g(i\omega)$ makes no net encirclements of $-k^{-1}$.*

Proof: Choose ω_o so that $C(\omega_o)$, defined as above, encloses all the points s_o in $\mathrm{Re}\{s\} > 0$ at which $1+kg(s_o) = 0$; this can be done because $g(s)$ is strictly proper. Now, the number of times which $g(s)$ encircles $-k^{-1}$ as s traverses $C(\omega_o)$ is the same as the number of times $1+kg(s)$ encircles zero as s goes around $C(\omega_o)$; by the Argument Principle (our Theorem 25.1), this is the same as the number of times which $1+kg(s)$ assumes the value zero in $\mathrm{Re}\{s\} > 0$, which in turn is the same as the number of poles in $\mathrm{Re}\{s\} > 0$ of the rational function $(1+kg(s))^{-1}$. This number, in turn, is the number of poles of $s \to g(s)(1+kg(s))^{-1}$ which lie in the closed right half-plane since $g(s)$ has no poles there and cannot, by assumption, be zero at any s_o where $1+kg(s_o) = 0$.

The last sentence in the theorem statement follows because the image under $s \to g(s)$ of the curve $C(\omega_o)$, as we noted earlier, approaches the Nyquist locus as $\omega_o \to \infty$. □

The Nyquist Criterion, at least in principle, gives a means by which the stability of a proposed feedback system may be determined from empirical data. Nyquist's Criterion is *robust* with respect to mis-modeling in the sense that a small change in $g(s)$, even a change in the degree of its numerator or denominator polynomial, will not change the answer to the question of whether the Nyquist locus encircles a certain point in the complex plane. We have presented Nyquist's Criterion as method for assessing when a *stable* system is *de-stabilized* by the addition of feedback. Given the effectiveness of feedback as a way of modifying system behavior, it is natural to ask whether feedback can be used to *stabilize* a system which is unstable to begin with.

Consider the single-input, single-output continuous-time system whose transfer function is

$$g(s) = \frac{1}{(s+1)(s-2)} .$$

This system is unstable, by the results of §24, because it's transfer function has a pole at 2, which is in $\text{Re}\{s\} > 0$. Consider a feedback compensation scheme as in Figure 25.2, using this particular $g(s)$ in the "forward" path. The closed-loop transfer function is

$$g_f(s) = \frac{1}{(s-2)(s+1)+k} ;$$

its poles are $\frac{1}{2} \pm \frac{1}{2}\sqrt{9-4k}$; there is no real value of k which places both poles in the "stable region" $\text{Re}\{s\} < 0$.

Accordingly, simple *constant-gain output feedback*, such as in Figure 25.2, cannot in general be employed as a means of stabilizing an unstable system. On the other hand, a glance at the unstable system whose transfer function is $g(s) = (s-1)^{-1}$ shows that constant-gain output feedback stabilization is sometimes possible. A great deal of work has been done by many researchers (see, for example, [Byrnes]) in an attempt to understand more precisely what can and cannot be done to a system, or to all systems in some specified class, by means of constant-gain output feedback.

The truth of the matter is that not very much *can* be done, in general; as k varies in Figure 25.2, the poles of the closed loop system's transfer function $g_f(s)$ follow certain paths in the complex plane known as *root loci*. Everything which may be accomplished for a given single-input, single-output system through constant-gain output feedback is exhibited, in some form or another, in the plot of the root loci. This plot may be approximated quite accurately without actually *solving* for the roots of any polynomials; see, for example, [Kuo] or [D'Azzo and Houpis]. Multi-input, multi-output systems may also be analyzed by means of suitable generalizations of Nyquist's Criterion and root loci; for a good survey of some of the applicable techniques, all of which involve quite a bit of multidimensional complex geometry, see [Byrnes and Brockett] or [MacFarlane].

In view of the foregoing discussion, it is evident that a system designer's options are somewhat limited if the only available feedback is of the constant-gain output variety, as in Figure 25.2. This fact is not surprising if one remembers that the "dynamics" of even a single-input, single-output system depend, at the bare minimum, on the locations of all of the poles and zeroes of the system's transfer function; if the system has a very large McMillan degree, say n, it is apparent that one shouldn't expect to be able influence the location of $2n$ complex numbers to any great extent by modifying a single real feedback gain k.

The limitations on output feedback may also be regarded as a by-product of having *too few measurements* (in the case of Figure 25.2, just a *single* one) to influence the system's behavior dramatically. One is prompted, for example, to attempt to make good use of one's only available measurement by allowing a *dynamic compensator* (i.e., a *transfer function*) in place of k in Figure 25.2. For a good discussion of dynamic output feedback compensation of single-input, single-output systems, see, for example, [Ghosh and Byrnes].

On the other hand, if we think of an input-output system as being associated with a state space system (via §21, Definition 18.18, or Definition 18.19), then the information about the system's behavior which is *missing* from a measurement of the output at some time t is merely information about the *state* of the system at time t. Perhaps feedback strategies predicated on the measurability of the *state* of such a system at each time could be designed so as to exert more effective control over the system than strategies which presume only the measurability of the system's *output*.

Keeping these remarks in mind, we approach our next major result, which was first proved for general (i.e., multi-input, multi-output) systems by Wonham in [Wonham]. The result states essentially that the "poles" of an m-input, p-output time-invariant continuous- or discrete-time *reachable* state space linear system may be placed *arbitrarily* by means of consatant-gain feedback of the system's *state*.

This result has evident importance for input-output linear systems, as well. Suppose that $(A, B, C, 0)$ is a minimal realization for a given m-input, p-output time-invariant input-output linear system. The state space system with state space $\mathbf{R}^n$ defined as in Example 6.5 or Example 6.6 using the matrices in $(A, B, C, 0)$ is a *reachable* system, since (A, B) is a reachable pair beacuse of Theorem 21.13. By being able to "place the poles" of such a state space system via "constant-gain state feedback" we shall mean being able to choose a constant $(m \times n)$ matrix F so as to make make the eigenvalues of $(A - BF)$ be essentially whatever we want them to be. Since the input-output system associated with the "compensated" realization $(A-BF, B, C, 0)$ has poles which are among the eigenvalues of $(A-BF)$ (c.f. the discussion at the end of §24), it is seen that the availability of state feedback would give us quite a bit of leeway in determining the pole locations of a compensated input-output linear system.

The state feedback construction we are about to describe is very similar in spirit to the constant-gain output feedback strategy depicted in Figure 25.2. Suppose that $(A, B, C, 0)$ is a (constant) realization for a time-invariant continuous-time state space linear system which is reachable (see §§6-7 and §13 for notation and terminology). The transfer function of the associated input-output linear system is $C(sI_n-A)^{-1}B$; the following equations reflect the system's behavior from the point of view of the choice of basis for the state space which led to $(A, B, C, 0)$:

$$\begin{aligned} \dot{x}(t) &= Ax(t)+Bu(t) \\ y(t) &= Cx(t). \end{aligned} \tag{I}$$

Controlling the system by means of state feedback entails letting

$$u(t) = -Fx(t)+v(t)$$

for each input $v: \mathrm{R} \to \mathrm{R}^m$. The idea is that we might want the system to respond in a certain way to some specific class of input functions v, and it might be that the system whose state evolution is governed by

$$\dot{x}(t) = (A-BF)x(t)+Bv(t)$$

responds in this desired fashion. For example, v could be identically zero, and our desired behavior could be the long-term decay of $t \to x(t)$ regardless of initial condition. Wonham's result on pole placement by constant gain state feedback says that the reachability of (A, B) enables one to specify the eigenvalues of $(A-BF)$ arbitrarily by choosing F; picking an F such that the eigenvalues of $(A-BF)$ have strictly negative real parts accomplishes the *asymptotic stabilization* in the simple example we have just described.

Let us consider first the case $m = 1$; that is, assume that A is a real $(n \times n)$ matrix, B is a real n-vector, and (A, B) is a reachable pair. Let $C = I_n$; by the method used in the proof of Theorem 21.9, we may construct a realization $(\hat{A}, \hat{B}, \hat{C}, 0)$

for the (continuous-time) m-input, n-output system with transfer function $C(sI_n - A)^{-1}B$ for which $\hat{A}$ and $\hat{B}$ take the form

$$\hat{A} = \begin{bmatrix} 0 & 1 & . & 0 & 0 \\ . & 0 & . & . & . \\ . & . & . & 1 & 0 \\ 0 & 0 & . & 0 & 1 \\ -q_n & -q_{n-1} & \cdot & -q_2 & -q_1 \end{bmatrix} ; \quad \hat{B} = \begin{bmatrix} 0 \\ 0 \\ . \\ 0 \\ 1 \end{bmatrix},$$

where the characteristic polynomial of A is

$$s^n + q_1 s^{n-1} + \ldots + q_{n-1}s + q_n .$$

Observe that the both realizations $(\hat{A}, \hat{B}, \hat{C}, 0)$ and $(A, B, C, 0)$ are *minimal*, since (A, B) is assumed reachable and (A, I_n) is obviously an observable pair, and $\hat{A}$ has the same size as A. As a consequence of Theorem 21.16, there exists a real invertible $(n \times n)$ matrix P such that $\hat{A} = P^{-1}AP$, $\hat{B} = P^{-1}B$, and $\hat{C} = P$. The formulas for $\hat{A}$ and $\hat{B}$ make it obvious that for any set $a_1, \ldots, a_n$ of real numbers, there exists a $(1 \times n)$ matrix $\hat{F}$ such that $\hat{A} - \hat{B}\hat{F}$ has the a_i's as its characteristic coefficients; simply set

$$\hat{F} = \left[(a_n - q_n) \; . \; . \; . \; (a_1 - q_1) \right] .$$

Setting $F = \hat{F}P^{-1}$ makes F an $(m \times n)$ matrix for which $(A - BF)$ has characteristic polynomial

$$s^n + a_1 s^{n-1} + \ldots + a_{n-1}s + a_n .$$

Since adjusting the characteristic polynomial is the same as adjusting the eigenvalues, the result follows in the case $m = 1$.

The general case (i.e., $m > 1$) is somewhat more difficult. The proof given below is not the most direct one possible; it hinges on the development in §22, and is a direct consequence of the construction which was used in proving Theorem 22.18.

25.3 Theorem (Wonham): *Let A and B be real matrices having respective sizes $(n \times n)$ and $(n \times m)$; suppose (A, B) is a reachable pair. Given any set of n complex numbers $\lambda_1, \lambda_2, \ldots, \lambda_n$, which are not necessarily distinct but come in complex conjugate pairs, there exists a real $(m \times n)$ matrix F such that the eigenvalues of $(A - BF)$ are precisely $\lambda_1, \lambda_2, \ldots, \lambda_n$.*

Proof: The constraint on $\{\lambda_i : 1 \leqslant i \leqslant n\}$ is only to guarantee that it may constitute the set of n eigenvalues of a real $(n \times n)$ matrix, each counted as many times as its algebraic multiplicity. Equivalently, the constraint is equivalent to saying that $\{\lambda_i : 1 \leqslant i \leqslant n\}$ is the set of roots of some polynomial of degree n which has real coefficients.

In any case, we prove the result by showing that for a reachable pair (A, B), the characteristic polynomial of $(A-BF)$ may be adjusted arbitrarily by choice of the real $(m \times n)$ matrix F. First, let $C = I_n$, and consider the m-input, p-ouput time-invariant (continuous-time) input-output linear system with transfer function $C(sI_n - A)^{-1}B$. The system has McMillan degree n, since (A, B) is assumed reachable and (A, I_n) is obviously an observable pair. By the construction leading up to Theorem 22.18, we may find another special minimal realization $(\hat{A}, \hat{B}, \hat{C}, 0)$ for this system, in which $\hat{A}$ and $\hat{B}$ are as follows.

Let $n_1, n_2, \ldots, n_m$ be the column Kronecker indices of $C(sI_n - A)^{-1}B$ (c.f. Definition 22.16 and the discussion which follows). Suppose that n_1 through n_ρ are nonzero, and that $n_{\rho+1}$ through n_m are zero. The $(n \times n)$ matrix $\hat{A}$ has ones in the $(i, i+1)$ positions (just above the main diagonal) and zeroes everywhere else *except* in the ρ rows which are indexed by $i = n_1 + \ldots + n_j$ for some $j \leqslant \rho$; each of these rows in $\hat{A}$ might be an arbitrary n-vector. The matrix $\hat{B}$ is of the form $\tilde{B}Q$, where Q is an invertible $(m \times m)$ matrix and $\tilde{B}$ has as its jth column, for $1 \leqslant j \leqslant \rho$, zeroes everywhere except for a one at the $n_1 + \ldots + n_j$ position; the columns of $\tilde{B}$ indexed by j for $j > \rho$ are zero.

Observe that because of the way that $\hat{A}$ and $\tilde{B}$ are structured, we may find, for any n real numbers $a_1, \ldots, a_n$, a real $(m \times n)$ matrix $\tilde{F}$ such that

$$\hat{A} - \tilde{B}\tilde{F} = \begin{bmatrix} 0 & 1 & . & 0 & 0 \\ . & 0 & . & . & . \\ . & . & . & 1 & 0 \\ 0 & 0 & . & 0 & 1 \\ -a_n & -a_{n-1} & . & -a_2 & -a_1 \end{bmatrix}. \qquad (*)$$

To see this, note that for each $j \leqslant \rho$, we may set the jth row of $\tilde{F}$ to be the $(n_1 + \ldots + n_j)$th row of $\hat{A}$ minus the jth row of the matrix on the right-hand side of (*). Since the ith row of $\tilde{B}\tilde{F}$ is zero unless i is of the form $(n_1 + \ldots + n_j)$, the other rows in $\hat{A}$ are unaltered when $\tilde{B}\tilde{F}$ is subtracted off.

Now, let $\hat{F} = Q^{-1}\tilde{F}$; then $\hat{A} - \hat{B}\hat{F}$ has the form (*). Finally, note that since $(A, B, I_n, 0)$ and $(\hat{A}, \hat{B}, \hat{C}, 0)$ are minimal realizations for the system with transfer function $(sI_n - A)^{-1}B$, there exists by Theorem 21.16 an invertible $(n \times n)$ matrix P such that $P^{-1}AP = \hat{A}$ and $P^{-1}B = \hat{B}$. Set $F = \hat{F}P^{-1}$; since

$$A - BF = P(\hat{A} - \hat{B}\hat{F})P^{-1} ,$$

it may be concluded that $(A - BF)$ has the same characteristic polynomial as $(\hat{A} - \hat{B}\hat{F})$, and the result follows since $a_1, \ldots, a_n$ were arbitrary real numbers. □

Note that although we have given a "continuous-time" proof of Theorem 25.2, the result is *purely algebraic*, and therefore applies equally well to discrete-time systems, since discrete- and continuous-time reachability of a pair (A, B) are equivalent properties. It should also be noted that much simpler proofs of Theorem 25.2 exist; perhaps the best-known of these is due to M. Heymann (see [Heymann]). Heymann's proof consists of reducing the case of arbitrary m (number of columns in B) to the case $m = 1$; we were able to argue that special case in a few lines before the statement of Theorem 25.2. His technique rests on a result which has become known as Heymann's Lemma.

25.4 Lemma (Heymann): *Let A and B be real matrices having respective sizes $(n \times n)$ and $(n \times m)$. If (A, B) is a reachable pair, and j is such that the jth column of B (call it b^j) is nonzero, then there exists an $(m \times n)$ matrix F such that $(A - BF, b^j)$ is a reachable pair.* □

The proof of Heymann's Lemma rests on some time-domain versions of the constructions in §22 which made up the proof of Theorem 22.18. We shall have more to say about this when we discuss the controllable canonical form.

25.5 Exercise: Complete the proof of Theorem 25.3 *given* Heymann's Lemma 25.4. Also, give a proof of Heymann's Lemma based on the construction of Theorem 22.18. □

25.6 Exercise: Show that reachability of the pair (A, B) is *equivalent* to the possibility of placing the poles arbitrarily as in Theorem 25.3. [*Suggestion:* Use Version I of the Canonical Structure Theorem (Theorem 17.5).] □

One might view Theorem 25.3 as saying that much can be done to a system, in terms of influencing its behavior, if state feedback is a possibility. Still, if input-output systems are to be appropriate models for real-world processes in which measurements of the "internal" (i.e., state) variables are unavailable, then one should not expect to be able to use state feedback to modify their behavior. Even in the case of state space systems, it is true in practice that if the state space dimension is sufficiently high, then a full measurement of the state will probably be impossible. The following question suggests itself: how might one *approximately* implement a state feedback control strategy for an input-output system or for a state space system in which one does not have full access to the system's state?

We consider the following basic setup. Let $(A, B, C, 0)$ be a realization of a m-input, p-output time-invariant continuous-time state space linear system having n-dimensional state space. The system's behavior is described by equations (I) above; we imagine that the initial condition $x(0)$, which we do *not* know, is some $x_o \in \mathbb{R}^n$. It is assumed that we know the matrices A, B, and C. We would like to set

$$u(t) = -Fx(t) + v(t), \quad t \geqslant 0,$$

but are allowed to observe only the output $y(t)$, $t \geqslant 0$. Our real interest is in compensating the system so that the transfer function from v to y is as close as possible to $C(sI_n - A + BF)^{-1}B$, which is what it *would* be if we could implement the state feedback strategy exactly.

Our first inclination is to build a *simulation* of the system; that is, build a system governed by

$$\begin{aligned} \dot{z}(t) &= Az(t) + Bu(t) \\ \hat{y}(t) &= Cz(t), \end{aligned}$$

where $t \rightarrow z(t)$, $t \geqslant 0$, is available for observation, and the initial condition $z(0)$ may be set arbitrarily. If we knew x_o, we could set $z(0) = x_o$, and $t \rightarrow z(t)$, $t \geqslant 0$, would be identical to $t \rightarrow x(t)$, $t \geqslant 0$. Since we don't know x_o, we would like to *compensate* our simulation in some fashion which guarantees that $z(t)$ "stays close" to $x(t)$ for $t \geqslant 0$; in other words, we would like the error in our initial setting of $z(0)$, namely, $x_o - z(0)$, to "damp out" over time.

Accordingly, let us add a *correction term* to the differential equation governing z; we might as well choose a correction term of the form $L(y(t)-\hat{y}(t))$, which is an instantaneous linear function of the deviation of our simulated system's output from the output of the true system. The differential equation for z becomes

$$\dot{z}(t) = (A-LC)z(t) + LCx(t) + Bu(t);$$

consider the resulting differential equation for $t \rightarrow x(t)-z(t)$:

$$\begin{aligned} \frac{d}{dt}(x(t)-z(t)) &= Ax(t)-Az(t)-LCx(t)+LCz(t) \\ &= (A-LC)(x(t)-z(t)). \end{aligned}$$

We see that $z(t)-x(t) \rightarrow 0$ as $t \rightarrow \infty$ for all $z(0)$ precisely when $(A-LC)$ has eigenvalues with strictly negative real parts (c.f. Theorem 23.15).

25.7 Lemma: *Let A and C be real matrices having respective sizes $(n \times n)$ and $(p \times n)$. If (A, C) is an observable pair, then there exists a real $(n \times p)$ matrix L such that the eigenvalues of $(A-LC)$ have strictly negative real parts.* □

The proof of Lemma 25.7 is left to the reader; the idea is that observability (A, C) is equivalent to the reachability of (A^T, C^T), and Theorem 25.3 therefore applies to the pair (A^T, C^T). Although Lemma 25.7 is all we need in the present context, the following version of Theorem 25.3 holds for observable pairs (A, C) by the same line of reasoning.

25.8 Theorem: *Let A and C be real matrices having respective sizes $(n \times n)$ and $(p \times n)$; suppose (A, C) is an observable pair. Given any set of n complex numbers $\lambda_1, \lambda_2, \ldots, \lambda_n$, which are not necessarily distinct but come in complex conjugate pairs, there exists a real $(n \times p)$ matrix L such that the eigenvalues of $(A-LC)$ are precisely $\lambda_1, \lambda_2, \ldots, \lambda_n$.* □

In any case, provided that $(A-LC)$ has eigenvalues with negative real parts, the state $t \rightarrow z(t)$ in the simulation of our "true" system will *track* the state $t \to x(t)$ of the true system asymptotically as $t \to \infty$. The simulated system with L chosen as above is called an *observer* for the system governed by (I). The idea of constructing observers for linear systems was first proposed by D. G. Luenberger in [Luenberger]. Recall that our original objective was to attempt to implement a state feedback strategy approximately by using only the outputs of our true system; since the observer state is available for observation, and since it follows the true state asymptotically, we are prompted to use a feedback strategy of the form

$$u(t) = -Fz(t) + v(t) .$$

Having choosen this u, let us think of the true system together with its observer as constituting a composite system with input v, output y, and $2n$-dimensional state vector obtained by stacking x and z. The differential equation which governs the evolution of the state of this composite system is

$$\begin{bmatrix} \dot{x}(t) \\ \dot{z}(t) \end{bmatrix} = \begin{bmatrix} A & -BF \\ LC & (A-BF-LC) \end{bmatrix} \begin{bmatrix} x(t) \\ z(t) \end{bmatrix} + \begin{bmatrix} B \\ B \end{bmatrix} v(t) ;$$

the output y is given in terms of the stacked state as

$$y(t) = \begin{bmatrix} C & 0 \end{bmatrix} \begin{bmatrix} x(t) \\ z(t) \end{bmatrix} .$$

Denote by $\tilde{A}$ and $\tilde{B}$ the matrices appearing in the differential equation, and let $\tilde{C} = [C \quad 0]$.

Recall that our original objective in approximating the state feedback strategy $u = -Fx + v$ by $u = -Fz + v$ was to make the transfer function from as close as possible to $C(sI_n - A + BF)^{-1}B$, which is what it would have been if we had been able to measure x. It turns out that the transfer function from v to y in the composite system above is *exactly equal* to this desired transfer function. To see this, define the invertible $(2n \times 2n)$ matrix P by

$$P = \begin{bmatrix} I_n & I_n \\ 0 & I_n \end{bmatrix},$$

and let $\hat{C} = \tilde{C}P$, $\hat{A} = P^{-1}\tilde{A}P$, and $\hat{B} = P^{-1}\tilde{B}$.

Since the transfer function of the composite system is $\tilde{C}(sI_{2n} - \tilde{A})^{-1}\tilde{B}$, it is also equal to $\hat{C}(sI_{2n} - \hat{A})^{-1}\hat{B}$; but

$$\hat{A} = \begin{bmatrix} (A-LC) & 0 \\ LC & (A-BF) \end{bmatrix}; \quad \hat{B} = \begin{bmatrix} 0 \\ B \end{bmatrix};$$

and $\hat{C} = [\,C \quad C\,]$. These formulas make it clear that $\hat{C}(sI_{2n}-\hat{A})^{-1}\hat{B}$ is the same as the desired transfer function $C(sI_n - A + BF)^{-1}B$.

The observer construction which we have just outlined has significant ramifications with regard to the feedback compensation of input-output linear systems. Suppose that $G(s) = C(sI_n - A)^{-1}B$ is the (strictly proper) transfer function of a real m-input, p-output time-invariant continuous-time input-output linear system and that $(A, B, C, 0)$ is a minimal realization for the system. The equations (I) govern the time-evolution of the "state variables" for the system which correspond to the realization $(A, B, C, 0)$ and show how these state variables determine the output. Observe that any "compensated" transfer function which we could achieve by means of a constant-gain *output feedback* strategy of the form

$$u(t) = -Ky(t) + v(t), \qquad t \in \mathbf{R},$$

where K is a real $(m \times p)$ matrix, may be achieved by means of the *state feedback* strategy

$$u(t) = -KCx(t) + v(t), \qquad t \in \mathbf{R}.$$

On the other hand, the observer construction enables us to compensate the system *dynamically* — using measurements of the inputs and outputs of the system alone — so that the transfer function of the compensated system is the same as that obtained by implementing any given *state feedback* strategy $u(t) = -Fx(t) + v(t)$, $t \in \mathbf{R}$. This *dynamic compensation* procedure, which is also known as a *controller-observer implementation* of a given state feedback strategy, is best illustrated by means of a block diagram.

First, note that the Laplace transforms of the various time functions appearing in the composite system-with-observer above are related as follows:

$$\begin{aligned} \tilde{u}(s) &= -F\tilde{z}(s) + \tilde{v}(s) \\ \tilde{z}(s) &= (sI_n - A + LC)^{-1}[B\tilde{u}(s) + L\tilde{y}(s)]\,. \end{aligned}$$

Thus $\tilde{u}(s)$ may be expressed as the sum of $\tilde{v}(s)$ and two other terms: one of them is a multiple of $\tilde{u}(s)$, and therefore represents the Laplace transform of the output of an input-output system driven by u, and the other is multiple of $\tilde{y}(s)$, and hence is the Laplace transform of the output of a system whose input is by y. The following block diagram should be illuminating.

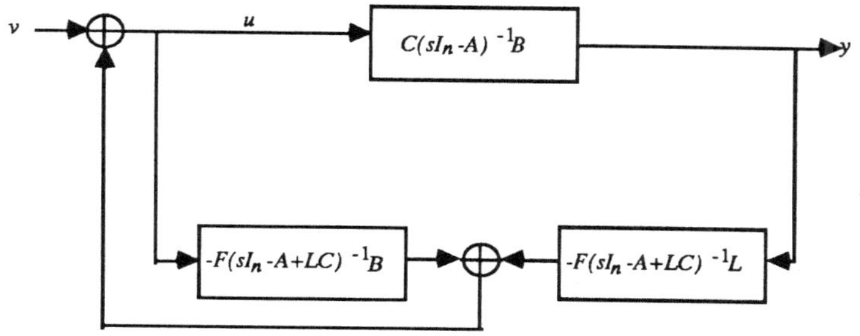

Figure 25.4 Controller-observer implementation

Recall once again that the transfer function from v to y is $C(sI_n - A + BF)^{-1}B$.

25.9 Exercise: Reformulate for discrete-time systems the entire discussion of observers which follows Exercise 25.6. □

We close this section with a brief introduction to the theory of *canonical forms* for input-output linear systems. The idea of a canonical form is fairly simple. Suppose that we are given some entity, perhaps an input-output or state space linear system; we think of the object in question as being specified *abstractly*, possibly by means of the definitions in §6 or §18. Since an abstract specification of this kind is not readily amenable to analysis, we would like to move to a more *concrete* description of the entity such as a realization or a weighting pattern. Generally, however, there are certain ambiguities inherent in such a representation procedure. For example, every choice of basis for the state space of a state space linear system leads to a different realization; likewise, in the case of time-invariant input-output systems, realizations (even minimal ones) are not uniquely specified.

Nonetheless, there are certain structural features of the entity which are *independent* of its concrete representation; it would be nice to be able to construct a special representation which not only exhibits these structural features but is also, in some sense, uniquely specified by the *manner* in which it displays them. Roughly speaking, a *canonical form* for the entity in question is just such a special representation.

We have seen a few examples of canonical forms already. Consider, for example, the development in §11; the "abstract entities" which we considered were the linear mappings ${}_N T : \mathbf{C}^n \to \mathbf{C}^n$ and ${}_A T \to \mathbf{C}^n \to \mathbf{C}^n$ defined, respectively, by a nilpotent $(n \times n)$ matrix N and a real $(n \times n)$ matrix A. The "concrete representations" were the matrices of these linear mappings with respect to various bases for $\mathbf{C}^n$; in Theorem 11.6, we showed how to construct for a given $(n \times n)$ nilpotent N the matrix of ${}_N T$ with respect to a special basis for $\mathbf{C}^n$ which exhibited clearly certain basis-independent structural features of ${}_N T$. Similarly, the Jordan Canonical form for A, which followed from Theorem 11.6 and the $M + N$ decomposition of §9, revealed a great deal about the way in which ${}_A T$ "acted" on vectors in A's generalized eigenspaces. In a brief discussion at the end of §11, we attempted to address at least partially some questions about uniqueness of the Jordan Canonical Form and the canonical form for nilpotent matrices given in Theorem 11.6.

The Canonical Structure Theorem in §17, especially Version II (Theorem 17.7), provided a canonical form of sorts for a given state space linear system. There, the abstract object was a time-invariant state space linear system and the concrete representations were realizations; the Canonical Structure Theorem said essentially that for any system, there is a special realization which exhibits a decomposition of the system's state space into pieces (e.g., the reachable subspace and the unobservable subspace) which have system theoretic relevance. We were careful to point out, however, that the realization whose existence is guaranteed by Theorem 17.7 is *not* uniquely determined; it is, however, sometimes referred to as a *Kalman canonical form* for the system in question.

In this section, the canonical forms in which we'll be interested are special realizations for time-invariant input-output linear systems. Our first canonical form, the *controllable canonical form*, has already appeared in another context. Recall from the discussion at the beginning of §22 that the single-input, single-output time-invariant continuous-time input-output linear system with transfer function

$$g(s) = \frac{p_1 s^{n-1} + \ldots + p_n}{s^n + q_1 s^{n-1} + \ldots + q_{n-1}s + q_n} + d$$

always has the realization $(\hat{A}, \hat{B}, \hat{C}, \hat{D})$, where $\hat{D} = d$;

$$\hat{A} = \begin{bmatrix} 0 & 1 & . & 0 & 0 \\ . & 0 & . & . & . \\ . & . & . & 1 & 0 \\ 0 & 0 & . & 0 & 1 \\ -q_n & -q_{n-1} & . & -q_2 & -q_1 \end{bmatrix}; \quad \hat{B} = \begin{bmatrix} 0 \\ 0 \\ . \\ 0 \\ 1 \end{bmatrix};$$

and

$$\hat{C} = \begin{bmatrix} p_n & p_{n-1} & . & . & p_1 \end{bmatrix}.$$

This special realization is the controllable canonical form for the given input-output system; as usual, the discrete-time system which has the "same" transfer function $g(z)$ is also realized by $(\hat{A}, \hat{B}, \hat{C}, \hat{D})$, and this special realization is also called the controllable canonical form for that discrete-time system.

The reason that $(\hat{A}, \hat{B}, \hat{C}, \hat{D})$ is called the controllable canonical form for the continuous-time system is illuminated by the following block diagram.

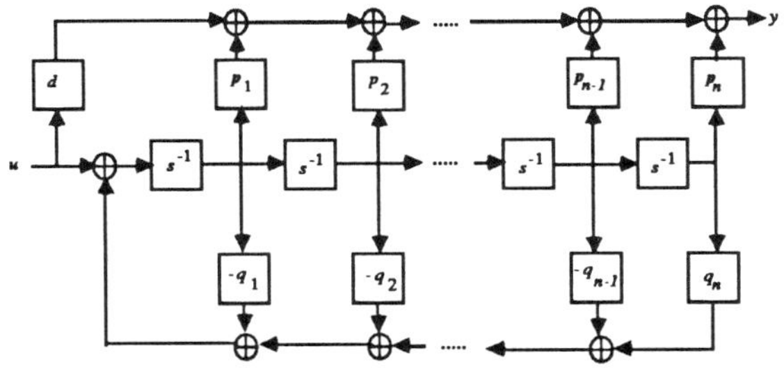

Figure 25.5 Controllable canonical form

Each box containing s^{-1} is meant to denote a single-input single-output system whose transfer function is s^{-1}; such a system plays the role of an *integrator*. The output of each integrator is one of the *state variables* of the system arising from the realization $(\hat{A}, \hat{B}, \hat{C}, \hat{D})$; the differential equation $\dot{x}(t) = \hat{A}x(t) + \hat{B}u(t)$ is shorthand for the n equations $\dot{x}_1(t) = x_2(t)$, $\dot{x}_2(t) = x_3(t)$, etc., $\dot{x}_{n-1}(t) = x_n(t)$, and finally

$$\dot{x}_n(t) = -q_n x_1(t) - q_{n-1}x_{n-1}(t) + \ldots + (-q_1)x_n(t) + u(t).$$

The diagram makes it clear how to choose u so as to make the state variables (and hence the output) behave in a prescribed fashion; simply note that if we demand that the state variables $x_1(t)$ through $x_n(t)$ follow certain paths $z_1(t)$ through $z_n(t)$, then a u which makes them do that is given by

$$u(t) = \dot{z}_n(t) + \sum_{i=1}^{n} q_i z_{n-i+1}(t);$$

ergo, the name controllable canonical form.

The same diagram with z^{-1} substituted for s^{-1} explains the controllable canonical form for a single-input, single-output time-invariant discrete-time input-output linear system; the z^{-1} blocks represent *delays*, and the differential equations for the continuous-time state variables metamorphose to the difference equations $x_1(k+1) = x_2(k)$, and so forth, down to $x_{n-1}(k+1) = x_n(k)$ and

$$x_n(k+1) = -\sum_{i=1}^{n} q_i x_{n-i+1}(k) + u(k).$$

Most of §22 was spent in *generalizing* the controllable canonical form for single-input, single-output systems to a similar construction for multi-input systems. The special realization constructed en route to Theorem 22.18 is exactly the controllable canonical form, or *standard controller realization*, for the m-input, p-output continuous-time input-output linear system with proper rational transfer function $G(s) = Z(s) + D$, with $Z(s)$ strictly proper. In the proof of the pole placement Theorem 25.3, we needed to consider that special realization again. The block diagram for this realization is quite complicated, so we won't include it here; instead, we'll attempt to describe it in words.

First, let's reconsider the block diagram for the single-input, single-output controllable canonical form; the diagram consists of a *bank of n integrators*, each of whose outputs is a state variable of the system corresponding to the realization. The input to the *first* integrator in the bank is a linear combination of the state variables and the input function u. The output is the sum of the "feed through" term du and a linear combination of the integrators' outputs.

As for the multi-input case, let $n_1, \ldots, n_\rho$ be the nonzero Kronecker column indices for $Z(s)$; see §22 or the proof of Theorem 25.3 for notation. The sum of the n_j is n, the McMillan degree of the system. To draw the block diagram for the controllable canonical form, it is probably best to begin by drawing ρ banks of integrators, one above the other, the first bank consisting of a chain of n_1 integrators, the second with n_2, and so on. The output of each integrator will be a state variable of the system; the output of the *leftmost* integrator in the *top* bank is $x_{n_1}(t)$, for example; the output of the *rightmost* integrator in the *second* chain from the top is $x_{n_1+1}(t)$.

The *input* to the *leftmost* integrator in each chain is a linear combination of the outputs of *all* the integrators and the input functions u_i, $1 \leqslant i \leqslant m$. Again in the notation of the proof of Theorem 25.3, if we define v_j as the linear combination of the u_i which feeds into the jth bank of integrators, for $1 \leqslant j \leqslant \rho$, then the vector v of v_j's is given by $v = Qu$. Q is the $(m \times m)$ matrix satisfying $\hat{B} = \tilde{B}Q$, where $\tilde{B}$ has zero jth column if $j > \rho$ and a single 1 in the $n_1 + \ldots + n_j$th position of the jth column if $j \leqslant \rho$.

The vector v may be regarded as a vector of inputs which is *equivalent to* u; since u and v are related by an invertible $(m \times m)$ matrix, anything that can be done to influence the behavior of the system by manipulating v can also be done through a manipulation of u. The way the v_j's enter the picture makes it at least somewhat transparent that the state variables may be made to behave in a prescribed fashion if v is chosen appropriately; see the discussion above of the single-input, single-output case.

Finally, each output of the system is a linear combination of the outputs of *all* the integrators (with coefficients coming from $\hat{C}$) along with a feed-through term which comes from Du.

Here is the block diagram for the controllable canonical form of the system in Example 22.21.

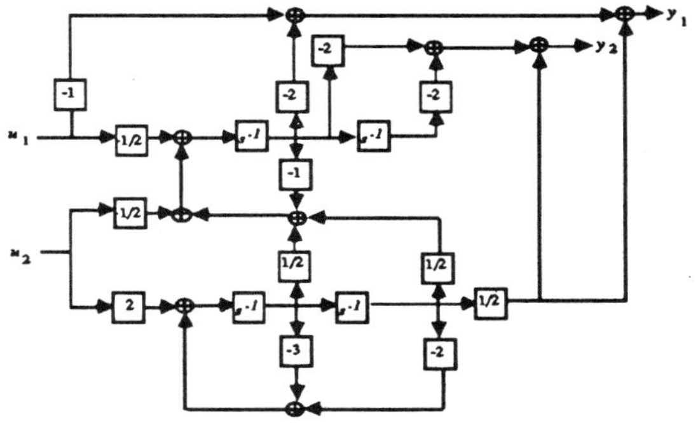

Figure 25.6 Controllable canonical form for Example 22.21

There is an *observable canonical form* for time-invariant input-output linear systems which is, in some sense, *dual* to the controllable canonical form. In the case of single-input, single-output systems, the observable canonical form is obtained simply by *transposing* the controllable canonical form $(\hat{A}, \hat{B}, \hat{C}, d)$ so as to obtain the realization $(\hat{A}^T, \hat{C}^T, \hat{B}^T, d)$. In the case of multi-input, multi-output systems, the observable canonical form may be obtained from a modification of the procedure which leads up to Theorem 23.18; the modified procedure works with *left* matrix fraction descriptions instead of *right* MFD's, and defines ordered and unordered *row* indices along with *Kronecker row indices* for $Z(s)$, the strictly proper part of the system's transfer function $G(s)$.

25.10 Exercise: Draw the block diagram for the observable canonical form of a single-input, single-output time-invariant continuous-time input-output linear system, and try to explain in terms of the diagram why it is called the observable canonical form. □

No treatment of the controllable canonical form would be complete without a more extensive discussion than we have given so far of the *Kronecker column indices* of an input-output system and some of their manifestations other than their rather arcane appearance in §22 as entries on "minimal lists of polynomial degrees." In order to obtain an alternative characterization of Kronecker column indices, which are also known as *Kronecker controllability indices*, we need to recall Definition 21.18 of the *N th order Hankel matrix* $\mathbf{H}_N$ of a given input-output linear system. The factorization of $\mathbf{H}_N$ given in the proof of Theorem 21.19 showed that if an input-output linear system has McMillan degree n, then H_N has rank n for every $N \geqslant n$.

Thus, for every $N \geqslant n$, we may choose n *linearly independent columns* from $\mathbf{H}_N$. This choice may be accomplished in many ways; one particular way of selecting the columns leads to a characterization of the Kronecker controllability indices in terms of $\mathbf{H}_n$.

25.11 Definition: Let H_n be the n th order Hankel matrix of a real m-input, p-output time-invariant continuous- or discrete-time input-output linear system which has McMillan degree n. A set of n linearly independent columns from H_n is called a *nice selection* of columns from H_n if and only if whenever $j > m$ and the jth column of H_n is a member of the set, then the $(j-m)$th column is also a member of the set. □

Definition 25.11 may seem somewhat obscure at first glance; it does, however, have an easy interpretation in terms of the columns of the *reachability matrix* $Q_r(A, B)$ coming from any minimal realization (A, B, C, D) for the system. Observe that the columns of H_n are simply products of the form $Q_o(A, C)$ times a column of $Q_r(A, B)$; hence, since $Q_o(A, C)$ has rank n, a set of columns in H_n is linearly independent if and only if the corresponding columns in $Q_r(A, B)$ are also linearly independent. The columns of $Q_r(A, B)$ have the form $A^k b^j$, where b^j is the jth column of B (j runs from 1 to m), while $0 \leqslant k \leqslant n-1$. A nice selection of columns from H_n is one for which the following assertion holds: for any j, $1 \leqslant j \leqslant m$, if $CA^k b^j$ is in the selected set and $k \geqslant 1$, then the set also contains $\{CA^l b^j : 0 \leqslant l < k\}$.

The last characterization of nice selections suggests a means by which they may be made whenever a minimal realization (A, B, C, D) for the system is available. We'll be speaking in terms of selecting columns from $Q_r(A, B)$, but keep in mind that we intend to build our nice selection out of the corresponding columns of H_n. Look at b^1, the first column of B; if $b^1 \neq 0$, select it, and continue selecting vectors of the form $A^k b^1$, $k > 0$, until the further selection of such a vector would result in a linearly dependent set. Now, do the same thing with b^2, and so on, until a selection of n linearly independent columns of $Q_r(A, B)$ has been made.

Another way of performing a nice selection is simply to pick columns from $Q_r(A, B)$, starting on the left, all the while skipping those columns which form a linearly dependent set when taken together with the columns already selected. The resulting selection is nice because if $A^k b^j$, for some k and j, depends on columns to the left of it in $Q_r(A, B)$, then $A^{k+1} b^j$ *also* depends on columns to its left. Proof of the last statement is left to the reader as a simple exercise in linear algebra.

In any case, suppose that a nice selection Σ of columns from H_n is made; for each j, $1 \leqslant j \leqslant m$, set $\hat{l}_j(\Sigma)$ equal to the highest value of l for which $A^{l-1} b^j$ is in the corresponding selection of columns from $Q_r(A, B)$. In a slight abuse of notation, we shall also denote by Σ this corresponding selection. Note that the $\hat{l}_j(\Sigma)$ depend only on the given selection Σ, and *not* on the realization (A, B, C, D). Observe that the $\hat{l}_j(\Sigma)$ sum to n; note also that since Σ is a nice selection, each $\hat{l}_j(\Sigma)$ is precisely the number of times that a column of the form $A^l b^j$ appears in Σ.

For each j, $1 \leqslant j \leqslant m$, let $l_1(\Sigma), \ldots, l_m(\Sigma)$ be the list of the $\hat{l}_j(\Sigma)$ arranged in decreasing order. The $l_j(\Sigma)$ are called the *ordered controllability indices* corresponding to the selection Σ. A certain number $\rho(\Sigma)$ of the ordered controllability indices corresponding to Σ are nonzero; thus $l_j(\Sigma) = 0$ if $j > \rho$. Recall from §22, particularly Definition 22.16 and the discussion preceding it, that there is a natural ordering on m-tuples of nonnegative integers which are arranged in decreasing order; specifically, we say that such an n-tuple $n_1, \ldots, n_m$ is smaller than another such m-tuple $l_1, \ldots, l_m$ if and only if there is a k such that $n_j = l_j$ for $j < k$ and $n_k < l_k$. Evidently, there must be some (or, possibly, more than one) nice selection Σ_o whose m-tuple of ordered controllability indices $l_j(\Sigma_o)$ is equal to or smaller than the m-tuple of ordered controllability indices coming from any other nice selection. As it happens, the last selection we described is a nice selection Σ_o which gives rise to such a *minimal list* of ordered controllability indices.

Once again, begin with the leftmost nonzero column of $Q_r(A, B)$; then move to the right, and adding a column to the selection if it is linearly independent of the ones already in the selected set. Eventually, n columns will have been selected, and no further columns from $Q_r(A, B)$ will form a linearly independent set when taken together with these. This selection procedure amounts essentially to a selection of the "first n linearly independent columns" from $Q_r(A, B)$.

25.12 Definition: Let $\mathbf{H}_n$ be the nth order Hankel matrix of a real m-input, p-output time-invariant continuous- or discrete-time input-output linear system which has McMillan degree n. The selection Σ_o just outlined of columns from $\mathbf{H}_n$ is called the *Kronecker column selection*, and the ordered column indices $l_1(\Sigma_o), \ldots, l_m(\Sigma_o)$ are called the *controllability Kronecker indices* of the system, and are denoted by $n_1(\mathbf{H}), \ldots, n_m(\mathbf{H})$. □

The reader should check that the Kronecker column selection, and hence the controllability Kronecker indices, as we have defined them, are indeed determined by **H** and not by any particular realization. Another straightforward but somewhat less trivial exercise is to show that the Kronecker selection is a *nice* selection. Perhaps the most interesting feature of the list of controllability Kronecker indices is that it is a minimal list of ordered column indices with respect to the ordering on such lists that we have defined above. In fact, the controllability Kronecker indices are precisely the same as the column Kronecker indices of the system which first appeared in see Definition 22.16 and later in the proof of Theorem 25.3.

25.13 Lemma: *In the above context, the list of controllability Kronecker indices $n_1(\mathbf{H}), \ldots, n_m(\mathbf{H})$ is minimal (with respect to the ordering defined above) among all lists of ordered controllability indices for nice selections of columns from* $\mathbf{H}_n$.

Proof: Suppose that the first k of the controllability Kronecker indices are the same, and that $n_{k+1}(\mathbf{H}) < n_k(\mathbf{H})$. We claim that that if $l_1, \ldots, l_m$ is the list of ordered controllability indices corresponding to any other nice selection Σ from the columns of $\mathbf{H}_n$, then either $l_1 > n_1$, or else the first k of the l_j must also be the same. To see this, first think of the selections as being from among the columns of $Q_r(A, B)$; then note note that $l_1 - 1$ denotes the power of A appearing in the rightmost $(n \times m)$ block of $Q_r(A, B)$ which contains a column belonging to Σ .

l_1 must be at least n_1, since there do not exist (by construction of the n_j) n linearly independent columns in the blocks to the left of the n_1th block. If $l_1 > n_1$, then the n_j-list is "less than" the l_j-list, with respect to our ordering on lists; if $l_1 = n_1$, then l_2 must also equal n_2 by the same argument, since $l_2 \leqslant l_1$, and there are not enough linearly independent columns in blocks strictly to the left of the n_2th block to fill out a nice selection along with the selected vectors corresponding with l_1. The point is that $n - l_1$ more columns need to be selected eventually; since $n_2 = l_1$, it follows that there are fewer than $n - l_1$ columns in blocks strictly to the left of the n_2th block which are linearly independent when taken together with the first n_1 already chosen. A continuation of this argument shows that if $l_1 = n_1$, then $l_j = n_j$ for all $j \leqslant k$.

The argument may be carried further, through the next run of equal n_j's. Assume that the first k of the l_j are equal to the first k of the n_j; conclude that $l_{k+1} \geqslant n_{k+1}$. If the inequality is strict, then we are done; if equality holds, then we may argue as above to conclude that a run of equal l_j's occurs, as well. The argument continues in this fashion until all the values of j have been exhausted. □

Our final result asserts the equality of the two kinds of Kronecker indices we have encountered.

25.14 Theorem: *Let there be given a real m-input p-output time-invariant continuous- or discrete-time input-output linear system which has McMillan degree n and nth order Hankel matrix H_n. The controllability Kronecker indices $\{n_j(\mathrm{H})\}$ of the system are the same as the Kronecker column indices $n_j(\mathrm{Z})$ of the strictly proper part $\mathrm{Z}(s)$ (or $\mathrm{Z}(z)$) of its transfer function $G(s)$ (or $G(z)$).*

Proof: A careful examination of the system's controllable canonical form defined before Exercise 25.10 should convince the reader that the Kronecker column indices of the system's transfer function, which are exhibited in the matrices $\hat{A}$ and $\tilde{B}$ in the controllable canonical form for the system, are the controllability Kronecker indices for the reachability matrix $Q_r(\hat{A}, \tilde{B})$. It remains to show that these controllability Kronecker indices are the same as those for $Q_r(\hat{A}, \hat{B})$, where $\hat{B} = \tilde{B}Q$ for an invertible $(m \times m)$ matrix Q. This is not difficult; the blocks in $Q_r(\hat{A}, \hat{B})$ are the products of the corresponding blocks in $Q_r(\hat{A}, \tilde{B})$ with the *same* invertible $(m \times m)$ matrix Q, and it is tedious but not hard to deduce from this fact that the Kronecker selection of columns from $Q_r(\hat{A}, \hat{B})$ has the same list of ordered controllability indices as the Kronecker selection from $Q_r(\hat{A}, \tilde{B})$. □

25.15 Exercise: Complete the proof of Theorem 25.14. [*Suggestion:* First observe that since $\hat{B} = \tilde{B}Q$, and Q is invertible, it follows that the same number of columns from $\hat{B}$ as from $\tilde{B}$ are selected when making the two Kronecker selections.] □

26. The Discrete-Time Linear Quadratic Regulator Problem

In this section and the next, we discuss what might be construed as one of the great triumphs of modern linear system theory. There is no doubt that the linear quadratic regulator problem has spawned a vast quantity of research, both theoretical and applied; the simplicity of its formulation, the elegance of its solution, and its evident relevance to applications in a variety of contexts make it the heart of perhaps the most widely understood and heavily applied body of system theoretic results. Moreover, the problem illustrates in a beautiful way the manner in which properties of a system such as reachability and observability manifest themselves as intuitively appealing and mathematically transparent sufficient conditions for the unraveling of a problem whose solution is by no means obvious at first glance.

The linear quadratic regulator problem is actually an *optimal control* problem for linear systems. We have not had much occasion to discuss control theory or control problems in this book; suffice it to say that the system theoretic constructions we have introduced constitute a large piece of the foundation of modern approaches control theory and control engineering. A control problem, roughly speaking, is a systems problem which is of a *prescriptive* as opposed to a *descriptive* kind (c.f. the discussion at the beginning of §13). By a descriptive problem, we mean a problem such as that of determining *what happens* in a system when a given input function is applied. A prescriptive problem, on the other hand, asks not what *does happen* under a given set of operating conditions, but rather what can be *made to happen* given certain assumptions about the system and our ability to *control* it.

One example of a prescriptive problem whose solution we have outlined is the problem of *pole placement* for state space linear systems; Theorem 25.3 states that the eigenvalues of the "A-matrix" in a realization of a reachable state space linear system may be placed essentially arbitrarily an appropriate *state feedback control* strategy is applied as the system's input. Another prescriptive problem we treated in §25 is the problem of implementing a state feedback control strategy in an input-output sense; the problem's solution employs an observer. Other prescriptive problems which we've mentioned only in passing are the problems of stabilization and pole placement by constant-gain output feedback.

The linear quadratic regulator problem is a prescriptive problem of a special kind. Instead of demanding a selection of control input in order to affect the *qualitative* behavior of a system, the problem requires the *minimization of a cost function* through choice of control. The cost function in the linear quadratic regulator problem does have its qualitative overtones, however, and it appeals to our instincts in the sense that it is a cost function whose minimization would seem to be a reasonable objective.

To set up the continuous-time version of the problem, which we'll consider in §27, let A, B, and C be real matrices having respective sizes $(n \times n)$, $(n \times m)$, and $(m \times m)$. Suppose that the vector functions $t \to x(t)$, $t \to y(t)$, and $t \to u(t)$ are related by

$$\begin{aligned} \dot{x}(t) &= Ax(t) + Bu(t) \quad , \quad t \in \mathbf{R} \\ y(t) &= Cx(t) \quad , \quad t \in \mathbf{R} \\ x(0) &= x_o \ . \end{aligned} \tag{I}$$

Let Q be an $(n \times n)$ nonnegative definite matrix. The "finite time horizon" version of the continuous-time linear quadratic regulator problem is to find $t \to u(t)$, $0 \leqslant t \leqslant t_1$, so as to minimize

$$J_0^{t_1} = \int_0^{t_1} u^T(t)u(t) + y^T(t)y(t)dt \ + \ x^T(t_1)Qx(t_1) \ ,$$

whereas the infinite-time, or *stationary* version of the problem is to find $u : [0, \infty) \to \mathbf{R}^m$ which minimizes

$$J_0^{\infty} = \int_0^{\infty} u^T(t)u(t) + y^T(t)y(t)dt \ .$$

The problem deals with a *linear* system, a cost which is *quadratic* in the system's input and output, and the *regulation* of the system's state so that the output remains or becomes small as time progresses. The $y^T y$ terms in the integrands reflect the fact that deviation of the output from zero is "costly;" the $u^T u$ terms are there because *control energy* may be expensive, as well. Besides, as we saw in §13, $t \to x(t)$ in (I) may be steered to zero *arbitrarily fast* if (A, B) is a reachable pair; such a quick zeroing out of the state, however, requires arbitrarily high input energy. Thus, minimizing $J_0^{t_1}$ or J_0^{∞} involves a tradeoff between high control energy with fast regulation on the one hand, and low energy control with slow regulation on the other.

The corresponding discrete-time linear quadratic regulator problems are stated similarly. With A, B, C, and Q as above, suppose that the vector discrete-time functions $k \to x(k)$, $k \to u(k)$, and $k \to y(k)$ are related as follows:

$$\begin{aligned} x(k+1) &= Ax(k)+Bu(k) \quad , \quad k \geqslant 0 \\ y(k) &= Cx(k) \quad , \quad k \geqslant 0 \\ x(0) &= x_o \ . \end{aligned} \tag{II}$$

The finite time horizon discrete-time linear quadratic regulator problem is to find $k \to u(k), 0 \leqslant k \leqslant k_1$, to minimize

$$J_0^{k_1} = \sum_{k=0}^{k_1-1} u^T(k)u(k) + y^T(k)y(k) + x^T(k_1)Qx(k_1) \ ,$$

whereas the infinite-time, or *stationary* version of the problem is to find u which minimizes

$$J_0^{\infty} = \sum_{k=0}^{\infty} u^T(k)u(k) + y^T(k)y(k) \ .$$

The solutions to the continuous- and discrete-time problems may both be derived in a number of ways; probably the most efficient is by means of the method of *dynamic programming*, which was first pioneered by Richard Bellman (see [Bellman]). There are a host of analytical details which attend the *derivation* of the dynamic programming solution to the continuous-time problems; for this reason, we won't derive these solutions, but instead will exhibit them and prove that they are indeed the correct answers. In the discrete-time case, however, no such analytical difficulties present themselves, and we are able to give a full-scale derivation of the solutions starting from first principles. Consequently, we treat the discrete-time problems first.

First, consider the finite time horizon problem of minimizing $J_0^{k_1}$. Since Q is nonnegative definite, and since every summand is nonnegative, $J_0^{k_1}$ is bounded below by zero, and hence it makes sense to try and minimize it. The dynamic programming approach to the problem begins by defining the *optimal cost to go* from time $k \leqslant k_1$, which is clearly a function of $x(k)$, as

$$J_k^o(x(k)) = \min_{u[k,k_1)} \sum_{l=k}^{k_1-1} u^T(l)u(l) + y^T(l)y(l) + x^T(k_1)Qx(k_1),$$

where the minimization is to be taken over all choices for $u(l)$, $l \leqslant k < k_1$. The *Principle of Dynamic Programming* asserts that for every k, $0 < k \leqslant k_1$, it is true that

$$J_{k-1}^o(x(k-1)) = \min_{u(k-1)} \{u^T(k-1)u(k-1)$$
$$+ y^T(k-1)y(k-1) + J_k^o(x(k))\} . \qquad \text{(DP)}$$

Equation (DP) states that the best one can do with regard to minimizing the cost function starting at time $k-1$ from some state $x(k-1)$ is to plan to proceed optimally from time k onward, no matter what $x(k)$ is reached because of $u(k-1)$, and then to choose $u(k-1)$ so as to minimize the sum of the cost due only to $u(k-1)$, the cost due only to $y(k-1)$, and the cost $J_k^o(x(k))$ associated with the "rest" of the optimal strategy, which depends, through $x(k)$, on $u(k-1)$. The Principle of Dynamic Programming is also known as the *Principle of Optimality* in optimal control theory.

We now apply (DP) to the solution of the finite time horizon problem. It is natural to assume, since all terms in the cost are quadratic, that the optimal cost J_0^o will be a quadratic function of x_o; more generally, it is reasonable to believe that for each k, $0 \leqslant k \leqslant k_1$, we'll have

$$J_k^o(x(k)) = x^T(k)P(k)x(k),$$

where $\{P(k)\}$, $0 \leqslant k \leqslant k_1$, is a sequence of nonnegative definite $(n \times n)$ matrices. It is required that $P(k_1) = Q$ since, as we have defined it, $J_{k_1}^o = x^T(k_1)Qx(k_1)$.

Under this assumption, write the quantity to be minimized in (DP) for $k \leqslant k_1$ as

$$u^T(k-1)u(k-1) + y^T(k-1)y(k-1) + x^T(k)P(k)x(k).$$

Observe that we can't do anything about $y^T(k-1)y(k-1)$ by choosing $u(k-1)$; we may invoke the system equations (II) and conclude that we need to choose $u(k-1)$ to minimize

$$[Ax(k-1) + Bu(k-1)]^T P(k)[Ax(k-1) + Bu(k-1)]$$
$$+ u^T(k-1)u(k-1).$$

We may either complete the square or differentiate with respect to the elements of $u(k-1)$ to find the best choice; if we do the latter, we obtain

$$u^T(k-1)+[Ax(k-1)+Bu(k-1)]^T P(k)B = 0 ,$$

from which it follows immediately that

$$u(k-1) = -[I_m+B^T P(k)B]^{-1}B^T P(k)Ax(k-1) .$$

Note that the inverse of the bracketed matrix exists because $P(k)$ is nonnegative definite by definition.

The last equation shows that the best input at time $k-1$ is of the form of an instantaneous linear feedback of the state at time $k-1$; although the feedback "gain" is time-varying in this case, the fact that the optimal control is a feedback strategy should come as no surprise in the light of §25. Readers who are uncomfortable about an optimal control presented in this way as a *feedback law* should observe that we could specify $u(k)$ equally well in *open-loop form* by first determining how $k \to x(k)$ evolves under the feedback control law and then plugging the result into the formula for $u(k)$. (See Theorem 26.2 below.)

In order to implement the optimal control, we need to know $P(k)$ at time $k-1$ for each k, $0 < k \leqslant k_1$. It turns out that the matrices $\{P(k)\}$ satisfy a simple recursion. Set

$$F(k) = [I_m+B^T P(k+1)B^T]^{-1}B^T P(k+1)A , \quad 0 \leqslant k < k_1 .$$

Thus, $F(k)$ is the negative of the feedback gain for the optimal control strategy at time k. Since $u(k) = -F(k)x(k)$, the optimal cost to go, $J_k^o(x(k)) = x^T(k)P(k)x(k)$ is given by

$$x^T(k)\{F^T(k)F(k)+C^T C+[A-BF(k)]^T P(k+1)[A-BF(k)]\}x(k) ,$$

where we have used (II). Since this formula must hold for all $x(k) \in \mathbf{R}^n$, we get

$$P(k) = F^T(k)F(k)+C^T C+[A-BF(k)]^T P(k+1)[A-BF(k)] ,$$

which, since $F(k)$ depends on $P(k+1)$, is a *difference equation* for $k \to P(k)$, $0 \leqslant k \leqslant k_1$, with "final condition" $P(k_1) = Q$ by definition of $J_{k_1}^o(x(k_1))$. We have solved the finite time horizon problem; the solution may be summarized as follows.

26.1 Theorem: *The minimum value of $J_0^{k_1}$ is equal to subject to the equations (II) is achieved by setting*

$$u(k) = -F(k)x(k), \quad 0 \leqslant k < k_1,$$

where for $0 \leqslant k < k_1$

$$F(k) = [I_m + B^T P(k+1)B]^{-1} B^T P(k+1)A ,$$

and $k \to P(k), 0 \leqslant k \leqslant k_1$, *satisfies* $P(k_1) = Q$ along with

$$P(k) = F^T(k)F(k) + C^T C + [A - BF(k)]^T P(k+1)[A - BF(k)] .$$

The minimum value of $J_0^{k_1}$ *is* $x_o^T P(0) x_o$.

The optimal control is given as a function of k by

$$u(k) = \Phi_{A-BF}(k, 0)x_o ,$$

where Φ_{A-BF} *is the state transition matrix (see §3) corresponding to* $k \to A - BF(k), 0 \leqslant k < k_1$. □

Buried in the argument leading up to Theorem 26.1 is a proof by induction that the matrices $P(k)$ are, indeed, nonnegative definite. The idea is that $J_{k_1}^o$ is equal to $x^T(k_1)Qx(k_1)$, where Q is nonnegative definite; and solving for $u(k-1)$ in (DP) shows that if J_k^o is a nonnegative quadratic in $x(k)$, then J_{k-1}^o is a nonnegative quadratic in $x(k-1)$. Since J_k^o evidently increases as k decreases, we see that each $P(k)$ must be a nonnegative definite $(n \times n)$ matrix.

Note that in solving the finite time horizon problem we made no assumptions about reachability, observability, and so on. In order to handle the minimization of J_0^∞, however, convergence requirements demand that such stipulations about the system (II) be made. For example, if the pair (A, B) is not discrete-time controllable (recall from §14 that this is a *weaker* property than reachability), then there might be initial conditions x_o for which the sum J_0^∞ does not converge for any choice of u. On the other hand, if (A, C) is not observable, then some initial conditions x_o will lead to zero cost if the zero input is applied; such an uncontrolled evolution of $x(k)$ might allow dangerous instabilities in x which are not revealed in $k \to y(k)$.

Accordingly, we shall soon be assuming that (A, B) and (A, C) from equations (II) are reachable and observable pairs, respectively. It turns out that these assumptions are not necessary, but the minor modifications which enable us to come up with "tight" sufficient conditions are not worth troubling over just yet (see Exercise 26.4 below, however). The problem, once again, is to find $u(0), u(1), \ldots$ so as to minimize

$$J_0^\infty = \sum_{k=0}^{\infty} u^T(k)u(k) + y^T(k)y(k)$$

where u and y are related by the equations (II). Assuming that (A, B) is a reachable pair guarantees that there exist input functions u which make J_0^∞ finite; in particular, any u which drives $x(k)$ in (II) to zero in finite time K and is zero for times larger than K will do the job.

It will be seen that the solution to the infintite time problem may be obtained as a suitably defined limit of the solution to the finite time problem. First, let us embellish the notation for the finite-time problem so as to exhibit more explicitly the dependence of its solution on the *time horizon* k_1 and the *final state cost matrix* Q.

For each $k_1 > 0$ and $(n \times n)$ nonnegative definite matrix Q, set $P_{k_1}(k; Q)$ denote the value at time $k \leqslant k_1$ of the solution to the matrix difference equation in Theorem 26.1 subject to the final condition $P_{k_1}(k_1; Q) = Q$. Also set $F_{k_1}(k; Q)$ equal to the corresponding $F(k)$; i.e.,

$$F_{k_1}(k; Q) = [I_m + B^T P_{k_1}(k+1; Q)B]^{-1} B^T P_{k_1}(k+1; Q)A .$$

In this new notation, the optimal cost in the finite-time problem is given by $x_o^T P_{k_1}(0; Q)x_o$. It turns out, but is somewhat difficult to show, that $x_o^T P_{k_1}(0; Q)x_o$ is an *increasing function of* k_1 for every $x_o \in \mathbf{R}^n$ and every nonnegative definite $(n \times n)$ matrix Q. In the special case where Q is the $(n \times n)$ zero matrix 0_n, it is clear that $x_o^T P_{k_1}(0; 0_n)x_o$ is increasing in k_1; simply note that for *any* input sequence $\{u(k)\}, 0 \leqslant k \leqslant k_1$, the $(k_1 + 1)$-fold sum

$$\sum_{k=0}^{k_1} u^T(k)u(k) + y^T(k)y(k) ,$$

whose smallest possible value subject to (II) is $x_o^T P_{k_1+1}(0; 0_n)x_o$, is at least as large as the sum of its first k_1 terms. The minimum possible value of the shorter sum is by definition $x_o^T P_{k_1}(0; 0_n)x_o$.

Having observed that $x_o^T P_{k_1}(0; 0_n)x_o$ increases monotonically in k_1, note also that it is *bounded from above* by any finite value of J_0^∞. Once again, any choice of u leads to a value of J_0^∞ which is bigger than or equal to the sum of its first k_1-1 terms, which in turn is at least $x_o^T P_{k_1}(0; 0_n)x_o$. We deduce the existence, for every $x_o \in \mathbf{R}^n$, of the limit

$$\lim_{k_1 \to \infty} x_o^T P_{k_1}(0; 0_n)x_o \ .$$

Moreover, for any symmetric matrix $(n \times n)$ R it is true for $1 \leqslant i, j \leqslant n$ that

$$[R]_{ij} = \tfrac{1}{4}\{(e^i+e^j)^T R(e^i+e^j) - (e^i-e^j)^T R(e^i-e^j)\} ,$$

where e^i and e^j are standard basis vectors for $\mathbf{R}^n$. Any symmetric R, for this reason, is determined by all the values of $x^T Rx$ for $x \in \mathbf{R}^n$; it follows from the existence of the above limit for all x_o that $[P_{k_1}(0; 0_n)]_{ij}$ converges for all i and j as $k_1 \to \infty$.

Define the $(n \times n)$ matrix P^* as

$$P^* = \lim_{k_1 \to \infty} P_{k_1}(0; 0_n) \ .$$

Observe first that P^* is a *nonnegative definite* $(n \times n)$ matrix; indeed, for every $k_1 > 0$ we have

$$x_o^T P^* x_o \geqslant x_o^T P_{k_1}(0; 0_n)x_o \geqslant 0,$$

where the last inequality holds because $P_{k_1}(0; 0_n)$ is nonnegative definite. Furthermore, P^* is an *equilibrium* of the difference equation in Theorem 26.1 which defines $k \to P_{k_1}(k ; Q)$. To see this, note that the constancy of A, B, and C implies that $P_{k_1}(0; 0_n)$ is the value at time $P(-k_1)$ of the solution $k \to P(k)$, $k \leqslant 0$, to the difference equation

$$P(k) = F^T(k)F(k) + C^T C + [A - BF(k)]^T P(k+1)[A - BF(k)] ,$$

with

$$F(k) = [I_m + B^T P(k+1)B]^{-1} B^T P(k+1)A , \qquad -\infty < k < 0 ,$$

subject to the condition $P(0) = 0_n$. Since $P(-k_1)$ converges to P^* as $k_1 \to \infty$, so does $P(-k_1+1)$, so that P^* is a solution to the matrix equation (regarding P as unknown)

$$P = F^T F + C^T C + [A - BF]^T P[A - BF] , \quad \text{(d.ARE)}$$

with

$$F = [I_m + B^T PB]^{-1} B^T PA \ .$$

Equation (d.ARE) is called a *discrete-time algebraic Riccati Equation*; equations resembling (d.ARE) play a central role in the theory of estimation and control for discrete-time linear systems. Because of the way in which F depends on P, (d.ARE) is a nonlinear matrix algebraic equation which is satisifed by the $(n \times n)$ matrix P^*. It turns out that there are many symmetric $(n \times n)$ matrices P which satisfy (d.ARE); we'll have more to say about this point later on.

In any case, we have shown so far that there exists an $(n \times n)$ nonnegative definite matrix P^* satisfying (d.ARE) such that

$$x_o^T P_{k_1}(0; 0_n) x_o \uparrow x_o^T P^* x_o$$

as $k_1 \to \infty$ for every $x_o \in \mathbf{R}^n$. Since the term on the left is the optimal cost for the finite time horizon linear quadratic regulator problem, it is natural to wonder whether $x_o^T P^* x_o$ is the optimal cost for the infinite time problem. Moreover, since the optimal *control* for the finite-time problem is $u(k) = -F_{k_1}(k; 0_n) x(k)$, $0 \leqslant k \leqslant k_1$, and since $F_{k_1}(k; 0_n)$ converges to

$$F^* = [I_m + B^T P^* B]^{-1} B^T P^* A ,$$

it might be asked whether $u(k) = -F^* x(k)$, $k \geqslant 0$, is the optimal control for the infinite-time problem. Both of these questions have affirmative answers if we introduce the assumption that (A, C) is an observable pair.

With F^* defined as above, set $A^* = A - BF^*$.

26.2 Lemma: *With notation as above, assume that the pair (A, C) in (II) is observable and that (A, B) is reachable. Then*

(a) P^* is a positive definite $(n \times n)$ matrix, and

(b) The eigenvalues of $A^* = A - BF^*$ all have magnitudes less than 1.

Proof: Observability implies that no initial condition $x_o \in \mathbf{R}^n$ for the difference equation in (II), when $u(k)$ is zero for all k, gives rise to an identically zero output $k \to y(k)$, $0 \leqslant k \leqslant k_1$, if $k_1 \geqslant n$; see §16. Thus, subject to (II),

$$\sum_{k=0}^{k_1-1} u^T(k)u(k)+y^T(k)y(k)>0$$

for *any* choice of input function u if $x_o \neq 0$. It follows that $P_{k_1}(0;0_n)$ is positive definite for every $k_1 > n$, since the the above sum can be made to equal $x_o^T P_{k_1}(0;0_n)x_o$ by appropriate choice of u. It follows that P^*, whose existence is guaranteed by the reachability assumption, is also positive definite, since $x_o^T P_{k_1}(0;0_n)x_o$ increases to $x_o^T P^* x_o$ for every $x_o \in \mathbf{R}^n$.

As for (b), we prove the result by constructing a Lyapunov function (see Theorems 23.16 and 23.22) for the equilibrium $\bar{x} = 0$ of the difference equation

$$x(k+1) = A^* x(k).$$

Define $V : \mathbf{R}^n \to \mathbf{R}$ by

$$V(x) = x^T P^* x, \quad x \in \mathbf{R}^n.$$

V is positive except at $x = 0$; furthermore,

$$V(A^*x) - V(x) = -x^T[(F^*)^T F^* + C^T C]x \leq 0 \qquad (*)$$

for all x. Thus, V is a Lyapunov function for $\bar{x} = 0$, and it follows that all the eigenvalues of A^* have magnitudes less than *or equal* to 1 by Theorems 23.20 and Theorem 23.26(b). To show that all the eigenvalues of A^* lie strictly inside the unit circle, suppose for the moment that some eigenvalue λ_o of A^* has magnitude 1. Let $z_o \in \mathbf{C}^n$ be a (possibly complex) corresponding eigenvector. Since $A^* z_o = \lambda_o z_o$, it follows that

$$V(A^* z_o) = |\lambda_o|^2 z_o^\dagger P^* z_o = V(z_o),$$

where we have extended the domain of definition of V from $\mathbf{R}^n$ to $\mathbf{C}^n$. It follows from (*) that

$$z^\dagger [F^{*T} F^* + C^T C] z_o = 0.$$

This, however, is impossible, since it implies both that $F^* z_o = 0$ and that $CA^{*k} z_o = 0$ for all $k \geq 0$, which together imply that $CA^k z_o = 0$ for all k, contradicting observability of (A, C). We conclude that all of the eigenvalues of A^* have magnitude strictly less than one. □

Lemma 26.2 provides the technical machinery which we require in order to complete the solution of the discrete-time stationary linear quadratic regulator problem. With notation as above, we know that for every $x_o \in \mathbb{R}^n$ and any choice of u for which the sum converges, it is true that

$$x_o^T P_{k_1}(0;0_n)x_o \leqslant \sum_{k=0}^{\infty} u^T(k)u(k)+y^T(k)y(k),$$

since the left-hand side is the smallest possible value of the sum of the first k_1 terms in the right. Taking the limit as $k_1 \to \infty$ of the left-hand side implies that

$$x_o^T P^* x_o \leqslant \sum_{k=0}^{\infty} u^T(k)u(k)+y^T(k)y(k)$$

for any choice of u which makes the sum converge. If we set

$$u(k) = -F^* x(k), \quad k \geqslant 0,$$

then the right-hand sum converges because $k \to x(k)$ obeys

$$x(k+1) = A^* x(k), \quad k \geqslant 0,$$

and because of Lemma 25.2(b), both $y(k)$ and $u(k)$ decrease exponentially as $k \to \infty$. Moreover, this choice of u makes the sum on the right-hand side equal to

$$\sum_{k=0}^{\infty} x^T(k)[F^{*T}F^* + C^T C]x(k),$$

which in turn, by (d.ARE), is the same as

$$\sum_{k=0}^{\infty} x^T(k)P^*x(k) - x^T(k+1)P^*x(k+1) = x_o^T P^* x_o .$$

Hence, the minimum value for J_0^∞ is precisely $x_o^T P^* x_o$, and is achieved if $u(k)$ is chosen as $-F^* x(k)$ for each $k \geqslant 0$.

We summarize the foregoing discussion as follows.

26.3 Theorem: *Suppose that (A, B) is reachable and (A, C) is observable. Then the minimum value of the discrete-time J_0^∞ subject to (II) is equal to $x_o^T P^* x_o$, and is achieved by choosing $u(k) = -F^* x(k)$, $k \geqslant 0$, where*

$$F^* = [I_m + B^T P^* B]^{-1} B^T P^* A ,$$

and

$$P^* = \lim_{k_1 \to \infty} P_{k_1}(0;0_n),$$

with $P_{k_1}(0;0_n)$ defined as in Lemma 26.2. P^* is positive definite, and satisfies the discrete-time algebraic Riccati *equation*

$$P^* = F^{*T} F^* + [A - BF^*]^T P^* [A - BF^*] + C^T C \ ,$$

with F^* *as above.*

The optimal control is given as a function of k *by*

$$u(k) = (A - BF^*)^k x_o \ , \quad k \geq 0 \ . \qquad \square$$

26.4 Exercise: Show that Theorem 26.3 holds, except for the assertion about the positive definiteness of P^*, if the reachability and observability assumptions are replaced respectively with the assumption that (A, B) is *discrete-time stabilizable* (i.e., there exists an $(m \times n)$ matrix F such that the eigenvalues of $A - BF$ have magnitudes less than 1) and the assumption that (A, C) is *discrete-time detectable* (i.e., (A^T, C^T) is discrete-time stabilizable.) $\square$

26.5 Exercise: Suppose that R_1 and R_2 are positive definite matrices having respective sizes $(m \times m)$ and $(p \times p)$. Derive the optimal control and optimal cost for

$$\hat{J}_0^{\infty} = \sum_{k=0}^{\infty} u^T(k) R_1 u(k) + y^T(k) R_2 y(k)$$

subject, once again, to equations (II). $\square$

Before moving on to the continuous-time linear quadratic regulator problem, it is worth discussing the equation (d.ARE) some more. We have shown, under the assumption that (A, B) is reachable and (A, C) is observable, that there exists a positive definite solution P^* satisfying (d.ARE), and that the eigenvalues of $A - BF^*$, with F^* given as above, lie inside the open unit disc in the complex plane. It happens that P^* is *unique* with respect to *each* of these properties. To prove these facts, we'll need to look back at the derivation preceding Theorem 26.1 of the solution to the finite time horizon regulator problem.

Suppose that $\hat{P}$ is a (real) nonnegative definite solution to

$$P = F^T F + [A - BF]^T P [A - BF] + C^T C \ , \quad \text{(d.ARE)}$$

with

$$F = [I_m + B^T P B]^{-1} B^T P A \ .$$

In this case, it is evident that $\hat{P}_{k_1}(k; \hat{P}) = \hat{P}$ for all k

between 0 and k_1, since $\hat{P}$ is an *equilibrium* of the difference equation defining $k \to P_{k_1}(k; Q)$. Moreover, by the argument leading up to Theorem 26.1, it follows that since $\hat{P}$ is nonnegative definite, the minimum value attained by

$$\sum_{k=0}^{k_1-1} u^T(k)u(k) + y^T(k)y(k) + x^T(k_1)\hat{P}x(k_1)$$

through choice of u subject to (II) is $x_o^T\hat{P}(0;\hat{P})x_o$, which in turn equals $x_o^T\hat{P}x_o$.

Now, for every choice of control sequence u, and for every $k_1 \geqslant 0$, it is true that

$$\sum_{k=0}^{k_1-1} u^T(k)u(k) + y^T(k)y(k) + x^T(k_1)\hat{P}x(k_1)$$

$$\geqslant \sum_{k=0}^{k_1-1} u^T(k)u(k) + y^T(k)y(k) . \qquad (**)$$

Because the right-hand side of (**) has minimum value $x_o^T P_{k_1}(0; 0_n)x_o$, it follows that

$$x_o^T\hat{P}x_o \geqslant x_o^T P_{k_1}(0; 0_n)x_o$$

for every $k_1 \geqslant 0$. Letting k_1 go to infinity yields

$$x_o^T\hat{P}x_o \geqslant x_o^T P^* x_o$$

for every $x_o \in \mathbf{R}^n$.

On the other hand, the choice of u which minimizes the left-hand side of (**) is

$$u(k) = -[I_m + B^T\hat{P}B]^{-1}B^T\hat{P}Ax(k) ,$$

giving rise to cost $x_o^T\hat{P}x_o$; hence, for all $k_1 \geqslant 0$,

$$x_o^T\hat{P}x_o \leqslant \sum_{k=0}^{k_1-1} (u^*(k))^T u^*(k) + (y^*(k))^T y^*(k) + (x^*(k_1))^T \hat{P}x^*(k_1),$$

where the starred quantities correspond to the inputs, outputs, and final state when the optimal strategy from Theorem 26.3 is used as input. Taking the limit as $k_1 \to \infty$, remembering from Lemma 26.2(b) that $x^*(k_1)$ goes to zero, we get

$$x_o^T\hat{P}x_o \leqslant x_o^T P^* x_o$$

for all $x_o \in \mathbf{R}^n$. Hence, $x_o^T\hat{P}x_o$ and $x_o^T P^* x_o$ are always equal, implying, since both are symmetric, that $\hat{P} = P^*$.

Thus, the only nonnegative definite solution to (d.ARE) is P^*. Furthermore, let us define

$$F(P) = [I_m + B^T PB]^{-1} B^T PA$$

whenever P is a real symmetric matrix for which the indicated inverse exists. We know from Lemma 26.2 that P^* satisfies (d.ARE) and that all the eigenvalues of $A - BF(P^*)$ have magnitudes strictly less than 1. We now show that P^* is the only real matrix having these properties. Suppose that $\hat{P}$ satisfies (d.ARE) and $A - BF(\hat{P})$ satisfies the eigenvalue condition. In this case, it is easy to show from (d.ARE) using the Lyapunov Lemma 23.23(b) that

$$\hat{P} = \sum_{k=0}^{\infty} [(A - BF(\hat{P}))^T]^k [F^T(\hat{P})F(\hat{P}) + C^T C][A - BF(\hat{P})]^k ,$$

from which it follows that $\hat{P}$ is a nonnegative definite solution of (d.ARE) and therefore must equal P^* by our earlier result.

We summarize these fundamental facts about (d.ARE) as follows.

26.6 Theorem: *P^* is the only nonnegative definite solution of (d.ARE). Moreover, $\hat{P}$ is a real symmetric $(n \times n)$ matrix for which $I_m + B^T \hat{P} B$ is invertible, and $\hat{P}$ satisfies (d.ARE) and is such that all the eigenvalues of*

$$A - B[I_m + B^T \hat{P} B]^{-1} B^T \hat{P} A$$

have magnitudes less than 1, then $\hat{P} = P^*$. □

27. *The Continuous-Time Linear Quadratic Regulator Problem*

We turn now to the continuous-time linear quadratic regulator problem(s). Recall from §26 that the finite time horizon problem is to choose $u : [0, t_1) \to \mathbf{R}^m$ so as to minimize

$$J_0^{t_1} = \int_0^{t_1} u^T(t)u(t) + y^T(t)y(t)dt + x^T(t_1)Qx(t_1) ,$$

subject to

$$\begin{aligned} \dot{x}(t) &= Ax(t)+Bu(t) \ , & t \in \mathbf{R} \\ y(t) &= Cx(t) \ , & t \in \mathbf{R} \\ x(0) &= x_0 \ , \end{aligned} \tag{I}$$

whereas the infinite-time, or *stationary* version of the problem is to find $u:[0,\infty) \to \mathbf{R}^m$ which minimizes

$$J_0^\infty = \int_0^\infty u^T(t)u(t)+y^T(t)y(t)dt \ .$$

The continuous-time dynamic programming approach to *deriving* the solution, as we noted in §26, involves some treacherous analytical details; most of the difficulties concern fundamental principles from the calculus of variations. We'll therefore pull the answers out of a hat, so to speak, and prove that they are, indeed,the solutions.

With A, B, and C as in (I), consider the following matrix differential equation.

$$\dot{P}(t) = -P(t)A-A^T P(t)+P(t)B^T BP(t)-C^T C \ . \tag{RDE}$$

Equation (RDE) is known as the *matrix Riccati differential equation*, and plays a part in continuous-time control theory which parallels the role in discrete time of the difference equation defining $P(k)$ in Theorem 26.1. Since (RDE) is a *nonlinear* differential equation, we are not guaranteed that there exist globally defined solutions $t \to P(t)$, $t \in \mathbf{R}$, as there would if (RDE) were a linear matrix differential equation such as the ones we encountered in §2. Fortunately, it turns out that for any $t_1 \in \mathbf{R}$ and any $(n \times n)$ nonnegative definite matrix Q, there exists a solution $t \to P(t)$ to (RDE) defined on the entire half-line $(-\infty, t_1]$. A proof of this assertion is contained in our argument below.

Suppose, for the moment, that $t \to P(t)$, $0 \leqslant t \leqslant t_1$, satisfies (RDE) along with "final condition" $P(t_1) = Q$. Then for any $u:[0, t_1] \to \mathbf{R}^m$, $J_0^{t_1}$ is given by

$$\int_0^{t_1} u^T u + x^T[-\dot{P}-PA-A^T P + PBB^T P]x \, dt \ + \ x^T(t_1)Qx(t_1) \ ,$$

where t-dependence of P, $\dot{P}$, u, and x in the integrand has been suppressed.

Without loss of generality, suppose that

$$u(t) = -B^T P(t)x(t)+v(t) \ , \quad 0 \leqslant t \leqslant t_1 \ .$$

In this notation,

$$\dot{x}(t) = (A - BB^T P(t))x(t) + Bv(t), \quad 0 \leqslant t \leqslant t_1;$$

thus,

$$\frac{d}{dt} x^T Px = x^T [\dot{P} + P(A - BB^T P) + (A - BB^T P)^T P] x,$$

and the integrand in $J_0^{t_1}$ becomes

$$u^T u - \frac{d}{dt} x^T Px + x^T PBB^T Px .$$

A quick "completing the squares" manipulation, using $u = -B^T Px + v$, yields

$$J_0^{t_1} = x_0^T P(0)x_0 - x^T(t_1)P(t_1)x(t_1) + x^T(t_1)Qx(t_1) + \int_0^{t_1} \|v(t)\|^2 dt .$$

Since $P(t_1) = Q$, it is obvious that the minimum value of $J_0^{t_1}$ occurs when $v : [0, t_1) \to \mathbf{R}^m$ is taken to be zero; that minimum value is $x_0^T P(0)x_0$, and is achieved by the feedback control law $u(t) = -B^T P(t)x(t)$, $t \in [0, t_1]$. Note that our argument implies that $P(0)$ must be nonnegative definite, since $x_0 \in \mathbf{R}^n$ is arbitrary.

The entire computation above was predicated on the existence of a solution P to (RDE), defined at least on $[0, t_1)$, which satisfied $P(t_1) = Q$. It turns out that the argument we have just given *ensures* that this solution exists. We know from the basic existence and uniqueness theorem for ordinary differential equations (see, for example, [Hirsch and Smale]) that for sufficiently small $a > 0$, there does exist a solution $t \to P(t)$, $t \in [t_1 - a, t_1]$, to (RDE) which satisfies $P(t_1) = Q$. If $t \to P(t)$ cannot be extended to the entire interval $[0, t_1)$, then it must be true that some of the entries of $P(t)$ become unbounded as t decreases toward 0 from t_1 (again, see [Hirsch and Smale]).

To see that this can't happen, observe that the minimum value of $J_0^{t_1}$ attatinable by choice of $u : [0, t_1] \to \mathbf{R}^m$ subject to the system constraints (I) is certainly bounded from above; we could let u be identically zero, for instance, and get such an upper bound for $J_0^{t_1}$. Similarly, by taking u identically zero on each $[t_0, t_1]$ and starting the differential equation off with x_0 at time t_0, a single upper bound $K \|x_0\|^2$ may be established for the minima of all the costs

$$J_{t_o}^{t_1} = \int_{t_o}^{t_1} u^T(t)u(t) + y^T(t)y(t)dt + x^T(t_1)Qx(t_1).$$

Now, suppose that $\hat{t} \in [0, t_1]$ is such that $P(t)$ is defined for $t \in (\hat{t}, t_1]$ but blows up as t approaches $\hat{t}$ from above. For $t_o \in (\hat{t}, t_1]$, we've seen that $x_o^T P(t_o)x_o \geqslant 0$ is the minimum value of $J_{t_o}^{t_1}$. Letting t_o approach $\hat{t}$ leads to a contradiction; since some element of $P(t_o)$ is becoming unbounded, $x_o^T P(t_o)x_o$ becomes unbounded for some $x_o \in \mathbf{R}^n$ satisfying $\|x_o\| = 1$. This, however, is impossible, since $x_o^T P(t_o)x_o \leqslant K \|x_o\|^2$.

Hence, for any nonnegative definite $(n \times n)$ matrix Q and any $t_1 \geqslant 0$ there exists a solution $t \to P(t)$ to (RDE) which is defined on the entire interval $[0, t_1]$. In fact, by considering the minimzation of $J_{t_o}^{t_1}$ for negative values of t_o, we may conclude that this solution P can be extended to the entire half-line $(-\infty, t_1]$. Furthermore, since $x_o^T P(t_o)x_o \geqslant 0$ for all $x_o \in \mathbf{R}^n$ and $t_o \leqslant t_1$, it follows that $P(t)$ is *nonnegative definite* for all $t \leqslant t_1$.

We summarize the foregoing discussion as follows.

27.1 Theorem: *Let Q be a real nonnegative definite $(n \times n)$ matrix; let $t_1 > 0$ be given. Let $t \to P(t)$, $t \leqslant t_1$, be the (necessarily nonnegative definite) solution to (RDE) satisfying the condition $P(t_1) = Q$. The minimum value of $J_0^{t_1}$ subject to (I) is attained by choosing $u(t) = -B^T P(t)x(t)$, $t \in [0, t_1]$, and is equal to $x_o^T P(0)x_o$. The optimal control u is given as a function of t by*

$$u(t) = \Phi_{(A-BB^TP)}(t, 0)x_o ,$$

where $\Phi_{(A-BB^TP)}$ is the state transition matrix (see §2) corresponding to $t \to A - BB^T P(t)$. □

Having solved the finite time horizon problem, we proceed as in §26 to find the solution to the stationary infinite time problem. First, we introduce some new, elaborate notation for the solution to the (RDE) whose existence we have proven above. Given t_1 and Q as in Theorem 27.1, set $P_{t_1}(t\,; Q)$ equal to the value at time $t \leqslant t_1$ of the solution to (RDE) with final $P(t) = Q$. We have shown that $P_{t_1}(t\,; Q)$ is defined for all $t \leqslant t_1$ and all nonnegative definite Q.

In particular, let us consider $P_{t_1}(0; 0_n)$. The quantity $x_o^T P_{t_1}(0; 0_n) x_o$ is the minimum value attainable by

$$\int_0^{t_1} u^T(t)u(t) + y^T(t)y(t)\,dt$$

through choice of u subject to (I). Evidently, this minimum value is *increasing* in t_1 for every $x_o \in \mathbb{R}^n$. Furthermore, it is *bounded from above* if (A, B) is a reachable pair, since in that case there exists a $u : [0, \infty) \to \mathbb{R}^m$ which zeroes $t \to x(t)$ in finite time. Thus, assuming reachability of (A, B), we conclude that

$$\lim_{t_1 \to \infty} x_o^T P_{t_1}(0; 0_n) x_o$$

exists; by an argument like the one in discrete-time, we deduce that there is an $(n \times n)$ nonnegative definite matrix P^* such that the above limit is equal to $x_o^T P^* x_o$ for every x_o.

We claim that the limiting matrix P^* is an *equilibrium* of (RDE); that is, P^* is a solution to

$$PA + A^T P - PBB^T P + C^T C = 0\,, \qquad \text{(ARE)}$$

which is called the *algebraic Riccati equation*. The fact that P^* is an equilibrium for (RDE), and hence a solution to (ARE), follows again from the theory of ordinary differential equations. It is a simple exercise in time-shifting to show that $P_{t_1}(0; 0_n)$ is the value at time $-t_1$ of the solution $t \to P(t)$, $t \leqslant 0$, to (RDE) along with the final condition $P(0) = 0_n$. With this notation, the convergence of $P(-t_1)$ to P^* as t_1 goes to infinity guarantees that P^* is an equilibrium of (RDE), since $\dot{P}(-t_1)$ must concurrently approach zero.

We now present the continuous-time analogue of Lemma 26.2. Once again, the additional assumption of observability, which has not yet been required, becomes the key.

27.2 Lemma: *With notation as above, assume that (A, B) is reachable and (A, C) is observable. Then*

(a) P^* is a positive definite $(n \times n)$ matrix, and

(b) The eigenvalues of $A^* = A - BB^T P^*$ have negative real parts.

Proof: The proof is very similar to the proof of Lemma 26.2. Reachability guarantees the existence of P^*; observability (see §15) guarantees that for every $t_1 > 0$, the only $x_o \in \mathbf{R}^m$ which gives rise to the identically zero output on $[t_o, t_1]$ is $x_o = 0$. Thus, the minimum value attained by

$$\int_0^{t_1} u^T(t)u(t) + y^T(t)y(t)dt$$

is positive for every $x_o \neq 0$; since this minimum value is $x_o^T P_{t_1}(0; 0_n)x_o$, it follows that $P_{t_1}(0; 0_n)$ is positive definite for all t_1, and hence that P^*, the "upper limit," is also positive definite.

Now define $V : \mathbf{R}^n \to \mathbf{R}$ by

$$V(x) = x^T P^* x \;, \qquad x \in \mathbf{R}^n \;.$$

We claim that V is a Lyapunov function for the equilibrium $\bar{x} = 0$ of the differential equation

$$\dot{x}(t) = (A - BB^T P^*)x(t) = A^* x(t) \;, \qquad t \in \mathbf{R} \;.$$

If $x_o \in \mathbf{R}^n$ is given, then for $t \geqslant 0$

$$\begin{aligned} \frac{d}{dt} V(x(t)) &= x^T(t)[P^*A^* + (A^*)^T P^*]x(t) \\ &= -x^T(t)[P^*BB^T P^* + C^T C]x(t) \;, \end{aligned}$$

where the last equality follows from (ARE). The right-hand side is always less than or equal to zero, and it follows that V is a Lyapunov function. By Theorems 23.15 and 23.20, the eigenvalues of A^* lie in $\mathrm{Re}\{s\} \leqslant 0$.

To show that the eigenvalues of A^* have strictly negative real parts, suppose that λ_o is a pure imaginary eigenvalue and that $z_o \in \mathbf{C}^n$ is a corresponding eigenvector. (ARE) implies that

$$z^\dagger[P^*A^* + (A^*)^T P^* + P^*BB^T P^* + C^T C]z_o = 0 \;;$$

The left-hand side, however, is just

$$z^\dagger[P^*BB^T P^* + C^T C]z_o$$

because $A^* z_o = \lambda_o z_o$ and λ_o is pure imaginary. We now have our contradiction of observability; the idea is that $B^T P^* z_o = 0$ implies that z_o is an eigenvalue of A as well as of A^*, and this fact coupled with $Cz_o = 0$ means that $CA^k z_o = 0$ for all k. □

With Lemma 27.2 in hand, we can show that the minimum value of J_0^∞ is, indeed, the limit as k_1 goes to infinity of the minimum value of $J_0^{k_1}$, and that the optimal control for the infinite time horizon problem is the limit of the optimal control for the finite time horizon problem. We know that if $u : [0, \infty) \to \mathbb{R}^m$ makes the integral converge, then for every $k_1 > 0$ it is true that

$$x_o^T P_{k_1}(0; 0_n) x_o \leqslant \int_0^\infty u^T(t)u(t) + y^T(t)y(t)dt ,$$

since the number on the left is the *minimum* value of the integral from 0 to k_1 of the integrand on the right. Take the limit as $k_1 \to \infty$; it follows that if u makes the integral converge, then

$$x_o^T P^* x_o \leqslant \int_0^\infty u^T(t)u(t) + y^T(t)y(t)dt .$$

On the other hand, if we set

$$u(t) = -B^T P^* x(t) , \quad t \geqslant 0 ,$$

then $t \to x(t)$ obeys the differential equation

$$\dot{x}(t) = (A - BB^T P^*)x(t) , \quad t \geqslant 0 ,$$

and by Lemma 27.2(b) $x(t)$, and therefore $y(t)$ and $u(t)$, decay exponentially as $t \to \infty$, implying that the integral converges. Note that for this choice of u, the fact that P^* satisfies (ARE) implies that

$$u^T(t)u(t) + y^T(t)y(t)$$

$$= -x^T(t)[P^*(A - BB^T P^*) + (A - BB^T P^*)^T]x(t)$$

for every $t \geqslant 0$; the right-hand side, however, is just minus the time derivative of $x^T(t)P^* x(t)$, and it therefore integrates to

$$x_o^T P^* x_o - \lim_{t \to \infty} x^T(t)P^* x(t) = x_o^T P^* x_o ,$$

where the last equality follows from the decay of $x(t)$.

Hence, the choice $u(t) = -BB^T P^* x(t)$, $t \geqslant 0$, *achieves* the lower bound on J_0^∞ which we established earlier, and therefore must be the optimal control for the infinite time horizon problem.

27.3 Theorem: *Suppose that (A, B) is reachable and (A, C) is observable. Then the minimum value of J_0^∞ subject to (I) is equal to $x_o^T P^* x_o$, and is achieved by choosing*

$u(t) = -B^T P^* x(t),\ t \geqslant 0$, where

$$P^* = \lim_{t_1 \to \infty} P_{t_1}(0; 0_n),$$

with $P_{t_1}(0; 0_n)$ defined as in Lemma 27.2. P^* is positive definite, and satisfies the algebraic Riccati equation

$$PA + A^T P - PBB^T P + C^T C = 0 .$$

The optimal control is given as a function of t by

$$u(t) = e^{t(A - BB^T P^*)} x_o , \quad t \geqslant 0 . \qquad \square$$

27.4 Exercise: Show that Theorem 27.3 holds, except for the assertion about positive definiteness of P, if the reachability and observability assumptions are weakened to the assumptions that (A, B) is *stabilizable* (i.e., there exists an $(m \times n)$ matrix F such that the eigenvalues of $(A - BF)$ have negative real parts) and that (A, C) is *detectable* (which means that (A^T, C^T) is stabilizable.) □

27.5 Exercise: Suppose that R_1 and R_2 are positive definite matrices having respective sizes $(m \times m)$ and $(p \times p)$. Find the optimal control and optimal cost for

$$\hat{J}_0^\infty = \int_0^\infty u^T(t) R_1 u(t) + y^T(y) R_2 y(t)\, dt$$

subject to conditions (I). □

As in the case of the discrete-time algebraic Riccati equation (d.ARE) of §26, equation (ARE) has associated with it a vast literature. There has probably been more written about (ARE) than about any other equation in the history of linear system theory; for an excellent survey of its control theoretic relevance, see [Willems] and the references therein. We shall spend the rest of the section proving a result about (ARE) which is the continuous-time analogue of Theorem 26.6. The result states essentially that P^*, as defined in Theorem 27.3, is the only nonnegative definite solution of (ARE), and that P^* is *also* the only "stabilizing" solution of (ARE), in the sense that if $\hat{P}$ is a real symmetric matrix for which $(A - BB^T \hat{P})$ has eigenvalues in $\mathrm{Re}\{s\} \leqslant 0$, then $\hat{P} = P^*$.

Suppose, for the moment, that $\hat{P}$ is a nonnegative definite solution of (ARE). Since $\hat{P}$ is an *equilibrium* of the Riccati differential equation (RDE), it follows that $P_{t_1}(t\,;\hat{P}) = \hat{P}$ for all $t_1 > 0$ and for all $t \in [0, t_1]$. Accordingly, by Theorem 27.1, $x_o^T\hat{P}x_o$ is the minimum value attained by

$$\int_0^{t_1} u^T(t)u(t) + y^T(y)y(t)\,dt + x^T(t_1)\hat{P}x(t_1) \qquad (*)$$

subject to (I), and that value is attained if u is chosen as

$$u(t) = -B^T\hat{P}x(t)\,, \qquad t \in [0, t_1]\,.$$

It is certainly true that for all t_1 and for any input u, the cost (*) is at least as large as

$$\int_0^{t_1} u^T(t)u(t) + y^T(y)y(t)\,dt\ ;$$

since this last expression is at least $x_o^TP_{t_1}(0; 0_n)x_o$ (again by Theorem 27.1), it follows by taking the limit as $t_1 \to \infty$ that

$$x_o^T\hat{P}x_o \geqslant x_o^TP^*x_o$$

for every x_o.

Conversely, suppose that we compute the cost (*) when the strategy $u^*(t) = -B^TP^*x(t)$, $t \in [0, t_1]$, is used as the input. Since the minimum value of (*) is $x_o^T\hat{P}x_o$, we get

$$x_o^T\hat{P}x_o \leqslant \int_0^{t_1} (u^*(t))^T u^*(t) + (y^*(t))^T y(t)\,dt + (x^*(t_1))^T\hat{P}x^*(t_1)$$

for every $t_1 > 0$. Taking the limit as $t_1 \to \infty$ yields, along with our earlier work,

$$x_o^T\hat{P}x_o = x_o^TP^*x_o$$

for every $x_o \in \mathbf{R}^n$. It follows as usual, from symmetry, that $P^* = \hat{P}$, and we conclude that there is only *one* nonnegative definite solution to (ARE)

Suppose now that $\hat{P}$ is a real symmetric $(n \times n)$ matrix which satisfies (ARE), and that the eigenvalues of $(A - BB^T\hat{P})$ have negative real parts. Let $\hat{A} = (A - BB^T\hat{P})$; it follows from (ARE) that

$$\hat{P}\hat{A} + \hat{A}^T\hat{P} = -\hat{P}BB^T\hat{P} - C^TC\ .$$

Since $\hat{A}$ has eigenvalues in the open left half-plane, we know from the Lyapunov Lemma (Theorem 23.23(a)) that

$$\hat{P} = \int_0^\infty e^{t\hat{A}^T}[\hat{P}BB^T\hat{P}+C^TC]e^{t\hat{A}}\,dt \ ,$$

which is clearly nonnegative definite. But we already know that the only nonnegative definite solution to (ARE) is P^*, so $\hat{P} = P^*$ once again.

The final result is as follows.

27.6 Theorem: *P^* is the only nonnegative definite solution of (ARE). Moreover, if $\hat{P}$ is a real symmetric $(n \times n)$ matrix satisfying (ARE), and the eigenvalues of $(A - BB^T\hat{P})$ have negative real parts, then $\hat{P} = P^*$.* □

REFERENCES

Baras, J. S., R. W. Brockett, and P. A. Fuhrmann, "State Space Models for Infinite-Dimensional Systems," *IEEE Transactions on Automatic Control 19* (1974), pp. 693-700.

Bellman, R. *Dynamic Programming*. Princeton: Princeton University Press, 1957.

Brockett, R. W. *Finite-Dimensional Linear Systems*. New York: John Wiley and Sons, Inc., 1970.

Brockett, R. W. and C. I. Byrnes, "Multivariable Nyquist Criteria, Root Loci, and Pole Placement: A Geometric Viewpoint," *IEEE Transactions on Automatic Control AC-26* (1981), pp. 271-284.

Byrnes, C. I. "Algebraic and Geometric Aspects of the Analysis of Feedback Systems," in *Geometrical Methods for the Theory of Linear Systems*. Dordrecht, Holland: D. Reidel Publishing Company, 1980.

Chen, C.-T. *Linear System Theory and Design*. New York: Holt, Rinehart, and Winston, 1984.

D'Azzo, J. J. and C. H. Houpis. *Linear Control System Analysis and Design: Conventional and Modern*. New York: McGraw-Hill, 1975.

Desoer, C. A. *Notes for a Second Course on Linear Systems*. New York: Van Nostrand Reinhold Company, 1970.

Desoer, C. A. and M. Vidyasagar. *Feedback Systems: Input-Output Properties*. New York: Academic Press, 1975.

Dym, H. and H. P. McKean. *Fourier Series and Integrals*. New York: Academic Press, 1972.

Forney, G. D. "Minimal Bases of Rational Vector Spaces, With Applications to Multivariable Linear Systems," *SIAM Journal on Control 13* (1975), pp. 493-520.

Ghosh, B. K. and C. I. Byrnes, "Simultaneous Stabilization and Simultaneous Pole Placement by Nonswitching Dynamic Compensators," *IEEE Transactions on Automatic Control AC-28* (1983), pp. 735-741.

Golub, G. H. and C. F. van Loan. *Matrix Computations*. Baltimore, MD: The Johns Hopkins University Press, 1983.

Hahn, W. *Stability of Motion*. New York: Springer-Verlag, 1967.

Halmos, P. R. *Finite-Dimensional Vector Spaces*. New York: Springer-Verlag, 1974.

Heymann, M. "Comments on 'On Pole Assignment in Multi-

Input Controllable Linear Systems'," *IEEE Transactions on Automatic Control 13* (1968), pp. 748-749.

Hirsch, M. W. and S. Smale. *Differential Equations, Dynamical Systems, and Linear Algebra*. New York: Academic Press, 1974.

Jacobson, N. *Basic Algebra I*. San Francisco: W. H. Freeman and Company, 1974.

Kailath, T. *Linear Systems*. Englewood Cliffs, NJ: Prentice Hall, Inc., 1980.

Kalman, R. E. "Mathematical Description of Linear Dynamical Systems," *SIAM Journal on Control, 1* (1963), pp. 152-292.

Kalman, R. E., Y.-C. Ho, and K. S. Narendra. "Controllability of Linear Dynamical Systems," *Contributions to the Theory of Differential Equations*. New York: Interscience, 1963.

Kalman, R. E., P. A. Falb, and M. Arbib. *Topics in Mathematical System Theory*. New York: McGraw-Hill, 1969.

Kuo, B. *Automatic Control Systems*. Englewood Cliffs, NJ: Prentice-Hall, Inc., 1982.

Lefschetz, S. *Differential Equations: Geometric Theory*. New York: Dover Publications, Inc., 1977.

Luenberger, D. G. "Observers for Multivariable Systems," *IEEE Transactions on Automatic Control 11* (1966), pp. 190-197.

Lyapunov, A. M. *Problème Général de la Stabilité du Mouvement*. Princeton, NJ: Princeton University Press, 1947. (Reproduction of a 1907 French translation of the 1892 Russian monograph)

MacFarlane, A. G. J. *Frequency Response Methods in Automatic Control*. New York: IEEE Press, 1979.

McMillan, B. "An Introduction to Formal Realizability Theory," *Bell System Technical Journal 31* (1952), pp. 217-219.

Padulo, L. and M. A. Arbib. *System Theory*. Philadelphia: Saunders, 1974.

Paley, R. E. A. C., and N. Wiener. *Fourier Transforms in the Complex Domain*. Providence, RI: American Mathematical Society, 1934.

Royden, H. L. *Real Analysis*. New York: MacMillan, 1963.

Rudin, W. *Real and Complex Analysis*. New York: McGraw-Hill, 1974.

Simmons, G. F. *Introduction to Topology and Modern*

Analysis. New York: McGraw-Hill, 1963.
Strang, G. *Linear Algebra and Its Applications*. New York: Academic Press, 1976.
Vidyasagar, M. *Nonlinear Systems Analysis*. Englewood Cliffs, NJ: Prentice-Hall, Inc., 1978.
Wiener, N. *Cybernetics*. New York: John Wiley, 1948.
Wonham, W. M. "On Pole Assignment in Multi-Input Controllable Linear Systems," *IEEE Transactions on Automatic Control 12* (1967), pp. 660-685.
Willems, J. C. "Least Squares Stationary Optimal Control and the Algebraic Riccati Equation," *IEEE Transactions on Automatic Control 16* (1971), pp. 621-634.
Zadeh, L. A. and C. A. Desoer. *Linear System Theory — A State Space Approach*. New York: McGraw-Hill, 1963.

INDEX